W0235295

Advances in Science, Technology & Innovation

IEREK Interdisciplinary Series for Sustainable Development

Editorial Board Members

Anna Laura Pisello, Department of Engineering, University of Perugia, Italy

Dean Hawkes, Cardiff University, UK

Hocine Bougdah, University for the Creative Arts, Farnham, UK

Federica Rosso, Sapienza University of Rome, Rome, Italy

Hassan Abdalla, University of East London, London, UK

Sofia-Natalia Boemi, Aristotle University of Thessaloniki, Greece

Nabil Mohareb, Beirut Arab University, Beirut, Lebanon

Saleh Mesbah Elkaffas, Arab Academy for Science, Technology, Egypt

Emmanuel Bozonnet, University of la Rochelle, La Rochelle, France

Gloria Pignatta, University of Perugia, Italy

Yasser Mahgoub, Qatar University, Qatar

Luciano De Bonis, University of Molise, Italy

Stella Kostopoulou, Regional and Tourism Development, University of Thessaloniki, Thessaloniki, Greece

Biswajeet Pradhan, Faculty of Engineering and IT, University of Technology Sydney, Sydney, Australia

Md. Abdul Mannan, Universiti Malaysia Sarawak, Malaysia

Chaham Alalouch, Sultan Qaboos University, Muscat, Oman

Iman O. Gawad, Helwan University, Egypt

Series Editor

Mourad Amer, Enrichment and Knowledge Exchange, International Experts for Research, Cairo, Egypt

Advances in Science, Technology & Innovation (ASTI) is a series of peer-reviewed books based on the best studies on emerging research that redefines existing disciplinary boundaries in science, technology and innovation (STI) in order to develop integrated concepts for sustainable development. The series is mainly based on the best research papers from various IEREK and other international conferences, and is intended to promote the creation and development of viable solutions for a sustainable future and a positive societal transformation with the help of integrated and innovative science-based approaches. Offering interdisciplinary coverage, the series presents innovative approaches and highlights how they can best support both the economic and sustainable development for the welfare of all societies. In particular, the series includes conceptual and empirical contributions from different interrelated fields of science, technology and innovation that focus on providing practical solutions to ensure food, water and energy security. It also presents new case studies offering concrete examples of how to resolve sustainable urbanization and environmental issues. The series is addressed to professionals in research and teaching, consultancies and industry, and government and international organizations. Published in collaboration with IEREK, the ASTI series will acquaint readers with essential new studies in STI for sustainable development.

More information about this series at http://www.springer.com/series/15883

Salma Alrasheed

Principles of Mechanics

Fundamental University Physics

OPEN

 Springer

Salma Alrasheed
Thuwal, Saudi Arabia

ISSN 2522-8714   ISSN 2522-8722   (electronic)
Advances in Science, Technology & Innovation
ISBN 978-3-030-15194-2   ISBN 978-3-030-15195-9   (eBook)
https://doi.org/10.1007/978-3-030-15195-9

Library of Congress Control Number: 2019934801

© The Editor(s) (if applicable) and The Author(s) 2019. This book is an open access publication.
Open Access This book is licensed under the terms of the Creative Commons Attribution 4.0 International License
(http://creativecommons.org/licenses/by/4.0/), which permits use, sharing, adaptation, distribution and reproduction in
any medium or format, as long as you give appropriate credit to the original author(s) and the source, provide a link to
the Creative Commons license and indicate if changes were made.
The images or other third party material in this book are included in the book's Creative Commons license, unless
indicated otherwise in a credit line to the material. If material is not included in the book's Creative Commons license
and your intended use is not permitted by statutory regulation or exceeds the permitted use, you will need to obtain
permission directly from the copyright holder.
The use of general descriptive names, registered names, trademarks, service marks, etc. in this publication does not
imply, even in the absence of a specific statement, that such names are exempt from the relevant protective laws and
regulations and therefore free for general use.
The publisher, the authors and the editors are safe to assume that the advice and information in this book are believed
to be true and accurate at the date of publication. Neither the publisher nor the authors or the editors give a warranty,
expressed or implied, with respect to the material contained herein or for any errors or omissions that may have been
made. The publisher remains neutral with regard to jurisdictional claims in published maps and institutional
affiliations.

This Springer imprint is published by the registered company Springer Nature Switzerland AG.
The registered company address is: Gewerbestrasse 11, 6330 Cham, Switzerland

Preface

This book is aimed at taking the reader step by step through the beautiful concepts of mechanics in a clear and detailed manner. Mechanics is considered to be the core of physics and a deep understanding of the concepts is essential for all branches of physics. Many proofs and examples are included to help the reader grasp the fundamentals fully, paving the way to deal with more advanced topics. This book is useful for undergraduate students majoring in physics or other science and engineering disciplines. It can also be used as a reference for more advanced levels.

I would like to express my deep gratitude to my parents Abdulkareem Alrasheed and Mona Alzamil for their encouragement and support. I am grateful to all of those who have contributed to this book and made its completion possible. In particular, I would like to thank Khalid Alzamil, Dr. Laila Babsail, and Abbie Clifford for their efforts in revising the book. My sincere thanks are also extended to Ardel Flavier and Rodolfo Rodriguez for their assistance in creating the figures and illustrations. Finally to my daughter Layla, words can't express my appreciation to you.

Jeddah, Saudi Arabia   Dr. Salma Alrasheed
January 2019

Contents

1.1 Introduction

Physics is an exciting adventure that is concerned with unraveling the secrets of nature based on observations and measurements and also on intuition and imagination. Its beauty lies in having few fundamental principles being able to reach out to incorporate many phenomena from the atomic to the cosmic scale. It is a science that depends heavily on mathematics to prove and express theories and laws and is considered to be the most fundamental of physical sciences. Astronomy, geology, and chemistry all involve applications of physics' principles and concepts. Physics doesn't only provide theories, but it also provides techniques that are used in every area of life. Modern physical techniques were the major contributors to the wealth of mankind's knowledge in the past century.

A simple law in physics can be used to explain a wide range of complex phenomena that may appear to be not related. When studying a complex physical system, a simplified model of the system is usually used, where the minor effects are neglected and the main features of the system are concentrated upon. For example, when dealing with an object falling near the earth's surface, air resistance can be neglected. In addition, the earth is usually assumed to be spherical and homogeneous. However, in reality, the earth is an ellipsoid and is not homogeneous. The difference between the calculations of these different models can be assumed to be insignificant.

Physics can be divided into two branches namely: classical physics and modern physics. This book focuses on mechanics, which is a branch of classical physics. Other branches of classical physics are: light and optics, sound, electromagnetism, and thermodynamics. Mechanics is the science of motion of objects and is the core of classical physics. On the other hand, modern branches of physics include theories that have been developed during the past twentieth century. Two main theories are the theory of relativity and the theory of quantum

mechanics. Modern physics explains many physical phenomena that cannot be explained by classical physics.

1.2 The SI Units

A physical quantity is a quantitative description of a physical phenomenon. For a precise description, one has to measure the physical quantity and represent this measurement by a number. Such a measurement is made by comparing the quantity with a standard; this standard is called a unit. For example, mass is a physical quantity that refers to the quantity of matter contained in an object. The unit kilogram is one of the units used to measure mass and is defined as the mass of a specific platinum–iridium alloy cylinder, kept at the International Bureau of Weights and Measures. Therefore, when we say that a block's mass is 300 kg, we mean that it is 300 times the mass of the cylindrical platinum–iridium alloy. All units chosen should obey certain properties such as being accurate, accessible, and should remain stable under varied environmental conditions or time.

In 1960, the International System of units (SI) (formally known as the Metric System MKS) was established. The abbreviation is derived from the French phrase "System International". As shown in Table 1.1, the SI system consists of seven base fundamental units, each representing a quantity assumed to be naturally independent. The system also includes two supplementary units, the radian which is a unit of the plane angle, and the steradian which is a unit of the solid angle. All other quantities in physics are derived from these base quantities. For example, mechanical quantities such as force, velocity, volume, and energy can be derived from the fundamental quantities length, mass, and time. Furthermore, the powers of ten are used to represent the larger and smaller values for a certain physical quantity as listed in Table 1.2. The most recent definitions of the units of length, mass, and time in the SI system are as follows:

© The Author(s) 2019
S. Alrasheed, *Principles of Mechanics*, Advances in Science,
Technology & Innovation, https://doi.org/10.1007/978-3-030-15195-9_1

Table 1.1 The SI system consists of seven base fundamental units, each representing a quantity assumed to be naturally independent

Quantity	Unit name	Unit symbol
Length	Meter	m
Mass	Kilogram	kg
Time	Second	s
Temperature	Kelvin	K
Electric Current	Ampere	A
Luminous Intensity	Candela	cd
Amount of Substance	mole	mol

Table 1.2 Prefixes for Powers of Ten

Factor	Prefix	Symbol
10^{-24}	yocto	y
10^{-21}	zepto	z
10^{-18}	atto	a
10^{-15}	femto	f
10^{-12}	pico	p
10^{-9}	nano	n
10^{-6}	micro	μ
10^{-3}	milli	m
10^{-2}	centi	c
10^{-1}	deci	d
10^{1}	deka	da
10^{2}	hecto	h
10^{3}	kilo	k
10^{6}	mega	M
10^{9}	giga	G
10^{12}	tera	T
10^{15}	peta	P
10^{18}	exa	E
10^{21}	zetta	Z

- The Meter: The distance that light travels in vacuum during a time of 1/299792458 s.
- The Kilogram: The mass of a specific platinum–iridium alloy cylinder, which is kept at the International Bureau of Weights and Measures.
- The Second: 9192631770 periods of the radiation from cesium-133 atoms.

1.3 Conversion Factors

There are two other major systems of units besides the SI units. The (CGS) system of units which uses the centimeter, gram and second as its base units, and the (FPS) system of units which uses the foot, pound, and second as its base units. The conversion factors between the SI units and other systems of units of length, mass, and time are

- $1 \text{ m} = 39.37 \text{ in} = 3.281 \text{ ft} = 6.214 \times 10^{-4} \text{ mi}$
- $1 \text{ kg} = 10^3 \text{ g} = 0.0685 \text{ slug} = 6.02 \times 10^{26} \text{ u}$
- $1 \text{ s} = 1.667 \times 10^{-2} \text{ min} = 2.778 \times 10^{-4} \text{ h} = 3.169 \times 10^{-8} \text{ yr}$

Example 1.1 If a tree is measured to be 10 m long, what is its length in inches and in feet?

Solution 1.1

$$10 \text{ m} = (10 \text{ m})\left(\frac{39.37 \text{ in}}{1 \text{ m}}\right) = 393.7 \text{ in}$$

$$10 \text{ m} = (10 \text{ m})\left(\frac{3.281 \text{ ft}}{1 \text{ m}}\right) = 32.81 \text{ ft}$$

Example 1.2 If a volume of a room is 32 m^3, what is the volume in cubic inches?

Solution 1.2

$$32 \text{ m}^3 = (32 \text{ m}^3)\left(\frac{39.37 \text{ in}}{1 \text{ m}}\right)^3 = 1.95 \times 10^6 \text{ in}^3$$

1.4 Dimension Analysis

The symbols used to specify the dimensions of length, mass, and time are L, M and T, respectively. Dimension analysis is a method used to check the validity of an equation and to derive correct expressions. Only the same dimensions can be added or subtracted, i.e., they obey the rules of algebra. To check the validity of an equation, the terms on both sides must have the same dimension. The dimension of a physical quantity is denoted using brackets [ ]. For example, the dimension of the volume is $[V] = \text{L}^3$, and that of acceleration is $[a] = \text{L}/\text{T}^3$.

Example 1.3 Show that the expression v$^2 = 2ax$ is dimensionally consistent, where v represents the speed, x represent the displacement, and a represents the acceleration of the object.

Solution 1.3

$$[\text{v}^2] = \text{L}^2/\text{T}^2$$

$$[xa] = (\text{L}/\text{T}^2)(\text{L}) = \text{L}^2/\text{T}^2$$

Each term in the equation has the same dimension and therefore it is dimensionally correct.

Fig. 1.1 A vector is represented geometrically by an arrow PQ 129 drawn to scale

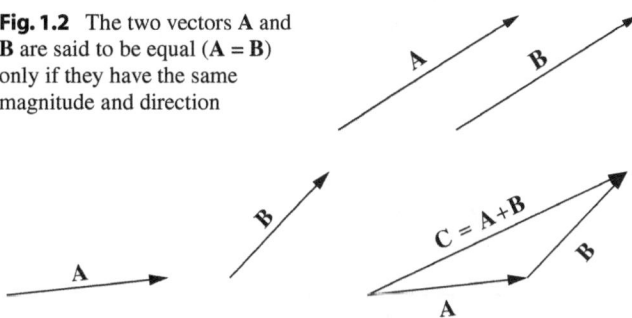

Fig. 1.2 The two vectors **A** and **B** are said to be equal (**A** = **B**) only if they have the same magnitude and direction

Fig. 1.3 To add two vectors **A** and **B** using the geometric method, place the head of **A** at the tail of **B** and draw a vector from the tail of **A** to the head of **B**

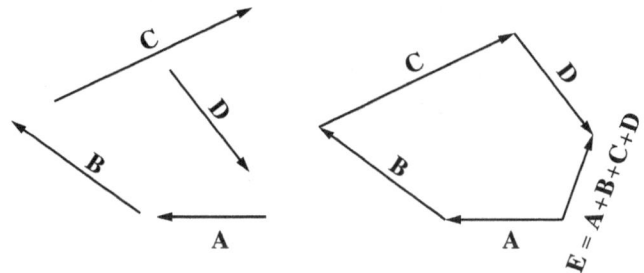

Fig. 1.4 Geometric method for summing more than two vectors

1.5 Vectors

When exploring physical quantities in nature, it is found that some quantities can be completely described by giving a number along with its unit, such as the mass of an object or the time between two events. These quantities are called scalar quantities. It is also found that other quantities are fully described by giving a number along with its unit in addition to a specified direction, such as the force on an object. These quantities are called vector quantities.

Scalar quantities have magnitude but don't have a direction and obey the rules of ordinary arithmetic. Some examples are mass, volume, temperature, energy, pressure, and time intervals by a letter such as $m, t, E. . .$, etc. Vector quantities have both magnitude and direction and obey the rules of vector algebra. Examples are displacement, force, velocity, and acceleration. Analytically, a vector is specified by a bold face letter such as **A**. This notation (as used in this book) is usually used in printed material. In handwriting, the designation $\overrightarrow{A}$ is used. The magnitude of **A** is written as $|\mathbf{A}|$ or A in print or as $|\overrightarrow{A}|$ in handwriting.

A vector is represented geometrically by an arrow PQ drawn to scale as shown in Fig. 1.1. The length and direction of the arrow represent the magnitude and direction of the vector, respectively, and is independent of the choice of coordinate system. The point P is called the initial point (tail of **A**) and Q is called the terminal point (head of **A**).

1.6 Vector Algebra

In this section, we will discuss how mathematical operations are applied to vectors.

1.6.1 Equality of Two Vectors

The two vectors **A** and **B** are said to be equal (**A** = **B**) only if they have the same magnitude and direction, whether or not their initial points are the same as shown in Fig. 1.2.

1.6.2 Addition

There are two ways to add vectors, geometrically and algebraically. Here, we will discuss the geometric method which is useful for solving problems without using a coordinate system. The algebraic method will be discussed later. To add two vectors **A** and **B** using the geometric method, place the head of **A** at the tail of **B** and draw a vector from the tail of **A** to the head of **B** as shown in Fig. 1.3. This method is known as the triangle method. An extension to sum up more than two vectors is shown in Fig. 1.4. An alternative procedure of vector addition using the geometric method is shown in Fig. 1.5. This is known as the parallelogram method, where **C** is the diagonal of a parallelogram with sides A and B. To find **C** analytically, Fig. 1.6 shows that

$$(DG)^2 = (DF)^2 + (FG)^2, \tag{1.1}$$

and that

$$DF = DE + EF = A + B\cos\theta,$$

Thus, Eq. 1.1 becomes

$$C^2 = (A + B\cos\theta)^2 + (B\sin\theta)^2 = A^2 + B^2 + 2AB\cos\theta,$$

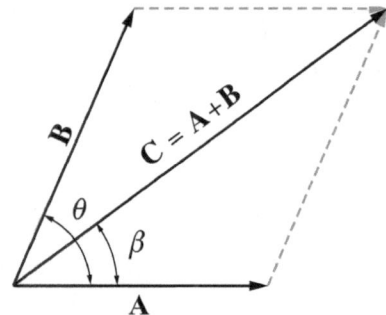

Fig. 1.5 The parallelogram method of adding two vectors

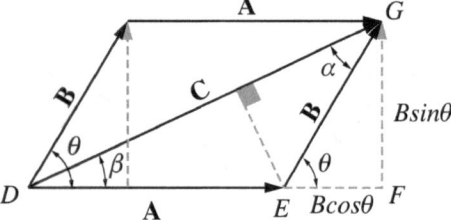

Fig. 1.6 Finding the magnitude and the direction of **C**

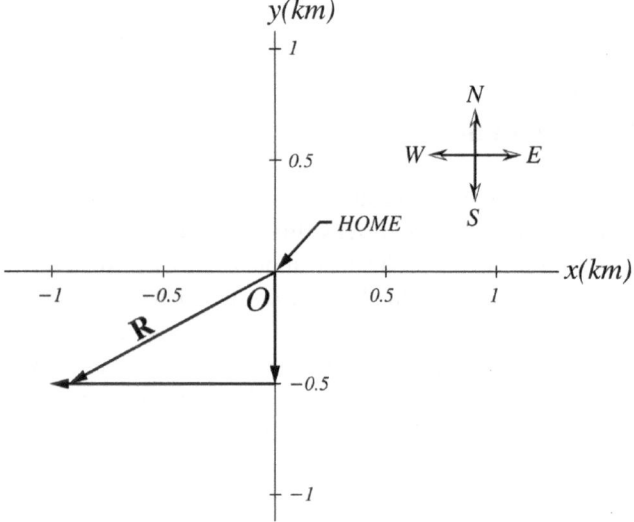

Fig. 1.7 The total displacement of the jogger is the vector **R**

or

$$C = \sqrt{A^2 + B^2 + 2AB\cos\theta},$$

The direction of **C** is

$$\tan\beta = \frac{GF}{DF} = \frac{GF}{DE + EF} = \frac{B\sin\theta}{A + B\cos\theta},$$

Note that only when **A** and **B** are parallel, the magnitude of the resultant vector **C** is equal to $A + B$ (unlike the addition of scalar quantities, the magnitude of the resultant vector **C** is not necessarily equal to $A + B$).

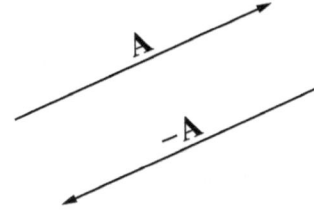

Fig. 1.8 The negative vector of **A** is a vector of the same magnitude of **A** but in the opposite direction

Example 1.4 **A** jogger runs from her home a distance of 0.5 km due south and then 1 km to the west. Find the magnitude and direction of her resultant displacement.

Solution 1.4 From Fig. 1.7, we can see that the magnitude of the resultant displacement is given by

$$R = \sqrt{(0.5 \text{ km})^2 + (1 \text{ km})^2} = 1.1 \text{ m}$$

The direction of R is

$$\theta = \tan^{-1}\frac{(0.5 \text{ m})}{(1 \text{ m})} = 26.6°$$

south of west.

1.6.3　Negative of a Vector

The negative vector of **A** is a vector of the same magnitude of **A** but in the opposite direction as shown in Fig. 1.8, and it is denoted by $-\mathbf{A}$.

1.6.4　The Zero Vector

The zero vector is a vector of zero magnitude and has no defined direction. It may result from $\mathbf{A} = \mathbf{B} - \mathbf{B} = \mathbf{0}$ or from $\mathbf{A} = c\mathbf{B} = \mathbf{0}$ if $c = 0$.

1.6.5　Subtraction of Vectors

The vector $\mathbf{A} - \mathbf{B}$ is defined as the vector that when added to **B** gives us **A**. Equivalently, $\mathbf{A} - \mathbf{B}$ can be defined as the vector **A** added to vector $-\mathbf{B}$ $(\mathbf{A} + (-\mathbf{B}))$ as shown in Fig. 1.9.

Fig. 1.9 Subtraction of two vectors

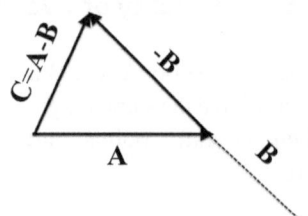

1.6.6 Multiplication of a Vector by a Scalar

The product of a vector **A** by a scalar q is a vector $q\mathbf{A}$ or $\mathbf{A}q$. Its magnitude is qA and its direction is the same as **A** if q is positive and opposite to **A** if q is negative, as shown in Fig. 1.10.

1.6.7 Some Properties

- $\mathbf{A} + \mathbf{B} = \mathbf{B} + \mathbf{A}$ (Commutative law of addition). This can be seen in Fig. 1.11.
- $(\mathbf{A} + \mathbf{B}) + \mathbf{C} = \mathbf{A} + (\mathbf{B} + \mathbf{C})$, as seen from Fig. 1.12 (Associative law of addition).
- $\mathbf{A} + \mathbf{0} = \mathbf{A}$
- $\mathbf{A} + (-\mathbf{A}) = \mathbf{0}$

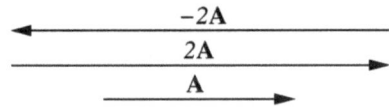

Fig. 1.10 The product of a vector by a scalar

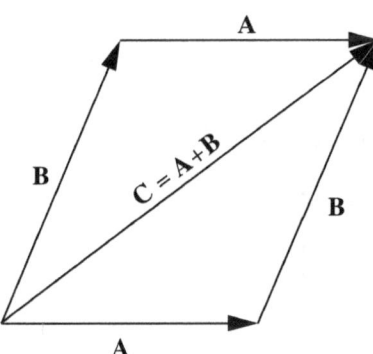

Fig. 1.11 Commutative law of addition

Fig. 1.12 Associative law of addition

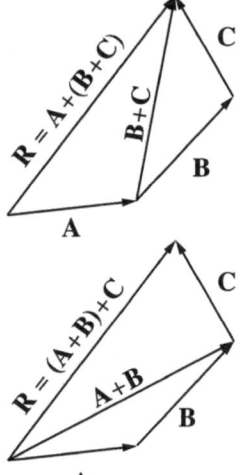

- $p(q\mathbf{A}) = (pq)\mathbf{A} = q(p\mathbf{A})$ (where p and q are scalars) (Associative law for multiplication).
- $(p + q)\mathbf{A} = p\mathbf{A} + q\mathbf{A}$ (Distributive law).
- $p(\mathbf{A} + \mathbf{B}) = p\mathbf{A} + p\mathbf{B}$ (Distributive law).
- $1\mathbf{A} = \mathbf{A}$, $0\mathbf{A} = \mathbf{0}$ (Here, the zero vector has the same direction as **A**, i.e., it can have any direction), $q\mathbf{0} = \mathbf{0}$

1.6.8 The Unit Vector

The unit vector is a vector of magnitude equal to 1, and with the same direction of **A**. For every $\mathbf{A} \neq 0$, $\mathbf{a} = \mathbf{A}/|\mathbf{A}|$ is a unit vector.

1.6.9 The Scalar (Dot) Product

The scalar product is a scalar quantity defined as $\mathbf{A} \cdot \mathbf{B} = AB \cos\theta$, where θ is the smaller angle between **A** and **B** ($0 \leq \theta \leq \pi$) (see Fig. 1.13).

1.6.9.1 Some Properties of the Scalar Product
- $\mathbf{A} \cdot \mathbf{B} = \mathbf{B} \cdot \mathbf{A}$ (Commutative law of scalar product).
- $\mathbf{A} \cdot (\mathbf{B} + \mathbf{C}) = \mathbf{A} \cdot \mathbf{B} + \mathbf{A} \cdot \mathbf{C}$ (Distributive law).
- $m(\mathbf{A} \cdot \mathbf{B}) = (m\mathbf{A}) \cdot \mathbf{B} = \mathbf{A} \cdot (m\mathbf{B}) = (\mathbf{A} \cdot \mathbf{B})m$, where m is a scalar.

1.6.10 The Vector (Cross) Product

The vector product is a vector quantity defined as $\mathbf{C} = \mathbf{A} \times \mathbf{B}$ (read A cross B) with magnitude equal to $|\mathbf{A} \times \mathbf{B}| = AB \sin\theta$, $(0 \leq \theta \leq \pi)$. The direction of **C** is found from the right-hand rule or of advance of a right-handed screw rotated from **A** to **B** as shown in Fig. 1.14. **C** is perpendicular to the plane formed by **A** and **B**.

1.6.10.1 Some Properties
- $\mathbf{A} \cdot \mathbf{A} = A^2$, $\mathbf{0} \cdot \mathbf{A} = 0$
- $\mathbf{A} \times \mathbf{B} = -\mathbf{B} \times \mathbf{A}$
- $\mathbf{A} \times (\mathbf{B} + \mathbf{C}) = \mathbf{A} \times \mathbf{B} + \mathbf{A} \times \mathbf{C}$ (Distributive law).
- $(\mathbf{A} + \mathbf{B}) \times \mathbf{C} = \mathbf{A} \times \mathbf{C} + \mathbf{B} \times \mathbf{C}$

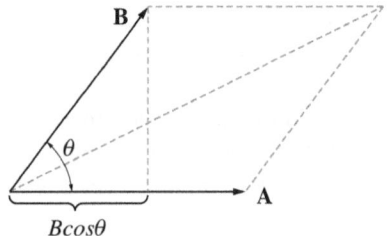

Fig. 1.13 The scalar product of two vectors

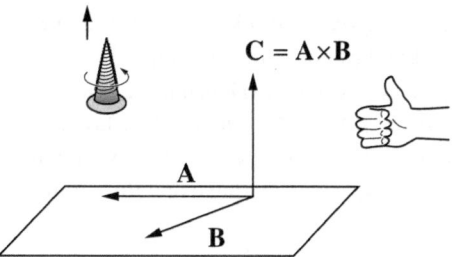

Fig. 1.14 The vector product of two vectors

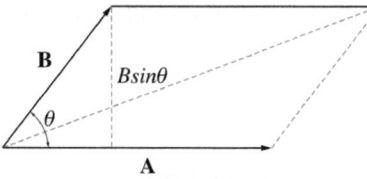

Fig. 1.15 The magnitude of the vector product $|\mathbf{A} \times \mathbf{B}| =$ is the area of a parallelogram with sides A and B

- $q(\mathbf{A} \times \mathbf{B}) = (q\mathbf{A}) \times \mathbf{B} = \mathbf{A} \times (q\mathbf{B}) = (\mathbf{A} \times \mathbf{B})q$, where q is a scalar.
- $|\mathbf{A} \times \mathbf{B}| =$ The area of a parallelogram that has sides A and B as shown in Fig. 1.15.

1.7   Coordinate Systems

To specify the location of a point in space, a coordinate system must be used. A coordinate system consists of a reference point called the origin O and a set of labeled axes. The positive direction of an axis is in the direction of increasing numbers, whereas the negative direction is opposite. Figures 1.16 and 1.17 show the rectangular (or Cartesian) coordinate system and the polar coordinates of a point, respectively The rectangular coordinates x and y are related to the polar coordinates r and θ by the following relations:

$$x = r \cos \theta$$

$$y = r \sin \theta$$

$$\tan \theta = y/x$$

$$r = \sqrt{x^2 + y^2}$$

In three dimensions, the cartesian coordinate system is shown in Fig. 1.18. Other used coordinate systems in three dimensions are the spherical and cylindrical coordinates (Figs. 1.19 and 1.20).

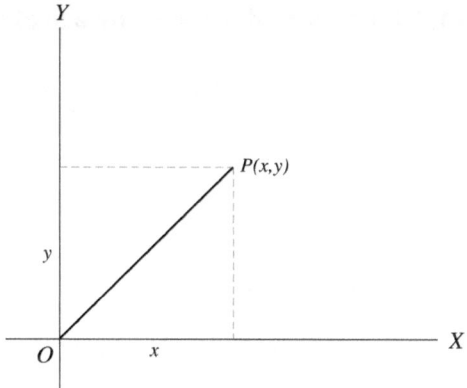

Fig. 1.16 The rectangular (cartesian) coordinate system

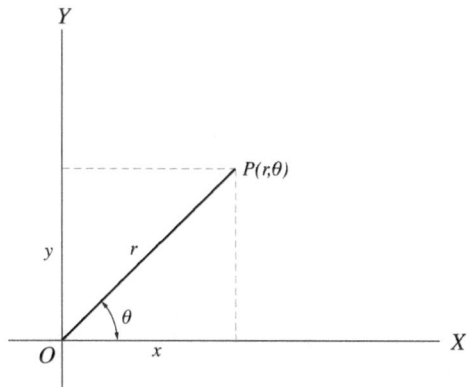

Fig. 1.17 The polar coordinate system

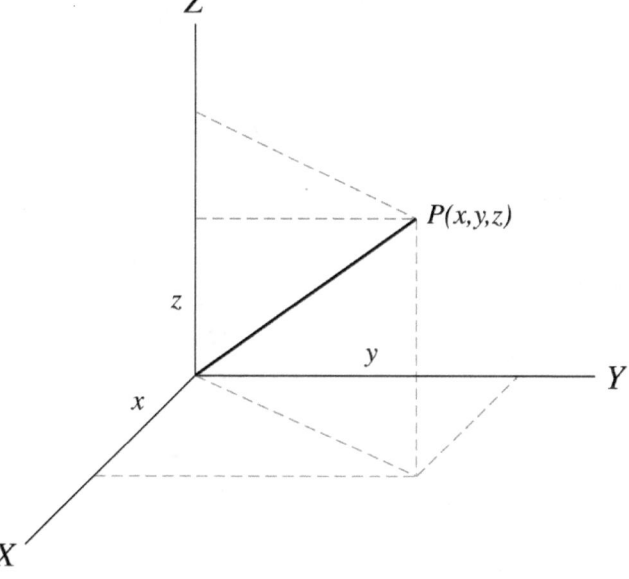

Fig. 1.18 The cartesian coordinate system in three dimensions

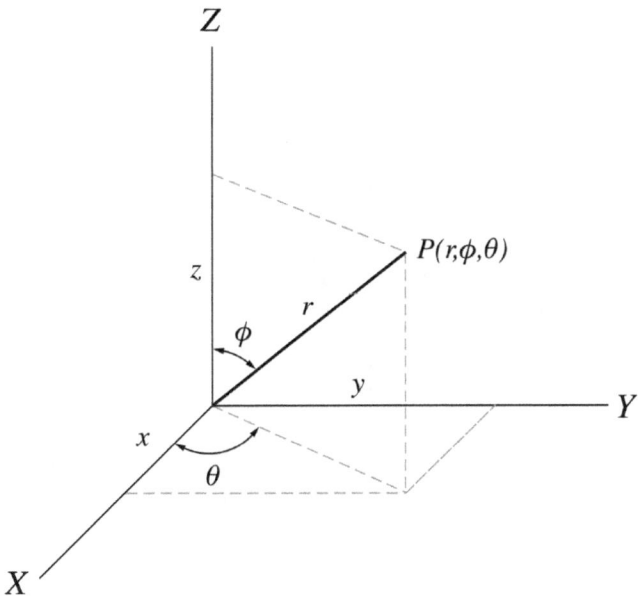

Fig. 1.19 The spherical coordinate system

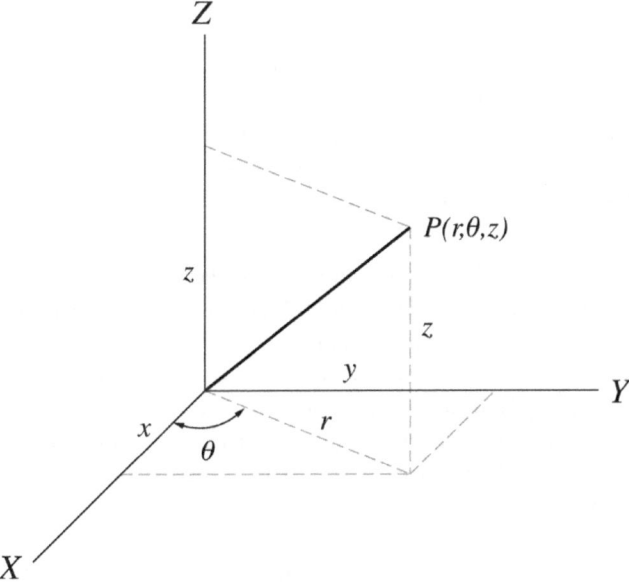

Fig. 1.20 The cylindrical coordinate system

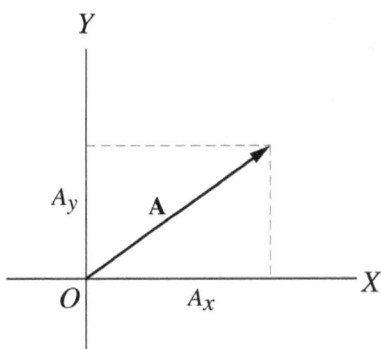

Fig. 1.21 In two dimensions, the vector **A** can be expressed as the sum of two other vectors $\mathbf{A} = \mathbf{A}_x + \mathbf{A}_y$, where $A_x = A\cos\theta$ and $A_y = A\sin\theta$

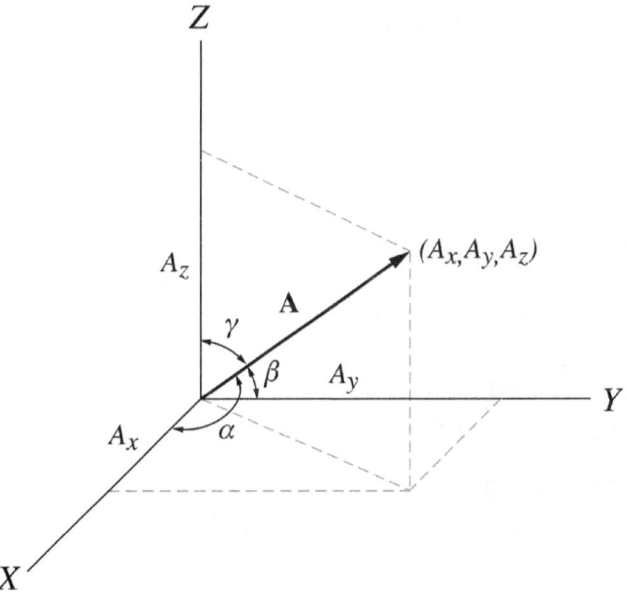

Fig. 1.22 In three dimensions the magnitude of A is $A = \sqrt{A_x^2 + A_y^2 + A_z^2}$

$$\tan\theta = A_y/A_x$$

In three dimensions (see Fig. 1.22), the magnitude of A is given by

$$A = \sqrt{A_x^2 + A_y^2 + A_z^2}$$

with directions given by

$$\cos\alpha = A_x/A, \quad \cos\beta = A_y/A, \quad \cos\gamma = A_z/A$$

1.8 Vectors in Terms of Components

In two dimensions, the vector **A** can be expressed as the sum of two other vectors $\mathbf{A} = \mathbf{A}_x + \mathbf{A}_y$, where $A_x = A\cos\theta$ and $A_y = A\sin\theta$ as shown in Fig. 1.21.

$\mathbf{A}_x$ and $\mathbf{A}_y$ are called the rectangular components, or simply components of **A** in the x and y directions respectively The magnitude and direction of **A** are related to its components through the expressions:

$$A = \sqrt{A_x^2 + A_y^2}$$

1.8.1 Rectangular Unit Vectors

The rectangular unit vectors **i**, **j**, and **k** are unit vectors defined to be in the direction of the positive x-, y-, and z-axes, respectively, of the rectangular coordinate system as shown in Fig. 1.23. Note that labeling the axes in this way forms a

Fig. 1.23 The rectangular unit vectors **i**, **j** and **k** are unit vectors defined to be in the direction of the positive x, y, and z axes respectively

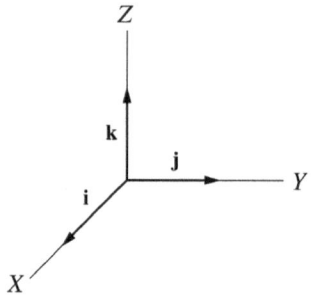

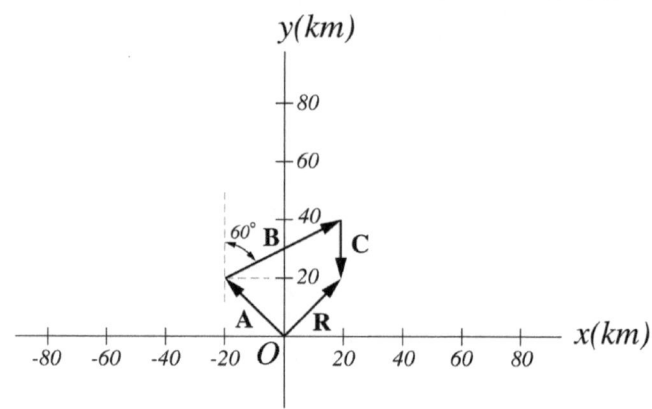

right-handed system. This name derives from the fact that a right- handed screw rotated through 90° from the x-axis into the y-axis will advance in the positive z-direction. (Note that throughout this book the right-handed coordinate system is used). In terms of unit vectors, vector A can be written as

$$\mathbf{A} = A_x\mathbf{i} + A_y\mathbf{j} + A_z\mathbf{k}$$

1.8.2 Component Method

Suppose we have $\mathbf{A} = A_x\mathbf{i} + A_y\mathbf{j}$ and $\mathbf{B} = B_x\mathbf{i} + B_y\mathbf{j}$

1.8.2.1 Addition
The resultant vector **C** is given by

$$\mathbf{C} = \mathbf{A} + \mathbf{B} = (A_x + B_x)\mathbf{i} + (A_y + B_y)\mathbf{j} = C_x\mathbf{i} + C_y\mathbf{j}$$

$$C_x = A_x + B_x$$

$$C_y = A_y + B_y$$

Thus, the magnitude of **C** is

$$C = \sqrt{C_x^2 + C_y^2}$$

with a direction

$$\tan\theta = \frac{C_y}{C_x} = \frac{A_y + B_y}{A_x + B_x}$$

in three dimensions

$$\mathbf{C} = (A_x+B_x)\mathbf{i}+(A_y+B_y)\mathbf{j} = (A_z+B_z)\mathbf{k} = C_x\mathbf{i}+C_y\mathbf{j}+C_z\mathbf{k}$$

the magnitude of **C** is

$$C = \sqrt{C_x^2 + C_y^2 + C_z^2}$$

And the directions are

$$\cos\alpha = C_x/C, \ \cos\beta = C_y/C, \ \cos\gamma = C_z/C$$

This component method is easy to use in adding any number of vectors.

Example 1.5 A truck travels northwest a distance of 30 km, and then 50 km at 30° north of east, and finally travels a distance of 20 km due south. Determine both graphically and analytically the magnitude and direction of the resultant displacement of the truck from its starting point.

Solution 1.5 Graphically, in Fig. 1.24 the displacements are drawn to scale with the head of **A** placed at the tail of **B** and the head of **B** placed at the tail of **C**. The resultant vector **R** is the vector that extends from the tail of **A** to the head of **C**. By using graph paper and a protractor, the magnitude of **R** is measured to have the value of 34.8 km and a direction of 49.8° from the positive x axis. Analytically, from Fig. 1.24, we have

$$A_x = A\cos 135° = (30\text{ km})(-0.707) = -21.2\text{ km}$$

$$A_y = A\sin 135° = (30\text{ km})(0.707) = 21.2\text{ km}$$

$$B_x = B\cos 30° = (50\text{ km})(0.866) = 43.3\text{ km}$$

$$B_y = B\sin 30° = (50\text{ km})(0.5) = 25\text{ km}$$

$$C_x = C\cos 270° = (20\text{ km})(0) = 0$$

$$C_y = C\sin 270° = (20\text{ km})(-1) = -20\text{ km}$$

$$\mathbf{R} = \mathbf{A}+\mathbf{B}+\mathbf{C} = (A_x+B_x+C_x)\mathbf{i}+(A_y+B_y+C_y)\mathbf{j}+(A_z+B_z+C_z)\mathbf{k} = 22.1\mathbf{i}+26.2\mathbf{j}$$

Thus, the magnitude of **R** is given by

Fig. 1.24 The displacements are drawn to scale with the head of **A** placed at the tail of **B** and the head of **B** placed at the tail of **C**. The resultant vector **R** is the vector that extends from the tail of **A** to the head of **C**

$$R = \sqrt{R_x^2 + R_y^2} = \sqrt{(221 \text{ km})^2 + (262 \text{ km})^2} = 34.3 \text{ km}$$

and its direction is

$$\theta = \tan^{-1}\left(\frac{26.2 \text{ km}}{22.1 \text{ km}}\right) = 49.9^\circ$$

north of east.

1.8.2.2 Subtraction

$$\mathbf{C} = \mathbf{A} - \mathbf{B} = (A_x - B_x)\mathbf{i} + (A_y - B_y)\mathbf{j} + (A_z - B_z)\mathbf{k}$$

The magnitude and direction of $\mathbf{C}$ are as in the case of addition except that the plus sign is replaced by the minus sign.

1.8.2.3 Scalar Product

$$\mathbf{A} \cdot \mathbf{B} = (A_x\mathbf{i} + A_y\mathbf{j} + A_z\mathbf{k}) \cdot (B_x\mathbf{i} + B_y\mathbf{j} + B_z\mathbf{k})$$

Using the definition of scalar product and by applying the distributive law we get nine terms: since $\mathbf{i} \cdot \mathbf{i} = \mathbf{j} \cdot \mathbf{j} = \mathbf{k} \cdot \mathbf{k}$, · and $\mathbf{i} \cdot \mathbf{j} = \mathbf{j} \cdot \mathbf{k} = \mathbf{j} \cdot \mathbf{k} = 0$, we get

$$\mathbf{A} \cdot \mathbf{B} = A_x B_x + A_y B_y + A_z B_z$$

The dot product of any vector (for example $\mathbf{A}$) by itself is

$$\mathbf{A} \cdot \mathbf{A} = A^2 = A_x^2 + A_y^2 + A_z^2$$

1.8.2.4 The Angle Between Two Vectors

$$\mathbf{A} \cdot \mathbf{B} = AB\cos\theta = A_x B_x + A_y B_y + A_z B_z$$

$$\cos\theta = \frac{A_x B_x + A_y B_y + A_z B_z}{AB}$$

Example 1.6 Two vectors $\mathbf{A}$ and B are given by $\mathbf{A} = \mathbf{i} + 5\mathbf{j} - 7\mathbf{k}$ and $\mathbf{B} = 6\mathbf{i} - 2\mathbf{j} + 3\mathbf{k}$. Find the angle between them.

Solution 1.6

$$\mathbf{A} \cdot \mathbf{B} = AB\cos\phi = A_x B_x + A_y B_y + A_z B_z$$

$$A = \sqrt{A_x^2 + A_y^2 + A_z^2} = \sqrt{1 + 25 + 49} = 8.7$$

$$B = \sqrt{B_x^2 + B_y^2 + B_z^2} = \sqrt{36 + 4 + 9} = 7$$

$$\cos\phi = \frac{A_x B_x + A_y B_y + A_z B_z}{AB} = \frac{6 - 10 - 21}{(8.7)(7)} = -0.4$$

$$\phi = 113.6^\circ$$

1.8.2.5 Perpendicular and Parallel Vectors

Nonzero vectors $\mathbf{A}$ and $\mathbf{B}$ are perpendicular if $\mathbf{A} \cdot \mathbf{B} = 0$ or $A_x B_x + A_y B_y + A_z B_z = 0$ and they are parallel if $\mathbf{A} \times \mathbf{B} = \mathbf{0}$. For any two parallel vectors $\mathbf{A}$ and $\mathbf{B}$, we have $\mathbf{A} = q\mathbf{B}$, where they have the same direction if $q > 0$, and are in opposite direction if $q < 0$. Also we can write

$$\frac{\mathbf{A}}{\mathbf{B}} = q$$

or

$$\frac{A_x}{B_x} = \frac{A_y}{B_y} = \frac{A_z}{B_z}$$

1.8.2.6 Vector Product

From the vector product definition, we can see that

$$\mathbf{i} \times \mathbf{i} = \mathbf{j} \times \mathbf{j} = \mathbf{k} \times \mathbf{k} = 0$$

$$\mathbf{i} \times \mathbf{j} = \mathbf{k}, \mathbf{j} \times \mathbf{k} = \mathbf{i}, \ \mathbf{k} \times \mathbf{i} = \mathbf{j}$$

$$\mathbf{j} \times \mathbf{i} = -\mathbf{k}, \ \mathbf{k} \times \mathbf{j} = -\mathbf{i}, \ \mathbf{i} \times \mathbf{k} = -\mathbf{j}$$

If we write the unit vectors around a circle as shown in Fig. 1.25, then reading counterclockwise gives the positive products and reading clockwise gives the negative products. Note that these results are for a right-handed coordinate system. We have

$$\mathbf{A} \times \mathbf{B} = (A_x\mathbf{i} + A_y\mathbf{j} + A_z\mathbf{k}) \times (B_x\mathbf{i} + B_y\mathbf{j} + B_z\mathbf{k})$$

using the distributive law and the above relations of unit vectors we get

$$\mathbf{A} \times \mathbf{B} = (A_y B_z - A_z B_y)\mathbf{i} + (A_z B_x - A_x B_z)\mathbf{j} + (A_x B_y - A_y B_x)\mathbf{k}$$

since a determinant of order 2 is defined as

$$\begin{vmatrix} a_1 & a_2 \\ b_1 & b_2 \end{vmatrix} = a_1 b_2 - a_2 b_1$$

Then, the above expression can be written as

Fig. 1.25 If we write the unit vectors around a circle, then reading counter clockwise gives the positive products and reading clockwise gives the negative products

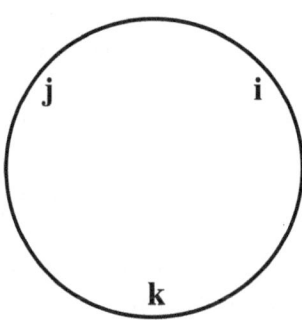

$$\mathbf{A} \times \mathbf{B} = \begin{vmatrix} A_y & A_z \\ B_y & B_z \end{vmatrix} \mathbf{i} - \begin{vmatrix} A_x & A_z \\ B_x & B_z \end{vmatrix} \mathbf{j} + \begin{vmatrix} A_x & A_y \\ B_x & B_y \end{vmatrix} \mathbf{k}$$

A determinant of order 3 is

$$\begin{vmatrix} c_1 & c_2 & c_3 \\ a_1 & a_2 & a_3 \\ b_1 & b_2 & b_3 \end{vmatrix} = \begin{vmatrix} a_2 & a_3 \\ b_2 & b_3 \end{vmatrix} c_1 - \begin{vmatrix} a_1 & a_3 \\ b_1 & b_3 \end{vmatrix} c_2 + \begin{vmatrix} a_1 & a_2 \\ b_1 & b_2 \end{vmatrix} c_3$$

Hence, the cross product can be expressed as

$$\mathbf{A} \times \mathbf{B} = \begin{vmatrix} \mathbf{i} & \mathbf{j} & \mathbf{k} \\ A_x & A_y & A_z \\ B_x & B_y & B_z \end{vmatrix} = (A_y B_z - A_z B_y)\mathbf{i} + (A_z B_x - A_x B_z)\mathbf{j} + (A_x B_y - A_y B_x)\mathbf{k}$$

Note that this is not a determinant since the elements in the first row are vectors and not scalars, but it is a convenient way to represent the cross product.

Example 1.7 Two vectors **A** and **B** are given by $\mathbf{A} = -\mathbf{i} + 3\mathbf{j}$ and $\mathbf{B} = 2\mathbf{i} + \mathbf{j}$. Find: (a) the sum of **A** and **B**, ·(b) $-\mathbf{B}$ and $3\mathbf{A}$, ·(c)$\mathbf{A} \cdot \mathbf{B}$ and $\mathbf{A} \times \mathbf{B}$.

Solution 1.7 (a)

$$\mathbf{R} = \mathbf{A} + \mathbf{B} = (A_x + B_x)\mathbf{i} + (A_y + B_y)\mathbf{j} = (-1+2)\mathbf{i} + (3+1)\mathbf{j} = \mathbf{i} + 4\mathbf{j}$$

$$R_x = 1$$

$$R_y = 4$$

(b)

$$-\mathbf{B} = -2\mathbf{i} - \mathbf{j}$$

$$3\mathbf{A} = -3\mathbf{i} + 9\mathbf{j}$$

(c)
$$\mathbf{A} \cdot \mathbf{B} = (-\mathbf{i} + 3\mathbf{j})(2\mathbf{i} + \mathbf{j}) = -\mathbf{i} \cdot 2\mathbf{i} - \mathbf{i} \cdot \mathbf{j} + 3\mathbf{j} \cdot 2\mathbf{i} + 3\mathbf{j} \cdot \mathbf{j} = -2 + 3 = 1$$

$$\mathbf{A} \times \mathbf{B} = (-\mathbf{i} + 3\mathbf{j}) \times (2\mathbf{i} + \mathbf{j}) = -\mathbf{i} \times \mathbf{j} + 3\mathbf{j} \times 2\mathbf{i} = -\mathbf{k} - 6\mathbf{k} = -7\mathbf{k}$$

Example 1.8 Find a vector of magnitude 1 that is perpendicular to each of the vectors $\mathbf{A} = 5\mathbf{i} + \mathbf{j} - 3\mathbf{k}$ and $\mathbf{B} = 3\mathbf{i} + 7\mathbf{j} - 2\mathbf{k}$.

Solution 1.8 By the definition of the unit vector, we have

$$\mathbf{c} = \frac{\mathbf{A} \times \mathbf{B}}{|\mathbf{A} \times \mathbf{B}|}$$

where c is a unit vector perpendicular to the plane formed by A and B. We have

$$\mathbf{A} \times \mathbf{B} = \begin{vmatrix} \mathbf{i} & \mathbf{j} & \mathbf{k} \\ 5 & 1 & -3 \\ 3 & 7 & -2 \end{vmatrix} = 19\mathbf{i} + \mathbf{j} + 32\mathbf{k}$$

$$|\mathbf{A} \times \mathbf{B}| = \sqrt{(19)^2 + (1)^2 + (32)^2} = 37.23$$

$$\mathbf{C} = \frac{19\mathbf{i} + \mathbf{j} + 32\mathbf{k}}{37.23} = 0.5\mathbf{i} + 0.027\mathbf{j} + 0.86\mathbf{k}$$

Example 1.9 Given that $\mathbf{A} = 2\mathbf{i} - 3\mathbf{j} - \mathbf{k}, \mathbf{B} = 3\mathbf{i} - \mathbf{j}$ and $\mathbf{C} = \mathbf{j} - 4\mathbf{k}$, find (a) $\mathbf{A} \times \mathbf{B}$ (b)$(\mathbf{A} \times \mathbf{B}) \times \mathbf{C}$ (c) $\mathbf{A} \cdot (\mathbf{B} \times \mathbf{C})$.

Solution 1.9 (a)

$$\mathbf{A} \times \mathbf{B} = \begin{vmatrix} \mathbf{i} & \mathbf{j} & \mathbf{k} \\ 2 & -3 & -1 \\ 3 & -1 & 0 \end{vmatrix} = -\mathbf{i} - 3\mathbf{j} + 7\mathbf{k}$$

(b)

$$\mathbf{A} \times (\mathbf{B} \times \mathbf{C}) = \begin{vmatrix} \mathbf{i} & \mathbf{j} & \mathbf{k} \\ -1 & -3 & 7 \\ 0 & 1 & -4 \end{vmatrix} = 5\mathbf{i} - 4\mathbf{j} - \mathbf{k}$$

(c)

$$\mathbf{B} \times \mathbf{C} = \begin{vmatrix} \mathbf{i} & \mathbf{j} & \mathbf{k} \\ 3 & -1 & 0 \\ 0 & 1 & -4 \end{vmatrix} = 4\mathbf{i} + 12\mathbf{j} + 3\mathbf{k}$$

$$\mathbf{A} \cdot (\mathbf{B} \times \mathbf{C}) = (2\mathbf{i} - 3\mathbf{j} - \mathbf{k}) \cdot (4\mathbf{i} + 12\mathbf{j} + 3\mathbf{k}) = 8 - 36 - 3 = -31$$

Example 1.10 Using vectors method, find the area of a triangle if the coordinates of its three vertices are $A(2, 1, 3)$, $B(2, 5, 7)$, $C(-1, 4, 2)$.

Solution 1.10

$$\mathbf{AB} = (2 - 2)\mathbf{i} + (5 - 1)\mathbf{j} + (7 - 3)\mathbf{k} = 4\mathbf{j} + 4\mathbf{k}$$

$$\mathbf{AC} = (-1 - 2)\mathbf{i} + (4 - 1)\mathbf{j} + (2 - 3)\mathbf{k} = -3\mathbf{i} + 3\mathbf{j} - \mathbf{k}$$

Area

$$= \frac{1}{2}|\mathbf{AB} \times \mathbf{AC}| = \frac{1}{2}|(4\mathbf{j} + 4\mathbf{k}) \times (-3\mathbf{i} + 3\mathbf{j} - \mathbf{k})| = \frac{1}{2}|4(-4\mathbf{i} - 3\mathbf{j} + 3\mathbf{k})|$$

$$= 2\sqrt{(-4)^2 + (-3)^2 + (3)^2} = 11.7$$

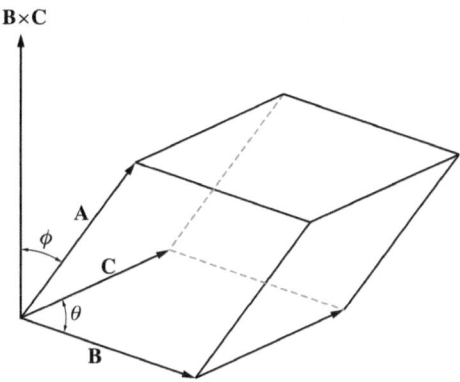

Fig. 1.26 The triple scalar product is equal to the volume of a parallepiped with sides **A**, **B**, and **C**

1.8.2.7 Triple Product
Scalar Triple Product

The triple scalar product is a scalar quantity defined as $\mathbf{A} \cdot (\mathbf{B} \times \mathbf{C})$. This quantity can be represented by a determinant that involves the components of the vectors,

$$\mathbf{A} \cdot (\mathbf{B} \times \mathbf{C}) = \begin{vmatrix} A_x & A_y & A_z \\ B_x & B_y & B_z \\ C_x & C_y & C_z \end{vmatrix}$$

where $\mathbf{A} = A_x\mathbf{i} + A_y\mathbf{j} + A_z\mathbf{k}$, $\mathbf{B} = B_x\mathbf{i} + B_y\mathbf{j} + B_z\mathbf{k}$, and $\mathbf{C} = C_x\mathbf{i} + C_y\mathbf{j} + C_z\mathbf{k}$. Furthermore, the triple scalar product is equal to the volume of a parallepiped with sides **A**, **B**, and **C** as shown in Fig. 1.26. Because any edges can be used, the triple scalar product can be written as $\mathbf{A} \cdot (\mathbf{B} \times \mathbf{C})$ or as $\mathbf{A} \cdot (\mathbf{C} \times \mathbf{B})$. These products are positive and negative for a right-handed coordinate system respectively. Therefore, there are 6 equal triple scalar products or 12 if you include the terms of the form $(\mathbf{B} \times \mathbf{C}) \cdot \mathbf{A}$. A. Three of these six products are positive and the rest are negative. By expanding the determinant, you can prove that

$$\mathbf{A} \cdot (\mathbf{B} \times \mathbf{C}) = \mathbf{B} \cdot (\mathbf{C} \times \mathbf{A}) = \mathbf{C} \cdot (\mathbf{A} \times \mathbf{B}) = -\mathbf{A} \cdot (\mathbf{C} \times \mathbf{B}) = -\mathbf{B} \cdot (\mathbf{A} \times \mathbf{C}) = -\mathbf{C} \cdot (\mathbf{B} \times \mathbf{A})$$

Vector Triple Product

The triple vector product is a vector quantity defined as $\mathbf{A} \times (\mathbf{B} \times \mathbf{C})$. You can prove by expanding this equation that

$$\mathbf{A} \times (\mathbf{B} \times \mathbf{C}) = (\mathbf{A} \cdot \mathbf{C})\mathbf{B} - (\mathbf{A} \cdot \mathbf{B})\mathbf{C}$$

Example 1.11 Given that $\mathbf{A} = A_x\mathbf{i}$, $\mathbf{B} = B_x\mathbf{i} + B_z\mathbf{k}$, and $\mathbf{C} = C_y\mathbf{j}$, show that the identity $\mathbf{A} \times (\mathbf{B} \times \mathbf{C}) = (\mathbf{A} \cdot \mathbf{C})\mathbf{B} - (\mathbf{A} \cdot \mathbf{B})\mathbf{C}$ is correct.

Solution 1.11

$$(\mathbf{B} \times \mathbf{C}) = \begin{vmatrix} \mathbf{i} & \mathbf{j} & \mathbf{k} \\ B_x & 0 & B_z \\ 0 & C_y & 0 \end{vmatrix} = -B_zC_y\mathbf{i} + B_xC_y\mathbf{k}$$

$$\mathbf{A} \times (\mathbf{B} \times \mathbf{C}) = \begin{vmatrix} \mathbf{i} & \mathbf{j} & \mathbf{k} \\ A_x & 0 & 0 \\ -B_zC_y & 0 & B_xC_y \end{vmatrix} = -A_xB_xC_y\mathbf{j}$$

$$(\mathbf{A} \cdot \mathbf{C})\mathbf{B} = 0$$

$$-(\mathbf{A} \cdot \mathbf{B})\mathbf{C} = -(A_xB_x)\mathbf{C} = -A_xB_xC_y\mathbf{j}$$

Hence, the identity is valid.

1.9 Derivatives of Vectors

If $\mathbf{A}(t)$ is a vector function of t, where t is a scalar variable such as

$$\mathbf{A}(t) = A_x(t)\mathbf{i} + A_y(t)\mathbf{j} + A_z(t)\mathbf{k}$$

Then

$$\frac{d\mathbf{A}(t)}{dt} = \frac{dA_x(t)}{dt}\mathbf{i} + \frac{dA_y(t)}{dt}\mathbf{j} + \frac{dA_z(t)}{dt}\mathbf{k}$$

1.9.1 Some Rules

If $\mathbf{A}(t)$ and $\mathbf{B}(t)$ are vector functions and $\phi(t)$ is a scalar function then

$$\frac{d}{dt}(\phi\mathbf{A}) = \phi\frac{d\mathbf{A}}{dt} + \frac{d\phi}{dt}\mathbf{A}$$

$$\frac{d}{dt}(\mathbf{A} \cdot \mathbf{B}) = \mathbf{A} \cdot \frac{d\mathbf{B}}{dt} + \frac{d\mathbf{A}}{dt} \cdot \mathbf{B}$$

$$\frac{d}{dt}(\mathbf{A} \times \mathbf{B}) = \mathbf{A} \times \frac{d\mathbf{B}}{dt} + \frac{d\mathbf{A}}{dt} \times \mathbf{B}$$

Example 1.12 Two vectors $\mathbf{r}_1$ and $\mathbf{r}_2$ are given by $\mathbf{r}_1 = 2t^2\mathbf{i} + \cos t\mathbf{j} + 4\mathbf{k}$ and $\mathbf{r}_2 = \sin t\mathbf{i} + \cos t\mathbf{k}$, find at $t = 0$ (a) $\dfrac{d^2\mathbf{r}_1}{dt^2}$ and (b) $\dfrac{d(\mathbf{r}_1 \cdot \mathbf{r}_2)}{dt}$.

Solution 1.12 (a)

$$\frac{d\mathbf{r}_1}{dt} = 4t\mathbf{i} - \sin t\mathbf{j}$$

$$\frac{d^2\mathbf{r}_1}{dt^2} = 4\mathbf{i} - \cos t\mathbf{j}$$

At $t = 0$

$$\frac{d^2\mathbf{r}_1}{dt^2} = 4\mathbf{i} - \mathbf{j}$$

(b)

$$\frac{d(\mathbf{r}_1 \cdot \mathbf{r}_2)}{dt} = \frac{d\{(2t^2\mathbf{i} + \cos t\mathbf{j} + 4\mathbf{k})(\sin t\mathbf{i} + \cos t\mathbf{k})\}}{dt} =$$

$$\frac{d(2t^2 \sin t + 4\cos t)}{dt} = 4t \sin t + 2t^2 \cos t - 4\sin t = 4(t-1)\sin t + 2t^2 \cos t$$

At $t = 0$

$$\frac{d(\mathbf{r}_1 \cdot \mathbf{r}_2)}{dt} = 0.$$

1.9.2 Gradient, Divergence, and Curl

If $\mathbf{A} = \mathbf{A}(x, y, z)$ is a vector function of x, y, and z then $\mathbf{A}(x, y, z)$ is called a vector field. Similarly, the scalar function $\phi(x, y, z)$ is called a scalar field.

1.9.2.1 Del
The vector differential operator *del* is defined as

$$\nabla = \mathbf{i}\frac{\partial}{\partial x} + \mathbf{j}\frac{\partial}{\partial y} + \mathbf{k}\frac{\partial}{\partial z}$$

1.9.2.2 Gradient
$$\nabla\phi = \left(\mathbf{i}\frac{\partial}{\partial x} + \mathbf{j}\frac{\partial}{\partial y} + \mathbf{k}\frac{\partial}{\partial z}\right)\phi = \mathbf{i}\frac{\partial\phi}{\partial x} + \mathbf{j}\frac{\partial\phi}{\partial y} + \mathbf{k}\frac{\partial\phi}{\partial z}$$

The vector $\nabla\phi$ is called the gradient of ϕ (written gradϕ).

1.9.2.3 Divergence
$$\nabla \cdot \mathbf{A} = \left(\mathbf{i}\frac{\partial}{\partial x} + \mathbf{j}\frac{\partial}{\partial y} + \mathbf{k}\frac{\partial}{\partial z}\right) \cdot (A_x\mathbf{i} + A_y\mathbf{j} + A_z\mathbf{k})$$

$$= \frac{\partial A_x}{\partial x} + \frac{\partial A_y}{\partial y} + \frac{\partial A_z}{\partial z}$$

$\nabla \cdot \mathbf{A}$ is called the divergence of A (written divA).

1.9.2.4 Curl
$$\nabla \times \mathbf{A} = \left(\mathbf{i}\frac{\partial}{\partial x} + \mathbf{j}\frac{\partial}{\partial y} + \mathbf{k}\frac{\partial}{\partial z}\right) \times (A_x\mathbf{i} + A_y\mathbf{j} + A_z\mathbf{k})$$

$$\begin{vmatrix} \mathbf{i} & \mathbf{j} & \mathbf{k} \\ \frac{\partial}{\partial x} & \frac{\partial}{\partial y} & \frac{\partial}{\partial z} \\ A_x & A_y & A_z \end{vmatrix} = \left(\frac{\partial A_z}{\partial y} - \frac{\partial A_y}{\partial z}\right)\mathbf{i} + \left(\frac{\partial A_x}{\partial z} - \frac{\partial A_z}{\partial x}\right)\mathbf{j} + \left(\frac{\partial A_y}{\partial x} - \frac{\partial A_x}{\partial y}\right)\mathbf{k}$$

$\nabla \times \mathbf{A}$ is called the curl of $\mathbf{A}$ (written curlA).

1.9.2.5 Some Identities
- divcurlA $= \nabla \cdot (\nabla \times \mathbf{A}) = 0$.
- curlgrad$\phi = \nabla \times (\nabla\phi) = \mathbf{0}$

.

Example 1.13 A vector field A and a scalar field B are given by $\mathbf{A} = 3xy\mathbf{i} + (2y^2 - x)\mathbf{j}$ and $B = 3x^2y$, Find at the point $(-1,1)$(a) $\nabla \cdot \mathbf{A}$ (b) $\nabla \times \mathbf{A}$ (c) ∇B.

Solution 1.13 (a)

$$\nabla \cdot \mathbf{A} = \frac{\partial A_x}{\partial x} + \frac{\partial A_y}{\partial y} = 3y + 4y = 7y$$

at $(-1, 1)$, $\nabla \cdot \mathbf{A} = 7$.
(b)

$$\nabla \times \mathbf{A} = \begin{vmatrix} \mathbf{i} & \mathbf{j} & \mathbf{k} \\ \frac{\partial}{\partial x} & \frac{\partial}{\partial y} & \frac{\partial}{\partial z} \\ 3xy & (2y^2 - x) & 0 \end{vmatrix} = (-3x - 1)\mathbf{k}$$

at $(-1, 1)$, $\nabla \times \mathbf{A} = 2\mathbf{k}$.
(c)

$$\nabla B = \frac{\partial B}{\partial x}\mathbf{i} + \frac{\partial B}{\partial y}\mathbf{j} + \frac{\partial B}{\partial z}\mathbf{k} = 6xy\mathbf{i} + 3x^2\mathbf{j}$$

at $(-1, 1)$, $\nabla B = -6\mathbf{i} + 3\mathbf{j}$.

1.10 Integrals of Vectors

If $\mathbf{A}(t) = A_x(t)\mathbf{i} + A_y(t)\mathbf{j} + A_z(t)\mathbf{k}$, where t is a scalar variable, the indefinite integral is defined as

$$\int \mathbf{A}(t)dt = \mathbf{i}\int A_x(t)dt + \mathbf{j}\int A_y(t)dt + \mathbf{k}\int A(t)dt$$

If $\mathbf{A}(t) = d\mathbf{B}(t)/dt$, then

$$\int \mathbf{A}(t)dt = \int \frac{d}{dt}\{\mathbf{B}(t)\}dt = \mathbf{B}(t) + \mathbf{C}$$

where $\mathbf{C}$ is an arbitrary constant vector. The definite integral between the limits $t = a$ and $t = b$ is defined as

$$\int_a^b \mathbf{A}(t)dt = \int_a^b \frac{d}{dt}\{\mathbf{B}(t)\}dt = \mathbf{B}(t) + \mathbf{C}|_a^b = \mathbf{B}(b) - \mathbf{B}(a)$$

1.10.1 Line Integrals

The line integral refers to an integral along a line or a curve. This curve may be open or closed. The line integral may

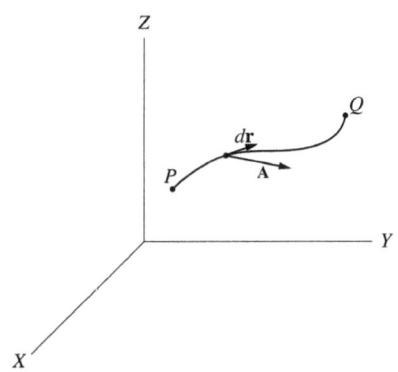

Fig. 1.27 The line integral

appear in three different forms shown by $\int_c \phi d\mathbf{r}$, $\int_c \mathbf{A}$. $d\mathbf{r}$, and $\int_c \mathbf{A} \times d\mathbf{r}$. The second is the most common one and it will be used throughout this book. Suppose the position vector of any point (x, y, z) on the curve C (see Fig. 1.27) that extends from $\mathbf{P}(x_1, y_1, z_1)$ at t_1 to $\mathbf{Q}(x_2, y_2, z_2)$ at t_2 is given by

$$\mathbf{r}(t) = x(t)\mathbf{i} + y(t)\mathbf{j} + z(t)\mathbf{k}$$

where t is a scalar variable, and suppose that $\mathbf{A} = \mathbf{A}(x, y, z) = A_x\mathbf{i} + A_y\mathbf{j} + A_z\mathbf{k}$ is a vector field, then the line integral of $\mathbf{A}$ is given by

$$\int_P^Q \mathbf{A} \cdot d\mathbf{r} = \int_C \mathbf{A} \cdot d\mathbf{r} = \int_C (A_x dx + A_y dy + A dz) \quad (1.2)$$

Note that $\mathbf{A} \cdot \mathbf{r}$ is the tangential component of $\mathbf{A}$ along C. If C is a simple closed curve (does not intersect with itself) then the line integral is written as

$$\oint_C \mathbf{A} \cdot d\mathbf{r} = \oint_C (A_x dx + A_y dy + A dz)$$

1.10.2 Independence of Path

The line integral in general depends on the path, but sometimes it does not. Instead, it depends only on the coordinates of the end points of the curve (path) but not on the curve itself. The line integral in Eq. 1.2 is independent of the path, joining the points P and Q if and only if $\mathbf{A} = \nabla\phi$, or equivalently $\nabla \times \mathbf{A} = \mathbf{0}$. The value of Eq. (1.2) is then given by

$$\int_P^Q \mathbf{A} \cdot d\mathbf{r} = \int_P^Q d\phi = \phi(P) - \phi(Q) = \phi(x_2, y_2, z_2) - \phi(x_1, y_1, z_1)$$

Note that $\phi(x, y, z)$ has continuous partial derivatives. Furthermore, if the line integral of $\mathbf{A}$ is independent of the path then the line integral of $\mathbf{A}$ about any closed path is equal to zero:

$$\oint_C \mathbf{A} \cdot d\mathbf{r} = 0$$

Example 1.14 A force field is given by $\mathbf{F} = (4xy^2 + z^2)\mathbf{i} + (4yx^2)\mathbf{j} + (2xz - 1)\mathbf{k}$
   (a) Show that $\nabla \times \mathbf{F}$,
   (b) Find a scalar function ϕ such that $\mathbf{F} = \nabla\phi$.

Solution 1.14 (a)

$$\nabla \times \mathbf{F} = \begin{vmatrix} \mathbf{i} & \mathbf{j} & \mathbf{k} \\ \frac{\partial}{\partial x} & \frac{\partial}{\partial y} & \frac{\partial}{\partial z} \\ (4xy^2 + z^2) & (4yx^2) & (2xz - 1) \end{vmatrix} = (2z - 2z)\mathbf{j} + (8xy - 8xy)\mathbf{k} = \mathbf{0}$$

(b)

$$\mathbf{F} \cdot d\mathbf{r} = \nabla\phi \cdot d\mathbf{r} = \frac{\partial\phi}{\partial x}dx + \frac{\partial\phi}{\partial y}dy + \frac{\partial\phi}{\partial z}dz = d\phi$$

$$d\phi = (4xy^2 + z^2)dx + (4yx^2)dy + (2xz - 1)dz$$

Hence

$$\phi = (2x^2y^2 + z^2x) + (2y^2x^2) + (z^2x - z)$$

Example 1.15 A vector $\mathbf{F}$ is given by $\mathbf{F} = 3x^2y\mathbf{i} - (4y + x)\mathbf{j}$. Compute $\int_c \mathbf{F} \cdot d\mathbf{r}$ along each of the following paths:
   (a) The straight lines from $(0, 0)$ to $(0, 1)$ and then to $(1, 1)$.
   (b) Along the straight line $y = x$. (c) Along the curve $x = t, y = t^2$.

Solution 1.15 (a) Along the straight line from $(0,0)$ to $(0,1)$ we have $x = 0$, and $dx = 0$, therefore

$$\int_C \mathbf{F} \cdot d\mathbf{r} = \int_C 3x^2y dx - (4y + x)dy = \int_{y=0}^1 -4y dy = -2y^2|_0^1 = -2$$

Along the straight line from $(0, 1)$ to $(1, 1)$ we have $y = 1$, $dy = 1$, hence

$$\int_C \mathbf{F} \cdot d\mathbf{r} = \int_{x=0}^1 3x^2 dx = x^3|_0^1 = 1$$

Thus, we have for the total path

$$\int_C \mathbf{F} \cdot d\mathbf{r} = -2 + 1 = -1$$

(b) Along the straight line $y = x$, we have $dy = dx$,

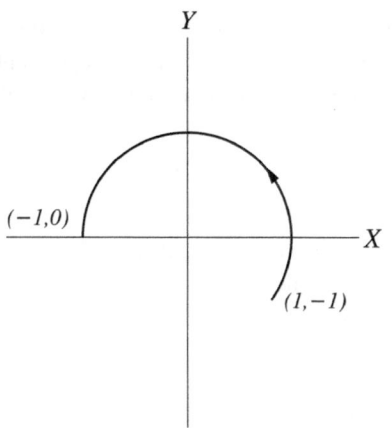

Fig. 1.28 The line integral along the curve using polar coordinates

$$\int_C \mathbf{F} \cdot d\mathbf{r} = \int_C 3x^2 y dx - (4y + x)dy = \int_{x=0}^{1} (3x^3 - 5x)dx$$

$$= 3/4x^4 - 5/2x^2|_0^1 = -3/2.$$

(c) Finally along the curve $x = t, y = t^2$, we have $dx = dt, dy = 2tdt$, furthermore the points $(0, 0)$ and $(1, 1)$ corresponds to $t = 0$ and $t = 1$, respectively. Hence

$$\int_C \mathbf{F} \cdot d\mathbf{r} = \int_C 3x^2 y dx - (4y + x)dy = \int_{t=0}^{1} 3t^4 dt - 2t(4t^2 + t)dt$$

$$= 3/5t^5 - 2t^4 - 2/3t^3|_0^1 = -31/15.$$

Example 1.16 If a vector $\mathbf{A}$ is given by $\mathbf{A} = xy\mathbf{i} - x^2\mathbf{j}$, find the line integral $\int_C \mathbf{A} \cdot d\mathbf{r}$ along the circular arc shown in Fig. 1.28.

Solution 1.16 By using the polar coordinates, we have $x = \cos\theta$ and $y = \sin\theta$ (since $r = 1$), $dx = -\sin\theta d\theta$ and $dy = \cos\theta d\theta$, also $x^2 + y^2 = r^2 = 1$, therefore we have

$$\int_c \mathbf{A} \cdot d\mathbf{r} = \int_{\theta=\pi}^{-\pi/4} -\cos\theta\sin^2\theta d\theta - \cos^3 d\theta = \int_{\theta=\pi}^{-\pi/4} -\cos\theta(\sin^2\theta + \cos^2\theta)d\theta$$

$$= \int_{\theta=\pi}^{-\pi/4} -\cos\theta d\theta = -\sin\theta|_\pi^{-\pi/4} = 0.71$$

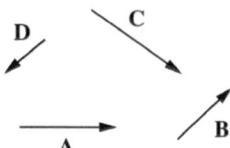

Fig. 1.29 Vectors A, B, C and D

Problems

1. Check if the relation $v = \sqrt{2GM_E/R_E}$ is dimensionally correct, where v represents the escape speed of a body, M_E and R_E are the mass and radius of the earth, respectively, and G is the universal gravitational constant.
2. If the speed of a car is 180 km/h, find its speed in m/s.
3. How many micrometers are there in an area of 3 km^2.
4. Figure 1.29 shows vectors **A**, **B**, **C**, and **D**. Find graphically the following vectors (a) $\mathbf{A} + 2\mathbf{B} - \mathbf{C}$ (b) $2(\mathbf{A} - \mathbf{B}) + \mathbf{C} - 2\mathbf{D}$ (c) show that $(\mathbf{A} + \mathbf{B}) + \mathbf{C} = \mathbf{A} + (\mathbf{B} + \mathbf{C})$.
5. A car travels a distance of 1 km due east and then a distance of 0.5 km north of east. Find the magnitude and direction of the resultant displacement of the car using the algebraic method.
6. Prove that $\mathbf{A} \cdot (\mathbf{B} + \mathbf{C}) = \mathbf{A} \cdot \mathbf{B} + \mathbf{A} \cdot \mathbf{C}$.
7. A parallelogram has sides **A** and **B**. Prove that its area is equal to $|\mathbf{A} \times \mathbf{B}|$.
8. If $\mathbf{A} = 2\mathbf{i} - 3\mathbf{j} + 4\mathbf{k}$ and $\mathbf{B} = \mathbf{i} + 5\mathbf{j} - 2\mathbf{k}$, find (a) $\mathbf{A} - 2\mathbf{B}$ (b) $\mathbf{A} \times \mathbf{B}$ (c) $\mathbf{A} \cdot \mathbf{B}$ (d) the length of **A** and the length of **B** (e) the angle between **A** and **B** (f) the scalar projection of **A** on **B** and the scalar projection of **B** on **A**.
9. Show that **A** is perpendicular to **B** if $|\mathbf{A} + \mathbf{B}| = |\mathbf{A} - \mathbf{B}|$.
10. Given that $\mathbf{A} = 2\mathbf{i} + \mathbf{j} + \mathbf{k}, \mathbf{B} = \mathbf{i} + 3\mathbf{j} - 5\mathbf{k}$ and $\mathbf{C} = 6\mathbf{i} + 3\mathbf{j} + 3\mathbf{k}$, determine which vectors are perpendicular and which are parallel.
11. Use the vectors $\mathbf{A} = \cos\theta\mathbf{i} + \sin\theta\mathbf{j}$ and $\mathbf{B} = \cos\phi\mathbf{i} - \sin\phi\mathbf{j}$ to prove that $\cos(\theta + \phi) = \cos\theta\cos\phi - \sin\theta\sin\phi$.
12. If $\mathbf{A} = 5x^2 y\mathbf{i} + yz\mathbf{j} - 3x^2 z^2\mathbf{k}, \mathbf{B} = 7y^3 z\mathbf{i} - 2zx\mathbf{j} + xz^2 y\mathbf{k}$ and $\phi(x, y, z) = 2z^2 y$, find at $(-1,1,1)$ (a) $\partial(\phi\mathbf{A})/\partial x$ (b) $\partial^2(\mathbf{A} \times \mathbf{B})/\partial z\partial y$ (c) $\nabla\phi$ (d) $\nabla \times (\phi\mathbf{A})$.
13. Evaluate $\nabla \times (r^2\mathbf{r})$ where $\mathbf{r} = x\mathbf{i} + y\mathbf{j} - z\mathbf{k}$ and $r = |\mathbf{r}|$.
14. If $\mathbf{r} = A\cos\omega t\mathbf{i} + A\sin\omega t\mathbf{j}$, show that $d^2\mathbf{r}/dt^2 + \omega^2\mathbf{r} = 0$.
15. A force field is given by $\mathbf{F} = -kx\mathbf{i} - ky\mathbf{j}$, find (a) $\nabla \times \mathbf{F}$ (b) a scalar field ϕ such that $\mathbf{F} = \nabla\phi$ (c) Calculate the line integral along the straight lines from $(0, 0)$ to $(1, 0)$ to $(1, 1)$ and from $(0, 0)$ to $(0, 1)$ to $(1, 1)$. Is the line integral independent of path?

Open Access This chapter is licensed under the terms of the Creative Commons Attribution 4.0 International License (http://creativecommons.org/licenses/by/4.0/), which permits use, sharing, adaptation, distribution and reproduction in any medium or format, as long as you give appropriate credit to the original author(s) and the source, provide a link to the Creative Commons license and indicate if changes were made.

The images or other third party material in this chapter are included in the chapter's Creative Commons license, unless indicated otherwise in a credit line to the material. If material is not included in the chapter's Creative Commons license and your intended use is not permitted by statutory regulation or exceeds the permitted use, you will need to obtain permission directly from the copyright holder.

Open Access This chapter is licensed under the terms of the Creative Commons Attribution 4.0 International License (http://creativecommons.org/licenses/by/4.0/), which permits use, sharing, adaptation, distribution and reproduction in any medium or format, as long as you give appropriate credit to the original author(s) and the source, provide a link to the Creative Commons license and indicate if changes were made.

The images or other third party material in this chapter are included in the chapter's Creative Commons license, unless indicated otherwise in a credit line to the material. If material is not included in the chapter's Creative Commons license and your intended use is not permitted by statutory regulation or exceeds the permitted use, you will need to obtain permission directly from the copyright holder.

Kinematics

2.1 Introduction

Mechanics is the science that studies the motion of objects and can be divided into the following:

1. Kinematics: Describes how objects move in terms of space and time.
2. Dynamics: Describes the cause of the object's motion.
3. Statics: Deals with the conditions under which an object subjected to various forces is in equilibrium.

This chapter is considered with kinematics which answers many questions such as: How long it takes for an apple to reach the ground when it falls from a tree? What is the maximum height reached by a baseball when thrown into air? What is the distance it takes an airplane to take off?

In physics, there are three types of motion: translational, rotational, and vibrational. A block sliding on a surface is in translational motion, a (Merry-go-Round) is an example of rotational motion, and a mass–spring system when stretched and released is in vibrational motion. From here until Chap. 7, the object studied will be treated as a particle (i.e., a point mass with no size). This assumption is possible only if the object moves in translational motion without rotating and by neglecting any internal motions that might exist in the object.

That is, an object can be treated as a particle only if all of its parts move in exactly the same way.

For example, if a man jumps into a pool without rotating by doing a somersault (freezing his body), he can be treated as a particle since all particles in his body will move in exactly the same way. Another example of an object that can be treated as a particle is the Earth in its motion about the Sun. Since the dimensions of the Earth are small compared to the dimensions of its path, it can be considered as a particle. The motion of an object is described either by equations or by graphs. Both ways provide information about the motion; however, equa-tions provide precise information while graphs give greater insight about the motion.

2.2 Displacement, Velocity, and Acceleration

This section will discuss the concepts of displacement, veloc-ity, and acceleration in one dimension. These concepts are essential in analyzing the motion of an object.

2.2.1 Displacement

Consider a car that is treated as a particle moving along the straight-line path shown in Fig. 2.1. The x-axis of a coordinate system is used to describe the position of the car with respect to the origin O, where the points P and Q correspond to the positions x_i at t_i and x_f at t_f, respectively. The position–time graph of this motion is shown in Fig. 2.2. The displacement of the truck is a vector quantity defined as the change in its position during the time interval from t_i to t_f and is given by

$$\Delta x = x_f - x_i$$

Hence displacement is a quantity that depends only on the initial and final positions of the object. The direction of the displacement in one dimension is specified by a plus or minus sign. It is positive if the particle is moving in the positive x direction and negative if the particle is moving in the negative x direction. In two or three dimensions, the displacement is represented by a vector. The SI unit of the displacement is the meter (m).

2.2.2 Average Speed

The average speed of an object is a scalar quantity defined as the total distance traveled divided by the total time:

© The Author(s) 2019
S. Alrasheed, *Principles of Mechanics*, Advances in Science,
Technology & Innovation, https://doi.org/10.1007/978-3-030-15195-9_2

Fig. 2.1 A car that is treated as a particle moving along the straight-line path

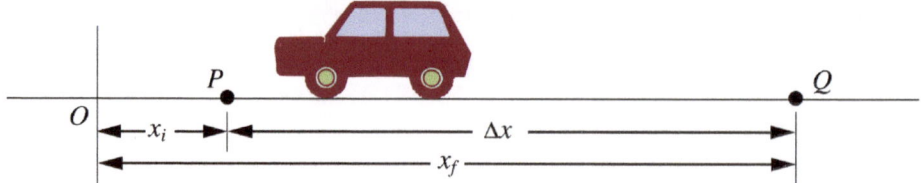

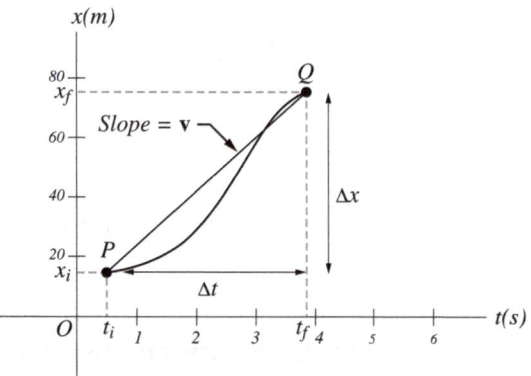

Fig. 2.2 The position time graph of the carõs motion

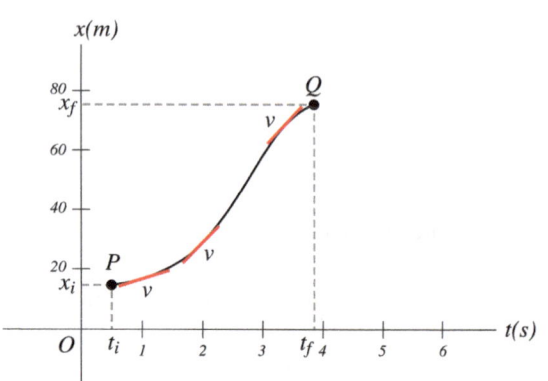

Fig. 2.3 Geometrically, the instantaneous velocity of a particle at a particular time on the position-time curve is the slope (the tangent) to the position-time curve at that point or instance

$$\text{Average speed} = \frac{\text{Total distance traveled}}{\text{Total time}}$$

The SI unit of the average speed is meter per second (m/s)

.

2.2.3 Velocity

The average velocity $\bar{v}$ of an object is a vector quantity defined in terms of displacement rather than the total distance traveled:

$$\bar{v} = \frac{\triangle x}{\triangle t}$$

$\bar{v}$ is positive if the motion is in the positive x-direction and negative if it is in the negative x-direction. On the position–time graph in Fig. 2.2, $\bar{v}$ is the slope of the straight line connecting the points P and Q. The average velocity helps in describing the overall motion of the particle in a certain time interval. To describe the motion in more detail, the instantaneous velocity is defined. This velocity corresponds to the velocity of a particle at a particular time. That involves allowing $\triangle t$ to approach zero:

$$v = \lim_{\triangle t \to \infty} \frac{\triangle x}{\triangle t} = \frac{dx}{dt}$$

Geometrically, the instantaneous velocity of a particle at a particular time on the position–time curve is the slope (the tangent) to the position–time curve at that point or instance (see Fig. 2.3). The SI unit of the velocity is m/s.

2.2.4 Speed

The speed of the particle is defined as the magnitude of its velocity. Note that speed and average speed are different since speed is defined in terms of displacement, whereas average speed is defined in terms of the total distance traveled.

2.2.5 Acceleration

If the particle's velocity changes with time, it is said to be accelerating. The average acceleration $\bar{a}$ of the particle is defined as the ratio of the change of its velocity $\triangle v$ to the time interval $\triangle t$:

$$\bar{a} = \frac{\triangle v}{\triangle t}$$

The SI unit of acceleration is m/s^2. The instantaneous acceleration is defined as

$$a = \lim_{\triangle t \to 0} \frac{\triangle v}{\triangle t} = \frac{dv}{dt}$$

The average acceleration is the slope of the line joining the points P and Q on the velocity–time graph, whereas the instantaneous acceleration is the slope of the curve at a particular point (see Fig. 2.4). Figure 2.5 shows the position, velocity, and acceleration for a particle simultaneously.

Example 2.1 A car travels along the path shown in Fig. 2.6, where it is located at $x_i = 3$ km at $t_i = 0$, and at $x_f = 19$ km

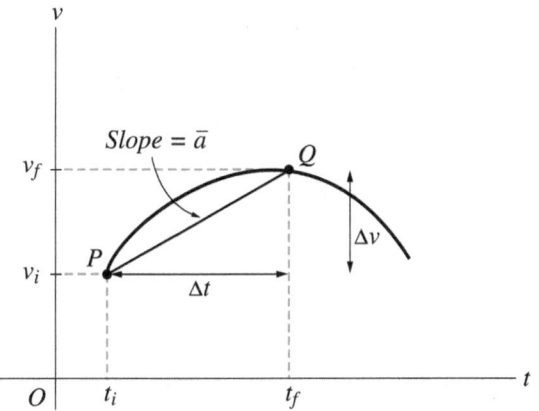

Fig. 2.4 The average acceleration is the slope of the line joining the points P and Q on the velocity-time graph, whereas the instantaneous acceleration is the slope of the curve at a particular point

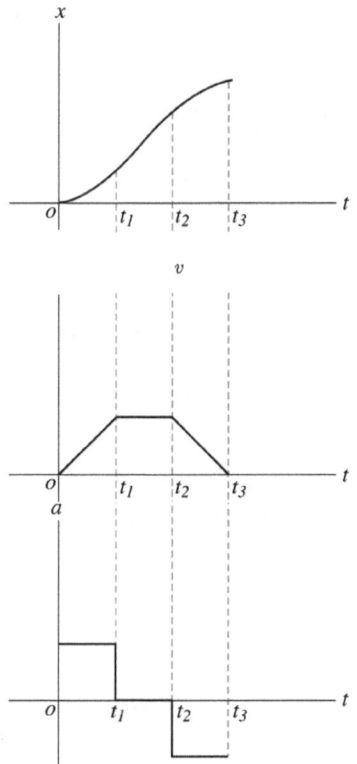

Fig. 2.5 This figure shows the position, velocity and acceleration as a function of time of a particle moving in one direction. The particle starts from rest, accelerates to a certain speed, is maintained at that speed for some time, then it decelerates back to rest

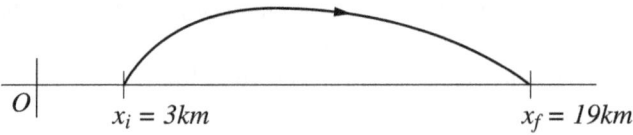

Fig. 2.6 A car moving along the curved path where it is located at $x_i = 3$ km at $t_i = 0$, and at $x_f = 19$ km at $t_f = 0.25$ hr

at $t_f = 0.25$ h. Find the displacement, average velocity, and average speed of the car during this time interval if the total distance traveled is 20 km.

Solution 2.1 The displacement of the car is

$$\Delta x = x_f - x_i = (19 \text{ km}) - (3 \text{ km}) = 16 \text{ km}$$

Its average velocity is

$$\bar{v} = \frac{\Delta x}{\Delta t} = \frac{x_f - x_i}{t_f - t_i} = \frac{(16 \text{ km})}{(0.25 \text{ h})} = 64 \text{ m/s}$$

$$\text{Average speed} = \frac{\text{Total distance traveled}}{\text{Total time}}$$

$$= \frac{(2.0 \text{ km})}{(025 \text{ h})} = 80 \text{ km/h}$$

Example 2.2 A particle moves along the x-axis according to the expression $x = 2t^2$. The plot of this equation is shown in Fig. 2.7. Find : (a) the displacement and average velocity of the particle during the time interval between $t = 1$ s and $t = 3$ s, ·(b) the instantaneous velocity of the particle as a function of time and at $t = 1$ s and $t = 3$ s.

Solution 2.2 (a)

$$x_i = 2t_i^2 = 2(1)^2 = 2 \text{ m}$$

$$x_f = 2t_f^2 = 2(3)^2 = 18 \text{ m}$$

The displacement of the particle is

$$\Delta x = x_f - x_i = (18 \text{ m}) - (2 \text{ m}) = 16 \text{ m}$$

The average velocity is

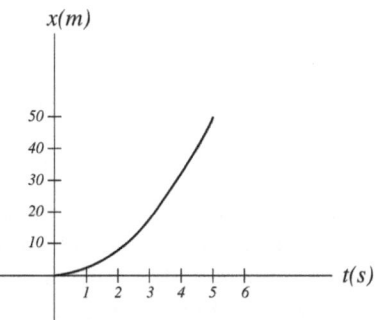

Fig. 2.7 A particle moves along the x-axis according to the expression $x = 2t^2$

$$\bar{v} = \frac{\Delta x}{\Delta t} = \frac{(16\,\text{m})}{(2\,\text{s})} = 8\,\text{m/s}$$

(b) The instantaneous velocity is given by

$$v = \frac{dx}{dt} = (4t)\,\text{m/s}$$

at $t = 1$ s, $v = 2$ m/s, and at $t = 3$ s, $v = 12$ m/s.

Example 2.3 **A** particle is moving along the x-axis. The position–time graph of its motion is shown in Fig. 2.8. Find: (a) the average velocity between a and b, ·(b) the instantaneous velocity at the points a, c and d.

Solution 2.3 (a)

$$\bar{v}_{ab} = \frac{\Delta x}{\Delta t} = \frac{(2\,\text{m}) - (-1.8\,\text{m})}{(3\,\text{s}) - (1\,\text{s})} = 1.9\,\text{m/s}$$

(b)

$$v_a = \frac{\Delta x}{\Delta t} = \frac{0 - (-2.5\,\text{m})}{(3\,\text{s}) - 0} = 0.83\,\text{m/s}$$

$$v_c = 0$$

$$v_d = \frac{\Delta x}{\Delta t} = \frac{0 - (3\,\text{m})}{(8.5\,\text{s}) - (4\,\text{s})} = -0.67\,\text{m/s}$$

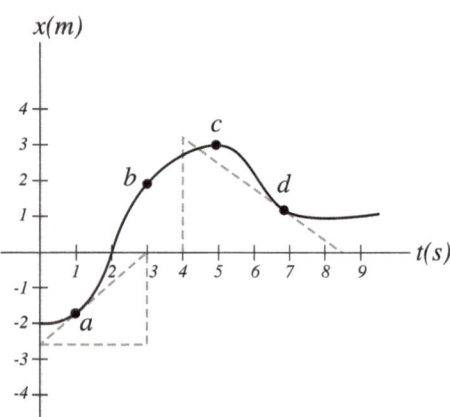

Fig. 2.8 The position-time graph of a particle moving along the x-axis

Example 2.4 **The** acceleration of an object is given by $a = (1-4t)$ m/s^2. If the object has an initial velocity of 3 m/s and an initial displacement of 2 m, determine (a) its velocity and displacement at any time; (b) the displacement of the object when it reaches its maximum speed.

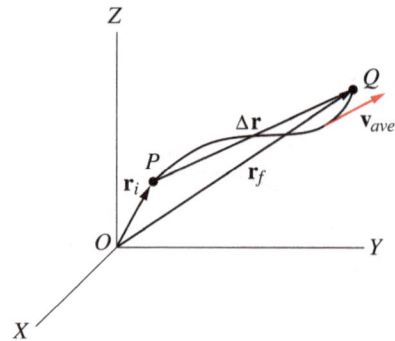

Fig. 2.9 A particle moving from point P to point Q along a path or curve C during a time interval $\Delta t = t_f - t_I$

Solution 2.4 (a)

$$v = \int a\,dt = \int (1 - 4t)dt = t - 2t^2 + c_1$$

At $t = 0$, $v = 3$ m/s and therefore $c_1 = 3$ m/s. Thus

$$v = (t - 2t^2 + 3)\,\text{m/s}$$

$$x = \int v\,dt = \int (t - 2t^2 + 3)dt = 0.5t^2 - 0.66t^3 + 3t + c_2$$

At $t = 0$, $x = 2$ m and $c_2 = 2$ m. Therefore

$$x = (0.5t^2 - 0.66t^3 + 3t + 2)\,\text{m} \qquad (2.1)$$

(b) When the object reaches its maximum speed $\dfrac{dv}{dt} = 0$ and hence $1 - 4t = 0$, that gives $t = 0.25$ s. Substituting into Eq. 2.1 gives

$$x = 1/2(0.25\,\text{s})^2 - 2/3(0.25\,\text{s})^3 + 3(0.25\,\text{s}) + 2 = 2.8\,\text{m}$$

2.3   Motion in Three Dimensions

Consider the particle moving from point P to point Q along a path or curve C during a time interval $\Delta t = t_f - t_i$ as shown in Fig. 2.9. To locate the particle at any point the position vector $\mathbf{r} = x\mathbf{i} + y\mathbf{j} + z\mathbf{k}$ is used. $\mathbf{r}_i$ and $\mathbf{r}_f$ corresponds to the position vectors of the particle at t_i and t_f respectively. A position vector should be drawn from a reference point (usually the origin of the coordinate system).

The displacement vector is then given by

$$\Delta \mathbf{r} = \mathbf{r}_f - \mathbf{r}_i$$

The average velocity is

$$\overline{\mathbf{v}} = \frac{\triangle \mathbf{r}}{\triangle t} = \frac{\mathbf{r}_f - \mathbf{r}_i}{t_f - t_i}$$

The instantaneous velocity at a particular time is defined as

$$\mathbf{v} = \lim_{\triangle t \to 0} \frac{\triangle \mathbf{r}}{\triangle t} = \frac{d\mathbf{r}}{dt}$$

As $\triangle t$ approaches zero, $\triangle \mathbf{r}$ becomes tangent to the path and it is replaced by $d\mathbf{r}$. The direction of y is in the direction of dr, hence, y is always tangent to the path at any point. In terms of components y is given by

$$\mathbf{v} = \frac{dx}{dt}\mathbf{i} + \frac{dy}{dt}\mathbf{j} + \frac{dz}{dt}\mathbf{k} = v_x\mathbf{i} + v_y\mathbf{j} + v_z\mathbf{k}$$

The magnitude of the instantaneous velocity is

$$|\mathbf{v}| = |\frac{d\mathbf{r}}{dt}| = v = \sqrt{\left(\frac{dx}{dt}\right)^2 + \left(\frac{dy}{dt}\right)^2 + \left(\frac{dz}{dt}\right)^2} = \frac{ds}{dt}$$

where ds is the infinitesimal arc length along the path and comes from the fact that as $\triangle t$ approaches zero, the distance traveled by the particle along the path becomes equal to the vector displacement $|\triangle \mathbf{r}|$. Figure 2.10 shows the instantaneous velocities along the path. The average acceleration is

$$\overline{\mathbf{a}} = \frac{\triangle \mathbf{v}}{\triangle t} = \frac{\mathbf{v}_f - \mathbf{v}_i}{t_f - t_i}$$

The direction of $\overline{\mathbf{a}}$ is of the same direction as $\triangle \mathbf{v}$. The instantaneous acceleration is then

$$\mathbf{a} = \lim_{\triangle t \to 0} \frac{\triangle \mathbf{v}}{\triangle t} = \frac{d\mathbf{v}}{dt}$$

In terms of components

$$\mathbf{a} = \frac{dv_x}{dt}\mathbf{i} + \frac{dv_y}{dt}\mathbf{j} + \frac{dv_z}{dt}\mathbf{k} = a_x\mathbf{i} + a_y\mathbf{j} + a_z\mathbf{k}$$

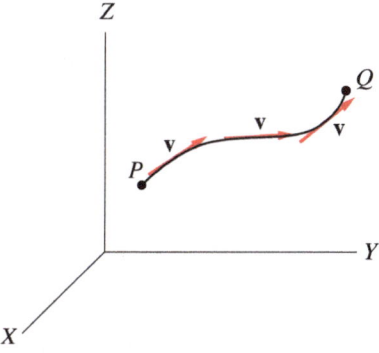

Fig. 2.10 The instantaneous velocity vectors along the path

Another way to describe motion in three dimensions is by using spherical or cylindrical coordinates. In this book, we will only use rectangular coordinates for three-dimensional motion.

2.3.1 Normal and Tangential Components of Acceleration

The acceleration describes the change in both the magnitude and direction of the velocity. That is, the acceleration is not necessarily produced due to the change in the magnitude of the velocity only. Sometimes, it is produced due to the change in the direction of the velocity even if its magnitude is unchanged, and sometimes due to the change in both the magnitude and direction. Furthermore, the direction of $\mathbf{a}$ is not necessarily in the direction of $\mathbf{v}$. If $\mathbf{v}$ is changed in magnitude only (motion along a straight line) then $\mathbf{a}$ is parallel to $\mathbf{v}$ if $\mathbf{v}$ is increasing, and antiparallel if $\mathbf{v}$ is decreasing. If $\mathbf{v}$ is changed in direction only (motion along a curved path with constant speed), then $\mathbf{a}$ is always perpendicular to $\mathbf{v}$ at any point (see Fig. 2.11). Finally, if $\mathbf{v}$ is changed in both magnitude and direction then $\mathbf{a}$ will be directed at some angle to $\mathbf{v}$ as in Fig. 2.12.

In this case, the acceleration can be resolved into parallel and perpendicular components. The parallel component corresponds to the change in the magnitude of $\mathbf{v}$, while the perpendicular component corresponds to the change in the direction of $\mathbf{v}$. These components can be viewed to be directed along a rectangular coordinate system that moves with the particle (as it moves in space), where the particle is located at the origin of this coordinate system. The parallel (or tangential) component of the acceleration is always tangent to the path while the perpendicular (or normal) component is normal to the path at each point as shown in Fig. 2.13.

Figure 2.14 shows the direction of the acceleration of a car moving down a ramp under the influence of gravity.

In terms of unit vectors, let $\mathbf{T}$ be the unit vector along the tangent axis, $\mathbf{N}$ is the unit vector along the normal axis (also called the principal unit normal vector) and $\mathbf{B}$ a third unit vector called the binormal vector defined by $\mathbf{B} = \mathbf{T} \times \mathbf{N}$. These unit vectors form a frame called the TNB frame, where it moves with the particle (see Fig. 2.15). Since $\mathbf{v}$ is always tangent to the path we may write

$$\mathbf{T} = \frac{\mathbf{v}}{|\mathbf{v}|} = \frac{d\mathbf{r}/dt}{|d\mathbf{r}/dt|} = \frac{d\mathbf{r}/dt}{ds/dt}$$

Because $\mathbf{T}$ is a unit vector we have $\mathbf{T} \cdot \mathbf{T} = 1$, differentiating this with respect to s gives

$$\mathbf{T} \cdot \frac{d\mathbf{T}}{ds} + \frac{d\mathbf{T}}{ds} \cdot \mathbf{T} = 2\mathbf{T} \cdot \frac{d\mathbf{T}}{ds} = 0$$

or

Fig. 2.11 If **v** is changed in magnitude only (motion along a straight line) then **a** is parallel to **v** if **v** is increasing, and antiparallel if **v** is decreasing. If **v** is changed in direction only (motion along a curved path with constant speed) then **a** is always perpendicular to **v** at any point

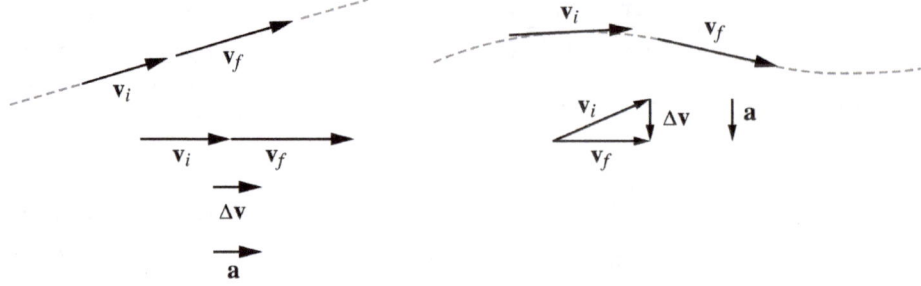

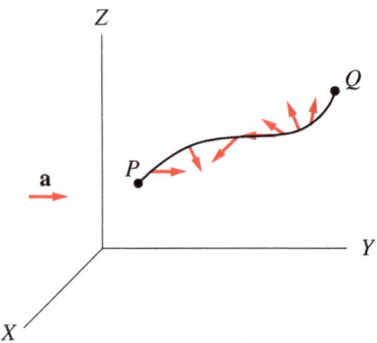

Fig. 2.12 If **v** is changed in both magnitude and direction then **a** will be directed at some angle to **v**

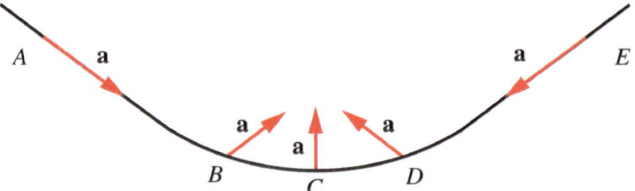

Fig. 2.14 At A the acceleration of a car is in the same direction of the velocity since the latter changes only in magnitude. As it moves its velocity is changed in both magnitude and direction. Therefore at B the direction of the acceleration is at some angle to the velocity. At C the speed reaches a maximum and therefore the instantaneous change of speed is zero at this point and the acceleration has only a perpendicular component. As the car moves up its velocity decreases and changes in direction also, thus the acceleration has both parallel and perpendicular components. Finally at E, the acceleration is in the opposite direction of the velocity since the velocity is decreasing but its direction is the same

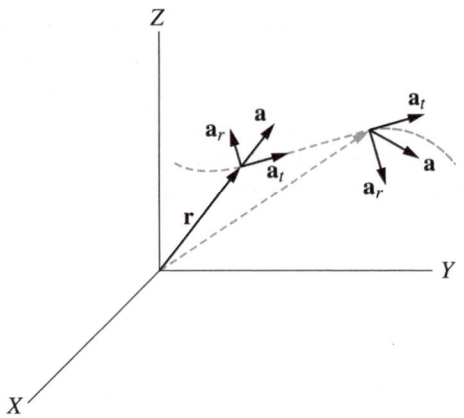

Fig. 2.13 The parallel (or tangential) component of the acceleration is always tangent to the path while the perpendicular (or normal) component is normal to the path at each point

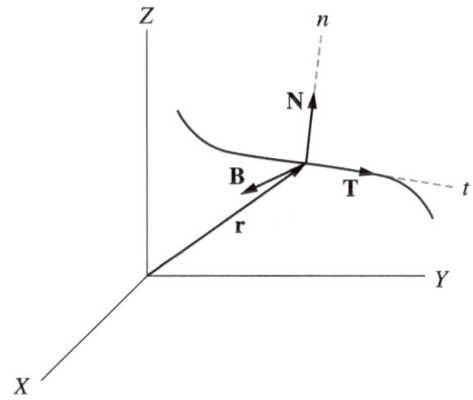

Fig. 2.15 The TNB frame moves with the particle

$$\mathbf{T} \cdot \frac{d\mathbf{T}}{ds} = 0$$

Hence, **T** is perpendicular to $d\mathbf{T}/ds$. Since **N** is also perpendicular to **T**, then we have

$$\mathbf{N} = \frac{d\mathbf{T}/ds}{|d\mathbf{T}/ds|} = \frac{1}{k}\frac{d\mathbf{T}}{ds}$$

k is called the curvature of C at a certain point and it has the value $k = |d\mathbf{T}/ds|$. The quantity $R = 1/k$ is the radius of curvature at that point. Thus, $\mathbf{N} = R(d\mathbf{T}/ds)$. The total

acceleration of the particle in terms of the unit tangent **T** vector and the principal unit normal vector **N** can be written as

$$\mathbf{a} = \frac{d\mathbf{v}}{dt} = \frac{d}{dt}(v\mathbf{T}) = \frac{dv}{dt}\mathbf{T} + v\frac{d\mathbf{T}}{dt} \qquad (2.2)$$

Furthermore,

$$\frac{d\mathbf{T}}{dt} = \frac{d\mathbf{T}}{ds}\frac{ds}{dt} = \frac{\mathbf{N}}{R}\frac{ds}{dt} = \frac{v\mathbf{N}}{R} \qquad (2.3)$$

Substituting Eq. 2.2 into Eq. 2.3 gives

$$\mathbf{a} = \frac{dv}{dt}\mathbf{T} + \frac{v^2}{R}\mathbf{N}$$

Therefore, $a_n = v^2/R$ and $a_t = dv/dt$. Note that unlike $d|\mathbf{v}|/dt$, $|d\mathbf{v}/dt|$ corresponds to the change in the magnitude of the velocity or in its direction or in both (as it represents the magnitude of the total acceleration vector), whereas $d|\mathbf{v}|/dt$ corresponds to the change in the magnitude only.

Example 2.5 A particle is moving in space according to the expression

$$\mathbf{r} = (5\cos t\mathbf{i} + 5\sin t\mathbf{j} + 7t\mathbf{k}) \text{ m}$$

Find the radius of curvature at any point on the space curve.

Solution 2.5

$$\frac{d\mathbf{r}}{dt} = (-5\sin t\mathbf{i} + 5\cos t\mathbf{j} + 7\mathbf{k}) \text{ m/s}$$

$$\frac{ds}{dt} = \left|\frac{d\mathbf{r}}{dt}\right| = \sqrt{(-5\sin t)^2 + (5\cos t)^2 + (7)^2} = 10 \text{ m/s}$$

Hence

$$\mathbf{T} = \frac{d\mathbf{r}/dt}{ds/dt} = \frac{(-5\sin t\mathbf{i} + 5\cos t\mathbf{j} + 7\mathbf{k})}{10} = -0.5\sin t\mathbf{i} + 0.5\cos t\mathbf{j} + 0.7\mathbf{k}$$

The radius of curvature is

$$R = \frac{1}{k} = \frac{1}{|d\mathbf{T}/ds|}$$

$$\frac{d\mathbf{T}}{ds} = \frac{d\mathbf{T}}{dt}\frac{dt}{ds} = \frac{d\mathbf{T}/dt}{ds/dt} = \frac{-0.5\cos t\mathbf{i} - 0.5\sin t\mathbf{j}}{10} = -0.05\cos t\mathbf{i} - 0.05\sin t\mathbf{j}$$

$$\left|\frac{d\mathbf{T}}{ds}\right| = \sqrt{(-005\cos t)^2 + (-005\sin t)^2} = 0.07$$

$$R = \frac{1}{0.07} = 14.3 \text{ m}$$

Example 2.6 A car moves with constant tangential acceleration down a ramp as shown in Fig. 2.16. If it starts from rest at A and reaches B after 4 s with a speed of 10 m/s, find the radius of curvature at B if the total acceleration of the car at that point is 3.2 m/s^2.

Solution 2.6 Since the tangential acceleration of the car is constant, it can be found from

$$a_t = \frac{v_B - v_A}{t} = \frac{(10 \text{ m/s}) - 0}{4 \text{ s}} = 2.5 \text{ m/s}^2$$

Fig. 2.16 A car moving with a constant tangential acceleration down a ramp

Since the total acceleration of the car at B is 2 m/s^2 then the normal acceleration is

$$a_n^2 = a^2 - a_t^2 = (3.2 \text{ m/s}^2)^2 - (2.5 \text{ m/s}^2)^2 = 4 \text{ (m/s}^2)^2$$

$$a_n = 2 \text{ m/s}^2$$

The radius of curvature is

$$R = \frac{v^2}{a_n} = \frac{(10 \text{ m/s})^2}{(2 \text{ m/s}^2)} = 50 \text{ m}$$

2.4 Some Applications

2.4.1 One-Dimensional Motion with Constant Acceleration

An acceleration that does not change with time is said to be a constant or uniform acceleration. In that case, the average and instantaneous accelerations are equal. This type of motion is more easily analyzed than when the acceleration is varied. Since the motion is in one dimension, it follows that the y and z components are zero. That is,

$$\mathbf{r} = x\mathbf{i}$$

$$\Delta\mathbf{r} = (x_f - x_i)\mathbf{i}$$

Hence, as we've mentioned earlier, the direction of the displacement can be specified with a plus or minus sign, as well as the directions of the velocity and acceleration. Let us assume that $t_i = 0$, $t_f = t$, $v_{xf} = v$, $v_{xi} = v_0$, $x_i = x_0$ and $x_f = x$. Since the acceleration is constant, the velocity will vary linearly with time, and thus the average velocity can be expressed as

$$\bar{v} = \frac{v_0 + v}{2}$$

$$a = -= \frac{v_f - v_i}{t_f - t_i} = \frac{v - v_0}{t}$$

$$v = v_0 + at \tag{2.4}$$

$$v = \frac{\Delta x}{\Delta t} = \frac{(v + v_0)}{2}$$

$$x - x_0 = \frac{1}{2}(v + v_0)t \qquad (2.5)$$

Furthermore,

$$x - x_0 = \frac{1}{2}(v + v_0)t = \frac{1}{2}(v_0 + v_0 + at)t$$

$$x - x_0 = v_0 t + \frac{1}{2}at^2 \qquad (2.6)$$

Finally,

$$x - x_0 = \frac{1}{2}(v + v_0)t = \frac{1}{2}(v + v_0)\left(\frac{v - v_0}{a}\right)$$

$$v^2 = v_0^2 + 2a(x - x_0) \qquad (2.7)$$

Equations 2.4, 2.5, 2.6, and 2.7 are called the kinematic equations for motion in a straight line under constant acceleration. The motion graphs for an object moving with constant acceleration in the positive x-direction are shown in Fig. 2.17.

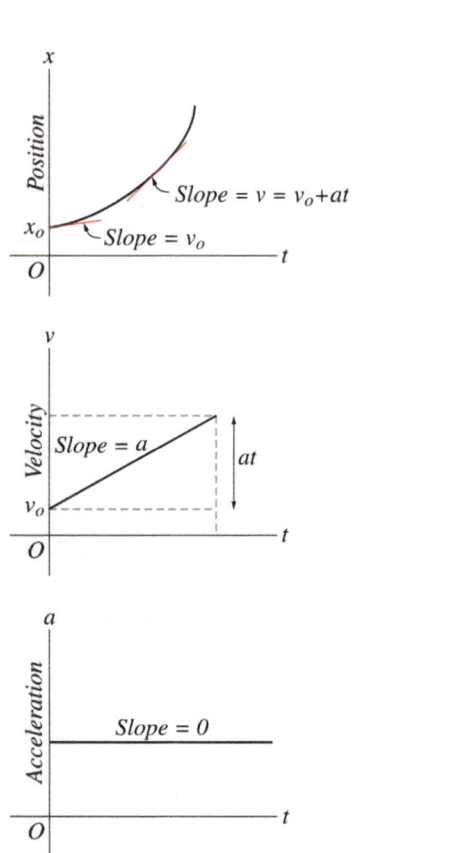

Fig. 2.17 The motion graphs for an object moving with constant acceleration in the positive x-direction

Example 2.7 A train accelerates uniformly from rest and travels a distance of 200 m in the first 8 s. Determine: (a) the acceleration of the train; (b) the time it takes the train to reach a velocity of 70 m/s, (c) the distance traveled during that time; (d) the velocity of the train 5 s later from the time calculated in (b).

Solution 2.7 (a)

$$x - x_0 = v_0 t - \frac{1}{2}at^2$$

Since $v_0 = 0$, we have

$$a = \frac{2(x - x_0)}{t^2} = \frac{2(200 \text{ m})}{(8 \text{ s})^2} = 6.25 \text{ m/s}^2$$

(b)

$$v = v_0 + at$$

$v_0 = 0$ and therefore

$$t = \frac{v}{a} = \frac{(70 \text{ m/s})}{(6.25 \text{ m/s}^2)} = 11.2 \text{ s}$$

(c)

$$x - x_0 = \frac{1}{2}at^2 = \frac{1}{2}(6.25)(11.2)^2 = 392 \text{ m}$$

(d)

$$v = v_0 + at = (70 \text{ m/s}) + (6.25 \text{ m/s}^2)(5 \text{ s}) = 101.25 \text{ m/s}$$

Example 2.8 An airplane accelerates uniformly from rest at a rate of 3 m/s^2 before taking off. If it is to take off at a speed of 100 m/s : (a) how much time is required for it to take off; (b) what distance will it have traveled before taking off?

Solution 2.8 (a)

$$v = v_0 + at$$

We have $v_0 = 0$, this gives

$$t = \frac{v}{a} = \frac{(100 \text{ m/s})}{(3 \text{ m/s}^2)} = 33.3 \text{ s}$$

(b)

$$x = \frac{1}{2}at^2 = \frac{1}{2}(3 \text{ m/s}^2)(33.3 \text{ s})^2 = 1.7 \times 10^3 \text{ m}$$

Example 2.9 A car moving at a constant velocity of 140 km/h passed a police car moving at a constant velocity of 80 km/h. 5 s after the car had passed the police car, the police vehicle begins to accelerate toward the car at a constant rate of 1.4×10^4 km/h^2 (a) How much time will it take the police car to catch the other car? (b) What is the distance traveled by both during that time? (c) How much time has passed from where the car passed the police car to where it was caught?

Solution 2.9 Let's assume that $x = 0$ at where the car passed the police car and that $t = 0$ at the instant the police car begins to accelerate. The velocity of the car is equal to 38.9 m/s, and the initial velocity and acceleration of the police car are 22.2 m/s and 1.1 m/s^2, respectively The police will catch the car when both their displacements from $x = 0$ are equal. (a) From the expression $x = x_0 + v_0 t + \frac{1}{2} a t^2$, the displacement of the car at any time is

$$x_c = x_{0c} + v_{0c} t = (194.5 \text{ m}) + (38.9 \text{ m/s})t$$

The displacement of the police car at any time is

$$x_p = x_{0p} + v_{0p} t + \frac{1}{2} a_p t^2 = (111 \text{ m}) + (22.2 \text{ m/s})t + \frac{1}{2}(1.1 \text{ m/s}^2)t^2$$

The police will catch the car when $x_c = x_p$, and therefore if

$(194.5 \text{ m}) + (38.9 \text{ m/s})t = (111 \text{ m}) + (22.2 \text{ m/s})t + \frac{1}{2}(1.1 \text{ m/s}^2)t^2$ or

$$t^2 - 30.4t - 151.8 = 0$$

Thus

$$t = \frac{(30.4) \pm \sqrt{(304)^2 + (4)(1518)}}{2}$$

That gives $t = 34.8$ s.

(b)

$$x_p = x_c = (111 \text{ m}) + (22.2 \text{ m/s})(34.8 \text{ s}) + \frac{1}{2}(1.1 \text{ m/s}^2)$$

$(34.8 \text{ s})^2 = 1.55 \times 10^3 \text{ m}$

(c)

$$t = (5 \text{ s}) + (34.8 \text{ s}) = 39.8 \text{ s}$$

2.4.2 Free-Falling Objects

Galileo Galilei (1564–1642) was an Italian scientist, who studied and experimented the acceleration of falling objects. By dropping various objects from the Leaning Tower of Pisa (or by releasing objects from inclined planes according to another story), Galileo discovered that when air resistance is neglected then all objects would fall with the same constant acceleration regardless of their mass or size. This acceleration, denoted by g, is known as the free-fall acceleration since air resistance is neglected and the object is assumed to be moving freely under gravity alone. The direction of the vector **g** is downwards toward the earth's center. However, g varies with

altitude as well as other factors which will be discussed in Chap. 9.

In solving problems involving objects falling near the surface of the earth, g can be assumed to be constant with a value of 9.8 m/s^2 and air resistance can be neglected. A free-falling motion is a motion along a straight line (for example along the y-axis) where objects may move upwards or downwards. The kinematics equations of the free-falling motion with constant acceleration can be found from Eqs. (2.4), (2.5), (2.6), and (2.7) by simply replacing x with y and a with g. If the positive direction of y is chosen to be upwards, then the acceleration is negative (downwards) and is given by $(a = -g)$. These substitutions give

$$v = v_0 - gt$$

$$y - y_0 = \frac{1}{2}(v + v_0)t$$

$$y - y_0 = v_0 t - \frac{1}{2}gt^2$$

$$v^2 = v_0^2 - 2g(y - y_0)$$

The displacement and velocity graphs are shown in Fig. 2.18. Note that it does not matter whether the object is falling or moving upward, it will experience the same acceleration g which is directed downwards. Figure 2.19 shows the important features of a free-falling object that is dropped from rest.

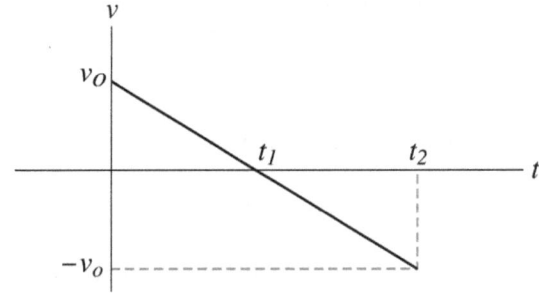

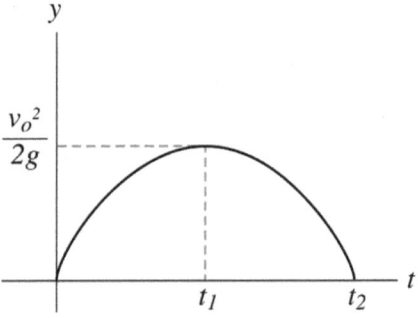

Fig. 2.18 The displacement and velocity graph for a free-falling object

Fig. 2.19 The important features of a free falling object that is dropped from rest

t(s)	y(m)	v(m/s)	a(m/s²)	The velocity vector	The acceleration vector
0	0	0	−9.8		↓
1	−4.9	−9.8	−9.8	↓	↓
2	−19.6	−19.6	−9.8	↓	↓
3	−44.1	−29.4	−9.8	↓	↓

Example 2.10 A ball is thrown directly upwards with an initial velocity of 15 m/s. On its way down, it was caught at a distance of 1m below the point from where it was thrown. Determine (a) the maximum height reached by the ball; (b) the time it takes the ball to reach that height; (c) the velocity of the ball when it is caught; (d) the total time elapsed from where the ball was thrown to where it was caught.

Solution 2.10 (a) First we take $y = 0$ at the position where the ball is thrown and positive y to be upwards. At the maximum height the velocity of the ball is zero,

$$v^2 = v_0^2 - 2g(y - y_0)$$

$$0 = (15 \text{ m/s})^2 - 2(9.8 \text{ m/s}^2)h_{max}$$

$$h_{max} = 11.5 \text{ m}$$

(b) Using the expression $v = v_0 - gt$ we have

$$0 = (15 \text{ m/s}) - (9.8 \text{ m/s}^2)t$$

$$t = 1.5 \text{ s}$$

(c) When the ball is caught its position is $y = -1$ m,

$$v^2 = v_0^2 - 2g(y - y_0)$$

taking the initial position of the ball at $y = 0$, we get

$$v^2 = (15 \text{ m/s})^2 - 2(9.8 \text{ m/s}^2)((-1 \text{ m}) - 0)$$

and

$$v = -15.6 \text{ m/s}$$

or if we take the initial position at $y = 11.5$ m we have

$$v^2 = 0 - 2(9.8 \text{ m/s}^2)((-1 \text{ m}) - (11.5 \text{ m}))$$

and

$$v = -15.6 \text{ m/s}.$$

(d) $v = v_0 - gt$, substituting for v and v_0 we have

$$(-15.6 \text{ m/s}) = (15 \text{ m/s}) - (9.8 \text{ m/s}^2)t$$

$$t = 3.1 \text{ s}$$

Example 2.11 A tennis ball is dropped from a building that is 30 m high. Find (a) its position and velocity 2 s later; (b) the total time it takes the ball to fall to the ground; (c) its velocity just before it hits the ground.

Solution 2.11 (a) Taking $y_0 = 0$ and $v_0 = 0$ at $t = 0$ we have

$$y - y_0 = v_0 t - \frac{1}{2}gt^2$$

at $t = 2$ s

$$y - 0 = 0 - \frac{1}{2}(9.8 \text{ m/s}^2)(2s)^2 = -19.6 \text{ m}$$

$$v = v_0 - gt = 0 - (9.8 \text{ m/s}^2)(2 \text{ s}) = -19.6 \text{ m/s}$$

(b)

$$y - y_0 = v_0 t - \frac{1}{2}gt^2$$

$$(-30 \text{ m}) - 0 = 0 - \frac{1}{2}(9.8 \text{ m/s}^2)t^2$$

$$t = 2.5 \text{ s}$$

(c)

$$v = v_0 - gt = 0 - (9.8 \text{ m/s}^2)(2.5 \text{ s})$$

$$v = -24.5 \text{ m/s}$$

Example 2.12 A ball is thrown vertically downwards from a 100 m high building with an initial speed of 1 m/s. 3 s later a second ball is thrown. What initial speed must the second ball have so that the two balls hit the ground at the same time?

Solution 2.12 The time it takes the first ball to hit the ground is found from

$$y - y_0 = v_0 t - \frac{1}{2} g t^2$$

$$0 - (100 \text{ m}) = (-1 \text{ m/s}) t_1 - \frac{1}{2} (9.8 \text{ m/s}^2) t_1^2$$

$$t_1 = 6.4 \text{ s}$$

The second ball must fall the same distance during a time of

$$t_1 - (3 \text{ s}) = (6.4 \text{ s}) - (3 \text{ s}) = 3.4 \text{ s}$$

and therefore

$$y - y_0 = v_0 t - \frac{1}{2} g t^2$$

$$0 - (100 \text{m}) = v_0 (3.4 \text{ s}) - \frac{1}{2} (9.8 \text{ m/s}^2)(3.4 \text{ s})^2$$

$$v_0 = -12.6 \text{ m/s}$$

2.4.3 Motion in Two Dimensions with Constant Acceleration

The position vector can be written as

$$\mathbf{r} = x\mathbf{i} + y\mathbf{j}$$

$$\mathbf{v} = v_x\mathbf{i} + v_y\mathbf{j}$$

$$\mathbf{a} = a_x\mathbf{i} + a_y\mathbf{j}$$

Because a is a constant both a_x and a_y are constants. Therefore, the kinematic in Sect. 2.4.1 applies in each direction:

$$v_x = v_{0x} + a_x t \tag{2.8}$$

$$x = x_0 + v_{0x} t + \frac{1}{2} a_x t^2 \tag{2.9}$$

$$v_y = v_{0y} + a_y t \tag{2.10}$$

$$y = y_0 + v_{0y} t + \frac{1}{2} a_y t^2 \tag{2.11}$$

$$\mathbf{r} = x\mathbf{i} + y\mathbf{j} = (x_0 + v_{0x} t + \frac{1}{2} a_x t^2)\mathbf{i} + (y_0 + v_{0y} t + \frac{1}{2} a_y t^2)\mathbf{j}$$

$$\mathbf{r} = \mathbf{r}_0 + \mathbf{v}_0 t + \frac{1}{2} \mathbf{a} t^2 \tag{2.12}$$

$$\mathbf{v} = v_x\mathbf{i} + v_y\mathbf{j} = (v_{0x} + a_x t)\mathbf{i} + (v_{0y} + a_y t)\mathbf{j}$$

$$= (v_{0x}\mathbf{i} + v_{0y}\mathbf{j}) + (a_x\mathbf{i} + a_y\mathbf{j})t$$

$$\mathbf{v} = \mathbf{v}_0 + \mathbf{a} t \tag{2.13}$$

Example 2.13 If the motion of a particle in a plane is described by $v_y = (-8t)$ m/s and $x = (5 - 2t^2)$ m : (a) plot the y component of the particle as a function of time if at $t = 0$, $y = 0$, ·(b) find the total speed and magnitude of the acceleration of the particle at $t = 2$ s.

Solution 2.13 (a) The y-component of position is

$$y = \int v_y dt = \int (-8t) dt = -4t^2 + c$$

since at $t = 0$, $y = 0$, then

$$y = (-4t^2) \text{ m}$$

The plot of y against t is shown in Fig. 2.20.

(b) The x-components of velocity and acceleration is

$$v_x = \frac{dx}{dt} = \frac{d(5 - 2t^2)}{dt}$$

$$v_x = (-4t) \text{ m/s}$$

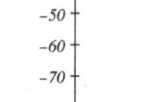

Fig. 2.20 The y component of the particle as a function of time

$$a_x = \frac{dv_x}{dt} = \frac{d(-4t)}{dt}$$

$$a_x = -4 \ \text{m/s}^2$$

The y-component of acceleration is

$$a_y = \frac{dv_y}{dt} = \frac{d(-8t)}{dt}$$

or

$$a_y = (-8) \ \text{m/s}^2$$

at $t = 2$ s, $v_x = -8$ m/s, $v_y = -16$ m/s and the velocity is

$$v = \sqrt{v_x + v_y} = \sqrt{(-8 \ \text{m/s})^2 + (-16 \ \text{m/s})^2} = 17.9 \ \text{m/s}$$

$$a_x = -4 \ \text{m/s}^2$$

and

$$a_y = (-8) \ \text{m/s}^2$$

Therefore, the acceleration of the particle is constant at any time and is given by

$$a = \sqrt{a_x + a_y} = \sqrt{(-4 \ \text{m/s}^2)^2 + (-8 \ \text{m/s}^2)^2} = 8.9 \ \text{m/s}^2$$

2.4.4 Projectile Motion

Projectile motion is the motion of an object thrown (projected) into the air at some angle with respect to the surface of the earth, such as the motion of a baseball thrown into the air or an object dropped from a moving airplane. In the simplified model where air resistance as well as other factors such as the Earth's curvature and rotation are neglected, and if the free-fall acceleration **g** is assumed constant in magnitude and direction throughout the motion of the object, then the path of the projectile is always a parabola that depends on the magnitude and direction of its initial velocity. Therefore, the projectile can be considered as a combination of a vertical motion with a constant acceleration directed downwards and a horizontal motion with zero acceleration (constant velocity). We can see from Fig. 2.21 that

$$\cos \theta_0 = v_{0x}/v_o$$

$$\sin \theta_0 = v_{0y}/v_o$$

At $t = 0$, we have $x_0 = y_0 = 0$ and $v_i = v_0$. Because $a_y = -g$ and $a_x = 0$ and by substituting in Eqs. 2.8, 2.9, 2.10, and 2.11 gives

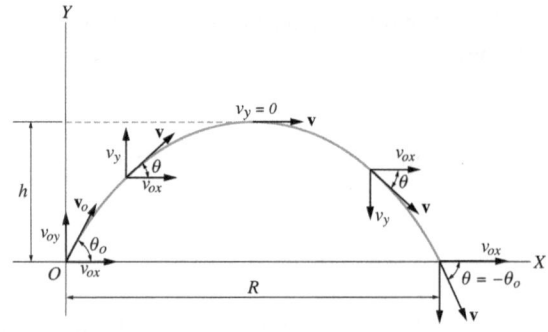

Fig. 2.21 The projectile motion

$$v_x = v_{0x} = v_0 \cos \theta_0 = \text{constant} \tag{2.14}$$

$$v_y = v_{y0} - gt = v_0 \sin \theta_0 - gt \tag{2.15}$$

$$x = v_{x0}t = (v_0 \cos \theta_0)t \tag{2.16}$$

$$y = v_{y0}t - \frac{1}{2}gt^2 = (v_0 \sin \theta_0)t - \frac{1}{2}gt^2 \tag{2.17}$$

Combining and eliminating t from Eqs. 2.16 and 2.17 we find that

$$y = (\tan \theta_0)x - \left(\frac{g}{2v_0^2 \cos^2 \theta_0}\right)x^2$$

$$\left(0 < \theta_0 < \frac{\pi}{2}\right)$$

This equation which is of the form $y = ax - bx2$ (a and b are constants), is the equation of a parabola. Therefore, when air resistance is neglected (when using the simplified model of the system), the trajectory of the projectile is always a parabola. At any instant, the velocity of the object is tangent to its trajectory Its magnitude and direction with respect to the positive x-direction are given by

$$v = \sqrt{v_x^2 + v_y^2}$$

and

$$\theta = \tan^{-1}(v_y/v_x)$$

respectively The maximum height h of the projectile, as in Fig. 2.22 , is found at $t = t_1$ by noting that at the peak h, $v_y = 0$. Substituting this in Eq. 2.15 gives

$$v_0 \sin \theta_0 = gt_1$$

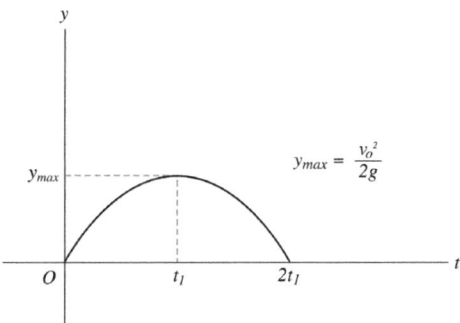

Fig. 2.22 The maximum height of a projectile

$$t_1 = \frac{v_0 \sin \theta_0}{g}$$

Substituting t_1 into Eq. 2.17 we get

$$y_{\max} = h = (v_0 \sin \theta_0)t_1 - \frac{1}{2}gt_1^2$$

$$h = (v_0 \sin \theta_0)\left(\frac{v_0 \sin \theta_0}{g}\right) - \frac{1}{2}g\left(\frac{v_0 \sin \theta_0}{g}\right)^2$$

$$h = \frac{v_0^2 \sin^2 \theta_0}{2g}$$

The maximum range R is at $t = 2t_1$. Substituting t into Eq. 2.16 gives

$$x = R = (v_0 \cos \theta_0)2t_1 = (v_0 \cos \theta_0)\frac{2v_0 \sin \theta_0}{g} = \frac{2v_0^2 \sin \theta_0 \cos \theta_0}{g}$$

$$R = \frac{v_0^2 \sin 2\theta_0}{g}$$

Example 2.14 A baseball is thrown at angle of $35°$ to the horizontal with an initial speed of $20\,\text{m/s}$. Neglecting air resistance, find: (a) the maximum height reached by the ball; (b) the time it takes the ball to hit the ground; (c) the range; and (d) the speed of the ball just before it strikes the ground.

Solution 2.14 (a) The maximum height reached by the ball is

$$h = \frac{v_0^2 \sin^2 \theta_0}{2g} = \frac{(20\ \text{m/s})^2 \sin^2(35°)}{2(9.8\ \text{m/s}^2)} = 6.7\ \text{m}$$

(b) The time it takes the ball to hit the ground is

$$t = 2t_1 = \frac{2v_0 \sin \theta_0}{g} = \frac{2(20\ \text{m/s})\sin(35°)}{(9.8\ \text{m/s}^2)} = 2.34\ \text{s}$$

(c) The range is

$$R = \frac{v_0^2 \sin 2\theta_0}{g} = \frac{(20\ \text{m/s})^2 \sin(70°)}{(9.8\ \text{m/s}^2)} = 38.4\ \text{m}$$

(d) The x-component of the velocity of the ball just before it hits the ground is

$$v_x = v_{0x} = v_0 \cos \theta_0 = (20\ \text{m/s})\cos(35°) = 16.4\ \text{m/s}$$

The y-component is

$$v_y = v_{0y} - gt = v_0 \sin \theta_0 - gt = (20\ \text{m/s})\sin(35°) - (9.8\ \text{m/s}^2)(2.34\ \text{s}) = -11.5\ \text{m/s}$$

Hence, the speed is

$$v = \sqrt{v_x^2 + v_y^2} = \sqrt{(164\ \text{m/s})^2 + (-11.5\ \text{m/s})^2} = 20\ \text{m/s}$$

Example 2.15 A boy throws a ball with a constant horizontal velocity of $1\,\text{m/s}$ at an altitude of $0.6\,\text{m}$. Find the horizontal distance between the releasing point to the point where the ball hits the ground.

Solution 2.15 Let the origin of the reference frame be the releasing point. Since $v_{0y} = 0$ we have

$$y = -\frac{1}{2}gt^2$$

and

$$x = v_{0x}t$$

Hence, when the ball reaches the ground, the elapsed time is

$$t = \sqrt{\frac{-2y}{g}} = \sqrt{\frac{-2(0.6\ \text{m})}{(-9.8\ \text{m/s}^2)}} = 0.34\ \text{s}$$

and

$$x = (1\ \text{m/s})(0.34\ \text{s}) = 0.34\ \text{m}$$

2.4.5 Uniform Circular Motion

A particle moving in a circular path with constant speed is said to be in uniform circular motion. The motion of the moon about earth, and the motion of clothes in a washing machine are examples of uniform circular motion. In this motion, the direction of the velocity of the particle is continuously changing but its magnitude is constant. As we have mentioned in Sect. 2.3.1, when only the direction of the velocity changes, the acceleration is then always perpendicular to the velocity

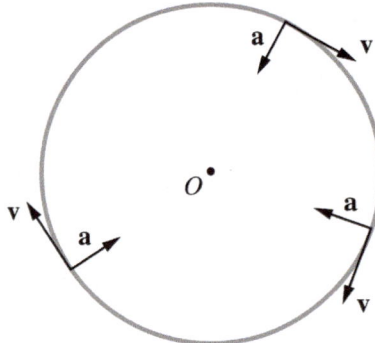

Fig. 2.23 The directions of y and a change continuously with time but their magnitudes are constant

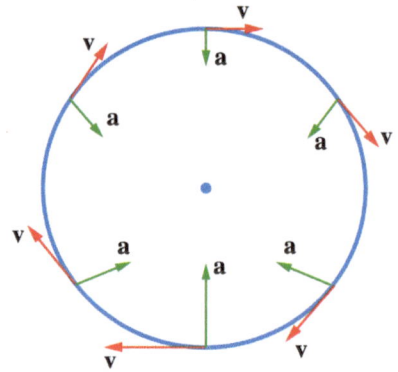

Fig. 2.24 The velocity and total acceleration vectors of a particle moving in a circular path with increasing speed (clockwise) until it reaches the maximum speed at the bottom, and then slows down as it goes back up. An example of this motion is in a roller coaster ride in a vertical circle

at any time. Therefore, we have only the normal component of the acceleration $a_n = v^2/R$, and the tangential component of the acceleration $a_t = dv/dt$ is zero. In the case of the circular path the radius of curvature R is constant, denoted by r, and the normal acceleration is directed along the radius of the circle

$$a_{rad} = \frac{v^2}{r}$$

The subscript *rad* is for radial. Thus, this radial or centripetal acceleration a_{rad} is always directed toward the center of the circle. Therefore, the directions of **v** and a change continuously with time but their magnitudes are constant (see Fig. 2.23). The time required for the particle to complete one revolution around the circle is called the period of revolution and is given by

$$T = \frac{2\pi r}{v}$$

Thus

$$a_{rad} = \frac{4\pi^2 r}{T^2}$$

Example 2.16 In a fun fair ride, the passengers rotate in a circle with a constant speed of 3 m/s. If the period of revolution is 1.5 s, find the total acceleration of the passenger.

Solution 2.16 Since the speed of the passenger is constant, it follows that the passenger's total acceleration is just the centripetal acceleration given by

$$a_{rad} = \frac{v^2}{r}$$

The radius of the circular path is

$$r = \frac{vT}{2\pi} = \frac{(3 \text{ m/s})(1.5 \text{ s})}{2(3.14)} = 0.7 \text{ m}$$

$$a_{rad} = \frac{v^2}{r} = \frac{(3 \text{ m/s})^2}{(0.7 \text{ m})} = 12.86 \text{ m/s}^2$$

2.4.6 Nonuniform Circular Motion

In nonuniform circular motion, the velocity of the particle varies in both magnitude and direction. As mentioned in Sect. 2.3.1, when both the magnitude and direction of the particle's velocity change then its acceleration is directed at some angle to **v**. Thus, in addition to the normal acceleration in uniform circular motion that corresponds to the change in the direction of **v**, there is a tangential component that corresponds to the change in the magnitude of **v**. Furthermore a_{rad} is not constant since **v** changes with time. Therefore, the resultant acceleration is

$$\mathbf{a} = \mathbf{a}_n + \mathbf{a}_t = \frac{v^2}{r}\mathbf{N} + \frac{d|\mathbf{v}|}{dt}\mathbf{T}$$

In Chap. 8, the concepts of angular velocity and acceleration and their vector relationship with the normal and tangential accelerations are introduced. Figure 2.24 shows the velocity and total acceleration vectors of a particle moving in a circular path with increasing speed (clockwise) until it reaches the maximum speed at the bottom, and then slows down as it goes back up. An example of this motion is in a roller coaster ride in a vertical circle.

Example 2.17 A car moving on a circular track of a 20 m radius accelerates uniformly from a speed of 30 km/h to a speed of 50 km/h in 3 s. Find the total acceleration of the car at the instant its speed is 40 km/s.

Solution 2.17 Since both the direction and the magnitude of the car's velocity change, its total acceleration is the vector sum of its tangential and radial accelerations. The tangential acceleration is

$$a_t = \frac{v - v_0}{t} = \frac{(13.8 \text{ m/s}) - (8.3 \text{ m/s})}{(3 \text{ s})} = 1.83 \text{ m/s}^2$$

When $v = 40 \text{ km/h} = 11.1 \text{ m/s}$ the radial acceleration is

$$a_{rad} = \frac{v^2}{r} = \frac{(11.1 \text{ m/s})^2}{(20 \text{ m})} = 6.2 \text{ m/s}^2$$

And the total acceleration is

$$a = \sqrt{(1.83 \text{ m/s}^2)^2 + (6.2 \text{ m/s}^2)^2} = 6.5 \text{ m/s}^2$$

2.5 Relative Velocity

In this section, we will see how observers moving relative to each other obtain different results when measuring the velocity of a moving body. Suppose two cars are moving besides each other at the same speed of 120 km/h with respect to earth. In this case, any of the two cars is at rest relative to the other. According to an observer who is stationary with respect to earth, each car is moving with a speed of 120 km/s. A second observer, in any of the cars, will see the stationary observer moving backwards at a speed of 120 km/h. In addition, if a third car is moving ahead of the two cars at a speed of 140 km/h relative to earth, then its speed relative to an observer in any of the two cars is 20 km/s. Thus, the displacement and velocities may have different values when measured relative to different observers. Therefore, the description of motion depends on the observer. By attaching a coordinate system to an observer together with an appropriate time scale, he or she are then said to be in a reference frame. In measuring quantities, it is essential to specify the reference frame. In most situations, the earth (the lab) is used as our frame of reference. To understand this, consider a particle moving in one dimension in the positive x-direction. Suppose two observers want to describe its motion, one is observer S who is stationary relative to the ground, and the other is observer S′, who is moving in the positive x-direction with a constant velocity relative to the ground (see Fig. 2.25). At any instant, the position of the particle relative to S is x_{PS}, and its position relative to S′ is $x_{PS'}$. The relation between these two observations is

$$x_{PS} = x_{PS'} + x_{S'S} \tag{2.18}$$

Therefore, the position of P relative to O_S is equal to the position of P relative to $O_{S'}$ plus the distance between O_S and $O_{S'}$. Differentiating Eq. 2.18 with respect to time we get

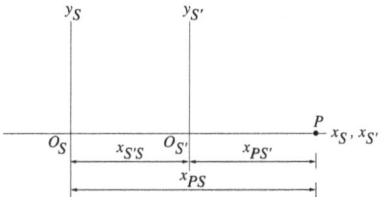

Fig. 2.25 Observer S is stationary relative to the ground, and observer S′ is moving in the positive x-direction with a constant velocity relative to the ground

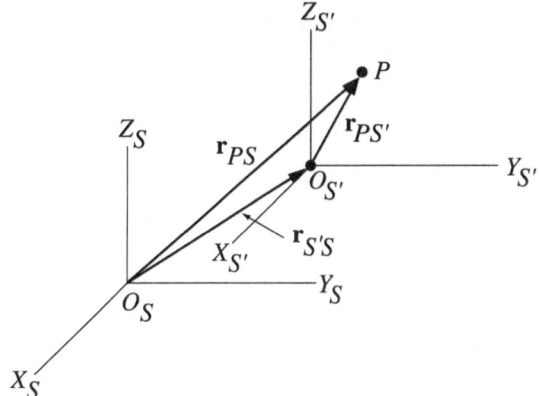

Fig. 2.26 The velocity of S′ with respect to S($v_{S'S}$) is constant in both magnitude and direction

$$\frac{dx_{PS}}{dt} = \frac{dx_{PS'}}{dt} + \frac{dx_{S'S}}{dt}$$

or

$$v_{PS} = v_{PS'} + v_{S'S}$$

We will extend this to three dimensions in the case where the velocity of S′ with respect to S($v_{S'S}$) is constant in both magnitude and direction (see Fig. 2.26). The position vector of the particle P relative to S is given by

$$\mathbf{r}_{PS} = \mathbf{r}_{PS'} + \mathbf{r}_{S'S} \tag{2.19}$$

Differentiating this with respect to time gives

$$\mathbf{v}_{PS} = \mathbf{v}_{PS'} + \mathbf{v}_{S'S} \tag{2.20}$$

Equations 2.19 and 2.20 are called the Galilean transformation equations. In addition, for any two frames of reference S and S we have

$$\mathbf{v}_{SS'} = -\mathbf{v}_{S'S}$$

Example 2.18 Two motor cyclists A and B are driving along the same road (See Fig. 2.27) with speeds 90 km/h and 50 km/s, respectively. Determine: (a) the velocity of motorcyclist A relative to B and of B relative to A?, · and (b) if the two motor cyclists approach each other along two parallel roads, (See Fig. 2.28), A moving at 80 km/s, and B moving at 60 km/s, what is the velocity of motorcyclist A relative to B and of B relative to A.

Fig. 2.27 Two motor cyclists A and B driving with speeds 90 km/h and 50 km/s respectively

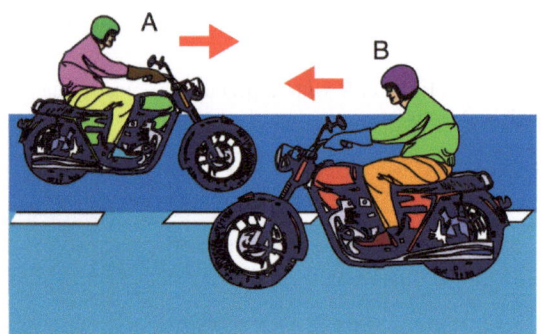

Fig. 2.28 A is moving at 80 km/s, and B moving at 60 km/s

Solution 2.18 Using the above discussion, consider S as the Earth's frame of reference denoted E, S′ as the frame of reference of motorcyclist B and the point P as the motor cyclist A

(a) The velocity of A relative to B is found from

$$v_{AB} = v_{AE} - v_{BE} = (90 \text{ km/h}) - (50 \text{ km/h}) = 40 \text{ km/h}$$

The velocity of B relative to A is

$$v_{BA} = -40 \text{ km/h}$$

(b)

$$v_{AB} = v_{AE} - v_{BE} = (80 \text{ km}) - (-60 \text{ km/h}) = 140 \text{ km/h}$$

$$v_{BA} = -140 \text{ km/h}$$

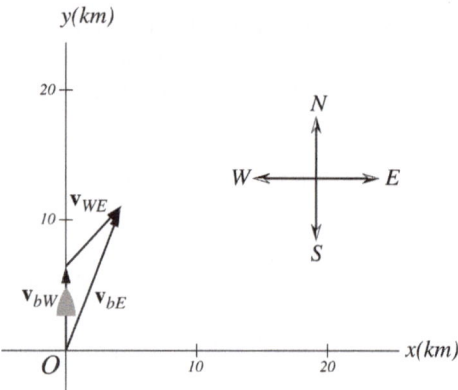

Fig. 2.29 A boat is traveling at 8 km/h north relative to the sea's waves, and the waves are traveling northeast relative to the earth at a constant speed of 4 km/h

Example 2.19 A boat is traveling at sea at 8 km/h north relative to the sea's waves, and the waves are traveling northeast relative to the earth at a constant speed of 4 km/h. What is the velocity of the boat relative to the earth?

Solution 2.19 Using Fig. 2.26, consider the Earth as S (denoted E), the waves as S′, and the boat as the point P. As we can see from Fig. 2.29, the velocity of the boat relative to the earth is given by $\mathbf{v}_{bE} = \mathbf{v}_{bw} + \mathbf{v}_{wE}$, where $\mathbf{v}_{bw}$ and $\mathbf{v}_{wE}$ are the velocities of the boat relative to the waves and the velocity of the waves relative to the earth respectively With the east as the direction of the positive x-axis we get

$$v_{(bE)y} = v_{(bw)y} + v_{(wE)y} = (8 \text{ km/h}) + (4 \text{ km/h}) \sin 45° = 10.83 \text{ km/h}$$

$$v_{(bE)x} = v_{(wE)x} = (4 \text{ km/h}) \cos 45° = 2.83 \text{ km/h}$$

Hence

$$v_{bE} = \sqrt{(v_{(bE)x})^2 + (v_{(bE)y})^2} = \sqrt{(10.83 \text{ km/h})^2 + (2.83 \text{ km/h})^2} = 11.2 \text{ km/h}$$

The direction of $\mathbf{v}_{bE}$ is

$$\theta = \tan^{-1} \frac{(v_{bE})_y}{(v_{bE})_x} = \tan^{-1} \frac{(10.83 \text{ km/h})}{(2.83 \text{ km/h})} = 75.35°$$

2.6   Motion in a Plane Using Polar Coordinates

Consider a particle moving in the x–y plane. A useful way to describe the position, velocity, and acceleration of the particle is by using its polar coordinates (r, θ). The relationship between the polar and rectangular coordinates is

Fig. 2.30 r_1 is a unit vector along the increasing r direction and θ_1 is a unit vector in the direction of increasing θ (anticlockwise direction)

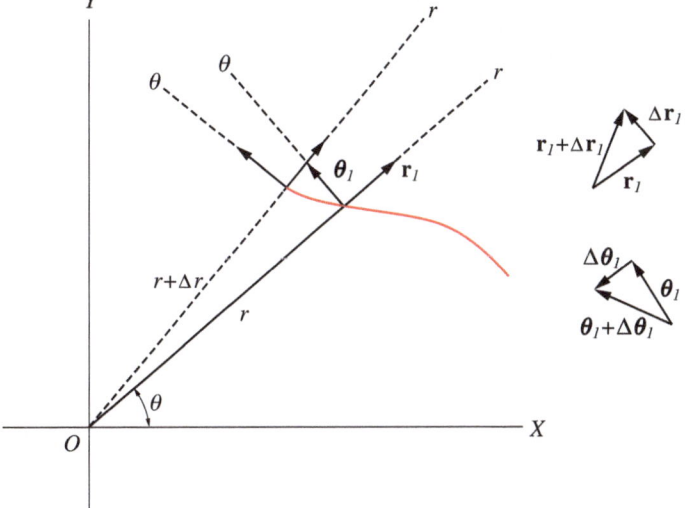

$$x = r\cos\theta$$

$$y = r\sin\theta$$

where θ is measured from the positive x- axis. Suppose a particle is located at (r, θ). If the particle moves in a straight line along the r direction, then θ is constant through the motion of the particle. If the particle moves in a circle, then r is constant. Let r_1 be a unit vector along the increasing r direction and θ_1 to be a unit vector in the direction of increasing θ (anticlockwise direction). From Fig. 2.30, we have

$$\mathbf{r_1} = \cos\theta\mathbf{i} + \sin\theta\mathbf{j}$$

and

$$\boldsymbol{\theta_1} = -\sin\theta\mathbf{i} + \cos\theta\mathbf{j}$$

Unlike the rectangular unit vectors, the polar unit vectors are not fixed in direction. Their direction changes as the particle moves along some path. Therefore, when finding the velocity and acceleration of a particle the derivatives of the polar unit vectors must be considered. The position vector of the particle is given by

$$\mathbf{r} = r\mathbf{r_1}$$

To find the velocity in terms of the polar unit vectors let us differentiate $\mathbf{r_1}$ and $\boldsymbol{\theta_1}$ with respect to time. That gives

$$\dot{\mathbf{r}}_1 = \frac{d\mathbf{r_1}}{dt} = -\sin\theta\frac{d\theta}{dt}\mathbf{i} + \cos\theta\frac{d\theta}{dt}\mathbf{j}$$

$$= \boldsymbol{\theta_1}\frac{d\theta}{dt} = \dot{\theta}\boldsymbol{\theta_1}$$

$$\dot{\boldsymbol{\theta}}_1 = \frac{d\boldsymbol{\theta_1}}{dt} = -\cos\theta\frac{d\theta}{dt}\mathbf{i} - \sin\theta\frac{d\theta}{dt}\mathbf{j}$$

$$= -\mathbf{r_1}\frac{d\theta}{dt} = -\dot{\theta}\mathbf{r_1}$$

The velocity of the particle is given by

$$\mathbf{v} = \frac{d\mathbf{r}}{dt} = \frac{d}{dt}(r\mathbf{r_1}) = \frac{dr}{dt}\mathbf{r_1} + r\frac{d\mathbf{r_1}}{dt}$$

$$= \dot{r}\mathbf{r_1} + r\dot{\mathbf{r}}_1 = \dot{r}\mathbf{r_1} + r\dot{\theta}\boldsymbol{\theta_1}$$

Hence, the velocity is (Fig. 2.31)

$$\mathbf{v} = \dot{r}\mathbf{r_1} + r\dot{\theta}\boldsymbol{\theta_1} \qquad (2.21)$$

We may write

$$\mathbf{v} = v_r\mathbf{r_1} + v_\theta\boldsymbol{\theta_1}$$

where $v_r = \dot{r}$ and $v_\theta = r\dot{\theta}$ and $v = \sqrt{v_r^2 + v_\theta^2}$. The total acceleration is

$$\mathbf{a} = \frac{d\mathbf{v}}{dt} = \frac{d}{dt}(\dot{r}\mathbf{r_1} + r\dot{\theta}\boldsymbol{\theta}_1) = \ddot{r}\mathbf{r_1} + \dot{r}\dot{\mathbf{r}}_1 + \dot{r}\dot{\theta}\boldsymbol{\theta}_1 + r\ddot{\theta}\boldsymbol{\theta}_1 + r\dot{\theta}\dot{\boldsymbol{\theta}}_1$$

$$= \ddot{r}\mathbf{r_1} + \dot{r}(\dot{\theta}\boldsymbol{\theta_1}) + \dot{r}\dot{\theta}\boldsymbol{\theta_1} + r\ddot{\theta}\boldsymbol{\theta_1} + r\dot{\theta}(-\dot{\theta}\mathbf{r_1})$$

$$\mathbf{a} = (\ddot{r} - r\dot{\theta}^2)\mathbf{r_1} + (r\ddot{\theta} + 2\dot{r}\dot{\theta})\boldsymbol{\theta_1} \qquad (2.22)$$

or

$$\mathbf{a} = a_r\mathbf{r_1} + a_\theta\boldsymbol{\theta_1}$$

where

$$a_r = (\ddot{r} - r\dot{\theta}^2)$$

Fig. 2.31 Unlike the rectangular
unit vectors, the polar unit vectors
are not fixed in direction. Their
direction changes as the particle
moves along some path

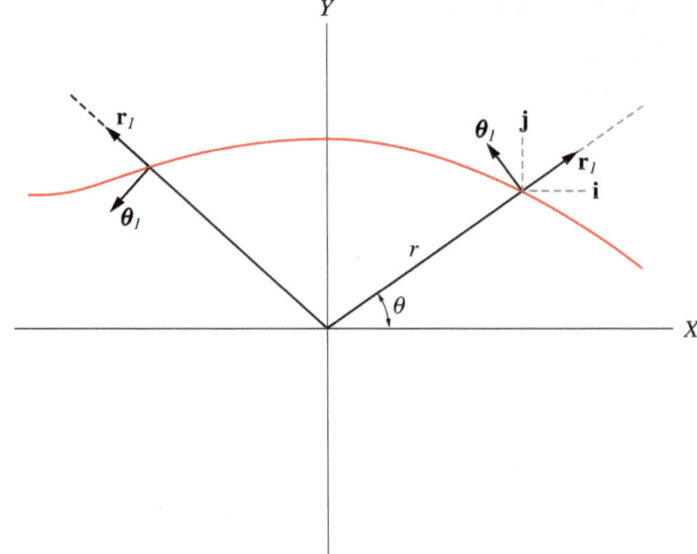

and

$$a_\theta = (r\ddot{\theta} + 2\dot{r}\dot{\theta})$$

and

$$a = \sqrt{a_r^2 + a_\theta^2}$$

Example 2.20 If a particle moves in a plane according to the
expressions $\theta = 0.3t + 0.2t^2$ and $r = 0.5t + 0.4t^2$. Find its
velocity and acceleration at $t = 2$ s

Solution 2.20 At $t = 2$ s, $\theta = 0.3t + 0.2t^2 = 1.4$ rad,
$\dot{\theta} = 0.3 + 0.4t = 1.1$ rad/s and $\ddot{\theta} = 0.4$ rad/s^2. Also
$r = 0.5t + 0.4t^2 = 2.6$ m, $\dot{r} = 0.5 + 0.8t = 2.1$ m/s and
$\ddot{r} = 0.8$ m/s^2. Therefore

$$v_r = \dot{r} = 2.1 \text{ m/s}$$

$$v_\theta = r\dot{\theta} = (2.6 \text{ m})(1.1 \text{rad/s}) = 2.9 \text{ m/s}$$

$$v = \sqrt{v_r^2 + v_\theta^2} = \sqrt{(2.1 \text{ m/s})^2 + (2.9 \text{ m/s})^2} = 3.6 \text{ m/s}$$

and

$$a_r = \ddot{r} - r\dot{\theta}^2 = (0.8 \text{ m/s}^2) - (2.6 \text{ m})(1.1 \text{rad/s})^2 = -2.35 \text{ m/s}^2$$

$$a_\theta = r\ddot{\theta} + 2\dot{r}\dot{\theta} = (2.6 \text{ m})(0.4 \text{ rad/s}^2) + 2(2.1 \text{ m/s})(1.1 \text{ rad/s}) = 5.7 \text{ m/s}^2$$

$$a = \sqrt{a_r^2 + a_\theta^2} = \sqrt{(-2.35 \text{ m/s}^2)^2 + (5.7 \text{ m/s}^2)^2} = 6.2 \text{ m/s}^2$$

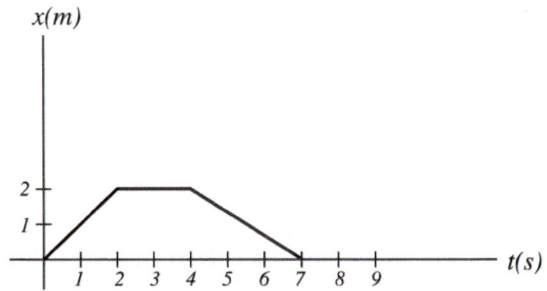

Fig. 2.32 An object moving in one dimension along the x-axis

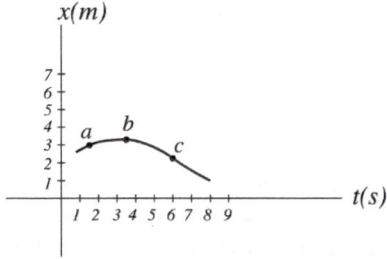

Fig. 2.33 The position-time graph of a particle moving along the x-axis

Problems

1. A sports car moves around a circular track of radius of
   100 m. If the car makes one round in 75 s, find the car's
   (a) average speed (b) average velocity.
2. An object is moving in one dimension along the x-
   axis according to Fig. 2.32. Describe the motion of the
   object.

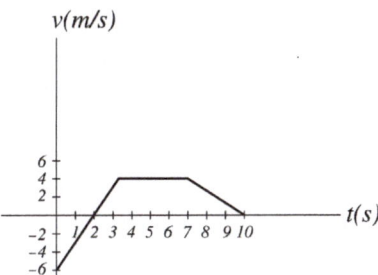

Fig. 2.34 The speed of a motorcyclist varying with time

Fig. 2.35 A car moves at a constant speed of 40 km/h along curved path

Fig. 2.36 An aircraft tracked by a radar coordinates

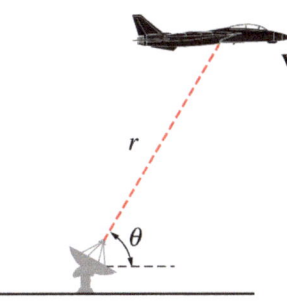

3. The position–time graph of a particle moving along the x-axis is shown in Fig. 2.33. Find (a) the average velocity between a and b(b) the instantaneous velocity at a, b, and c.

4. A motorist drives along a straight-line road. His speed varies with time according to Fig. 2.34. Sketch the position versus time and acceleration versus time graphs of the motorist.

5. A particle moves along the curve defined by $x = 5e^{-t}$ and $y = \sin 5t$. Find the position, velocity and acceleration of the particle at any time.

6. A car moves at constant speed of 40 km/h along the road shown in Fig. 2.35. If the radius of curvature at A is 350 m and the total acceleration of the car at B is 1m/s², find (a) the total acceleration of the car at A and C(b) the radius of curvature at B.(Hint: the radius of curvature at C is infinite).

7. A body with initial speed of 15 m/s undergoes a uniform acceleration of -2 m/s². Find the elapsed time and the distance it traveled when it reaches a speed of 3 m/s.

8. A stone is thrown downwards from a height of 10 m. Find its initial speed if it reaches the ground after 1s.

9. A block is thrown horizontally from the top of a cliff that is 30 m high with a speed of 10 m/s. Find (a) the block's magnitude of displacement from the origin and its velocity after 1.5 s, (b) the horizontal distance from the releasing point to where the block hits the ground.(Hint: the magnitude of displacement from the origin is $d = \sqrt{x^2 + y^2}$).

10. A river has a uniform speed of 0.5 m/s due east. If a boat travels east at a speed of 3 m/s relative to the water, find the time it takes the boat to travel a distance of 1100 km and return to its starting point.

11. An aircraft is tracked by a radar (see Fig. 2.36). If at a certain instant the radar measurements give $r = 7 \times 10^4$ m, r = 1000 m/s, $\ddot{r} = 7$ m/s², $\theta = 45^\circ$, $\dot{\theta} = 0.6$ deg /s, and $\ddot{\theta} = 0.02$ deg /s². Find the velocity and acceleration of the airplane at that instant.

Open Access This chapter is licensed under the terms of the Creative Commons Attribution 4.0 International License (http://creativecommons.org/licenses/by/4.0/), which permits use, sharing, adaptation, distribution and reproduction in any medium or format, as long as you give appropriate credit to the original author(s) and the source, provide a link to the Creative Commons license and indicate if changes were made.

The images or other third party material in this chapter are included in the chapter's Creative Commons license, unless indicated otherwise in a credit line to the material. If material is not included in the chapter's Creative Commons license and your intended use is not permitted by statutory regulation or exceeds the permitted use, you will need to obtain permission directly from the copyright holder.

3.1 Introduction

In this chapter, dynamics which is a branch of mechanics will be discussed. Dynamics is concerned with the cause behind the motion of objects and answers questions such as: Why does a skydiver float in air? What makes an apple fall from a tree? Why a block connected to a spring oscillates when the spring is stretched? We will find that these motions occur when objects interact with each other, i.e., the apple is interacting with earth, the skydiver is interacting with air, and the block is interacting with the spring.

3.1.1 The Concept of Force

The interaction between one object and another or between the object and its environment defines a quantity called force. A force is a pull or a push in a certain direction that may cause the object to move or deform. However, motion does not always occur if the force is not large enough to overcome other forces such as friction or gravity But whether or not an object moves due to a force, there is always some deformation. In this book, it is assumed that objects remain undeformed under the influence of any forces. Experimentally, a force is found to be a vector quantity The net external force acting on an object (the vector sum of all forces acting on the object) causes the object to accelerate where the direction of the acceleration is in the direction of that force.

Hence, acceleration is a measure of force. If the net force equals zero, the acceleration of the object is zero, and the velocity of the object remains unchanged (constant). Forces in nature are one of two:

1. *Contact forces* resulting from direct contact between two objects (e.g., kicking a ball or punching a bag);
2. *Field forces* that can act through empty space and in which physical contact is not necessary (e.g., gravitational force between two objects and the electric force between two electric charges).

3.1.2 The Fundamental Forces in Nature

The following fundamental forces are all field forces:

1. The gravitational force between any two objects;
2. The electromagnetic force between two electric charges;
3. The strong nuclear force between subatomic particles which is responsible for the stability of the nuclei;
4. The weak nuclear force which produces certain kinds of radioactive decay and is responsible for the instability of some nuclei.

The first two fundamental forces are examples of long-range forces, which act over a great distance. The second two are examples of short-range forces, which are forces that act over a very short distance. Note that contact forces are fundamentally electromagnetic since they involve electromagnetic forces between the atoms of the surfaces in contact.

3.2 Newton's Laws

Sir Isaac Newton (1642–1727) formulated his three famous laws of motion describing the relationship between the force acting on an object and the acceleration of that object. Newtonian or classical mechanics which is based mainly on Newton's three laws of motion, deals only with objects that are

- Large compared to the size of an atom ($\approx 10^{-10}$ m).
- Moving at speeds much less than the speed of light ($\approx 3 \times 10^8$ m/s).

Einstein's special theory of relativity replaces Newtonian mechanics when an object's speed approaches the speed of light. On the other hand, quantum mechanics replaces Newtonian mechanics when the object's dimensions are close to atomic scale.

© The Author(s) 2019
S. Alrasheed, *Principles of Mechanics*, Advances in Science,
Technology & Innovation, https://doi.org/10.1007/978-3-030-15195-9_3

3.2.1 Newton's First Law

It was believed long ago that a force is necessary to keep an object moving and that any object's natural state is to be at rest. Later, these statements were proved to be incorrect. To understand this, suppose a block resting on a surface is given a push and is released. As a result, the block will slide for sometime before coming to rest. The time elapsed between pushing the block until it comes to rest will increase as the surface gets smoother. If the surface becomes so smooth, such that friction is almost negligible, the block will continue to move along a straight line with constant speed for a greater distance before coming to rest.

An example of frictionless motion is the motion of the puck in the air-hockey table. The puck floats on a thin column of air that is used as the lubricant. In situations where there is no friction at all, the object will continue to move along a straight line with constant speed without requiring any force to keep it moving. However, a force is required to initiate motion. This concept was formulated by Newton and became his first law of motion: *An object at rest remains at rest and an object in motion will continue in motion with constant velocity (constant speed in a straight line) unless acted upon by a net external force.* That is, if

$$\Sigma \mathbf{F} = 0$$

$$\mathbf{a} = 0$$

A body's tendency to stay at rest or maintain uniform motion in a straight line is called inertia. Thus, Newton's first law is often referred to as the law of inertia, where it defines specific kinds of reference frames called inertial reference frames. An inertial reference frame is a frame in which Newton's first law is valid. That is, in an inertial frame of reference, an object has no acceleration if there is no net force acting upon it. Any reference frame moving with constant velocity relative to an inertial frame is also inertial. Observers in different inertial frames measure the same acceleration for a moving object. To prove this, consider the two inertial reference frames S and S′ mentioned in Sect. 2.5, where S is stationary and S′ is moving with constant velocity relative to S. By differentiating Eq. 2.20, we have

$$\frac{d\mathbf{v}_{PS}}{dt} = \frac{d\mathbf{v}_{PS'}}{dt} + \frac{d\mathbf{v}_{S'S}}{dt}$$

Because $\mathbf{v}_{S'S}$ is constant we have

$$\mathbf{a}_{PS} = \mathbf{a}_{PS'}$$

That is, the acceleration of the particle P measured from both inertial reference frames S and S′ is the same. To show that

Newton's first law is only valid when applied with respect to an inertial frame of reference; consider a girl named Mia that is at rest while watching her friend Lea driving a car moving at constant velocity. Lea has her seatbelt fastened and put her suitcase in the seat right next to her without restraining it. Now, suppose that Lea steps on the brakes, which would cause her vehicle to decelerate, her suitcase will start to move forward. According to Lea, who is in an accelerated frame, the suitcase moved from rest even though there was no apparent net external force acting on it. Therefore, in Lea's frame, Newton's first law seems to be incorrect.

The situation, however, is different to Mia, who is in an inertial frame of reference. In her perspective, the suitcase was initially moving with constant velocity and the net force on it was zero. When the car started to decelerate, the net force on the suitcase is still equal to zero and thus the suitcase must continue to move forward with constant velocity and stop by friction or impact with the inside of the car. Therefore, it is apparent to Mia that Newton's first law is valid. From the previous example, we conclude that Newton's first law (and in general Newton's laws) is not valid in all kinds of reference frames; it is only valid when applied with respect to inertial frames. That is, Lea must not apply Newton's first law in her reference frame.

The same situation would be observed by Lea if she were to turn her car while moving. When the car turns, the suitcase will start to move in the direction opposite to the turn. Once again, Lea observes that the suitcase has moved from rest without any apparent force acting on it which contradicts Newton's first law in her opinion. Mia sees no contradiction with Newton's first law because when the car turns, the suitcase tends to continue its initial uniform straight line motion, and thus it moves toward the direction opposite to the turn. Therefore Newton's laws are obeyed by objects when observed from inertial frames of reference (see Fig. 3.1).

To apply classical (Newtonian) mechanics with respect to a noninertial reference frame, new forces named as pseudo forces are introduced. In this book, only inertial frames are used, and all laws are stated with respect to those frames. One convenient inertial frame of reference, used throughout this book, is the surface of the earth. The earth can be considered as an inertial frame since its motion about its axis and about the sun has a small effect on calculations and thus can be neglected.

3.2.2 The Principle of Invariance

Some quantities such as mass, force, time, and acceleration are invariant, which means that they have the same numerical values when measured in different inertial frames of reference. Other quantities such as velocity, kinetic energy, and work have different values in different inertial frames. How-

Fig. 3.1 The boy is throwing the water out by pitching the bucket forward. If he stops, the water will continue its motion along a straight line. However, because of the force of gravity, it follows a parabolic path

ever, the laws of physics have the same form in all inertial frames of reference. This is called the principle of invariance.

3.2.3 Mass

As mentioned earlier, the tendency of an object to resist any change in its motion (i.e., to remain at rest or maintain uniform motion along a straight line) is called inertia. From experiments and everyday experience, it is observed that a certain force produces different accelerations when applied to different bodies. This variation in the produced acceleration depends upon the quantity of matter contained in the body Such quantity is known as the mass of the body Therefore, mass is a measure of inertia. Objects with large masses have less acceleration when exposed to the same force. Thus, mass is a quantity that relates the acceleration of the body to the force acting on it. The SI unit of mass is the kilogram (kg). Experimentally, it is found that the ratio of the masses of any two bodies (say m_1 and m_2) is equal to the inverse of the ratio of the magnitudes of their accelerations if both are acted upon by the same force. That is, we have

$$\frac{m_1}{m_2} = \frac{a_2}{a_1}$$

The mass of any body can be found by comparing its acceleration to the acceleration of a 1 kg mass when both bodies are acted upon by the same force. This leads to the conclusion that

mass is independent of force; it is an inherent characteristic of matter. Furthermore, it has been experimentally proved that when two masses m_1 and m_2 are attached together, the combined body behaves as a single body of mass $m_1 + m_2$. Thus, mass is a scalar quantity and obeys the rules of ordinary arithmetic.

3.2.4 Newton's Second Law

Unlike Newton's first law, Newton's second law describes the situation in which the net force acting on an object is not zero. It was found that when different forces act on an object, the object undergoes different accelerations. The magnitude of the acceleration is directly proportional to the magnitude of the applied force, and its direction is in the direction of that force. Furthermore, this acceleration is inversely proportional to the mass of the object for a certain applied force. These observations are summarized in Newton's second law of motion: *The acceleration of an object produced by a net external force is directly proportional to the force in a direction parallel to that force and is inversely proportional to its mass* That is,

$$\Sigma \mathbf{F} = m\mathbf{a}$$

In terms of components, the vector equation $\Sigma \mathbf{F} = m\mathbf{a}$ can be written as

$$\Sigma F_x = ma_x$$

$$\Sigma F_y = ma_y$$

$$\Sigma F_z = ma_z$$

where $a_x = \ddot{x}, a_y = \ddot{y}$ and $a_z = \ddot{z}$ In terms of normal and tangential coordinates, the net normal force is

$$\Sigma F_n = ma_n$$

and the net tangential force is

$$\Sigma F_t = ma_t$$

where $a_n = v^2/R$ and $a_t = dv/dt$. Finally, in terms of polar coordinates we have

$$\Sigma F_r = ma_r$$

and

$$\Sigma F_\theta = ma_\theta$$

where $a_r = (\ddot{r} - r\dot{\theta}^2)$ and $a_\theta = (r\ddot{\theta} + 2\dot{r}\dot{\theta})$. The unit of force in the SI system is the Newton (N). One Newton (1N)

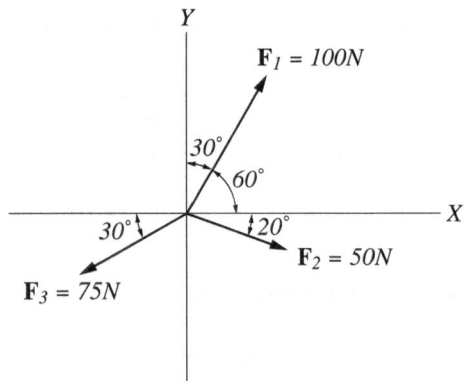

Fig. 3.2 A body is exposed to three forces acting in different directions

is defined as the force that gives a 1kg mass an acceleration of $1 \, \text{m/s}^2$.

$$1 \, \text{N} = 1 \, \text{kg.m/s}^2$$

Example 3.1 A body is exposed to three forces acting in different directions as shown in Fig. 3.2. Find the magnitude and direction of the resultant force acting on the body and the corresponding acceleration.

Solution 3.1 The net force in the x-direction is

$$\sum F_x = F_{1x} + F_{2x} + F_{3x} = F_1 \cos 60° + F_2 \cos 20° - F_3 \cos 30°$$

$$= (100 \, \text{N})(0.5) + (50 \, \text{N})(0.94) - (75 \, \text{N})(0.866) = 32.1 \, \text{N}$$

The net force in the y-direction is

$$\sum F_y = F_{1y} + F_{2y} + F_{3y} = F_1 \sin 60° - F_2 \sin 20° - F_3 \sin 30°$$

$$= (100 \, \text{N})(0.866) - (50 \, \text{N})(0.34) - (75 \, \text{N})(0.5) = 32.1 \, \text{N}$$

The magnitude of the net force is

$$F_{net} = \sqrt{(32.1)^2 + (32.1)^2} = 45.4 \, \text{N}$$

The direction of F_{net} relative to the x-axis is

$$\theta = \tan^{-1} \frac{F_{nety}}{F_{netx}} = \tan^{-1} \frac{(32.1 \text{N})}{(32.1 \, \text{N})} = 45°$$

The acceleration of the body is, therefore,

$$a = \frac{F}{m} = \frac{(45.4 \, \text{N})}{(50 \, \text{kg})} = 0.91 \, \text{m/s}^2$$

and its direction is the same as that of the force.

Example 3.2 If a man pushes a 60 kg box with a constant horizontal force of 100 N : (a) how far will the container be moved when its speed reaches a value of 3 m/s; (b) how far will the container be moved when its speed reaches a value of 3 m/s if the same force is applied at 30° to the horizontal; (c) find the normal force acting on the block in (a) and (b).

Solution 3.2 (a) The acceleration of the container is given by

$$a = \frac{F}{m} = \frac{(100 \, \text{N})}{(60 \, \text{kg})} = 1.67 \, \text{m/s}^2$$

the distance moved when its speed reaches 3 m/s is found from

$$v^2 - v_0^2 = 2a(x - x_0)$$

By taking $x_0 = 0$ at the starting point we have

$$(3 \, \text{m/s})^2 - 0 = 2(1.67 \, \text{m/s}^2)(x - 0)$$

That gives $x = 2.7 \, \text{m}$.

(b) If the force is at 30° to the horizontal, the acceleration is

$$a = \frac{F}{m} = \frac{(100 \, \text{N}) \cos 30°}{(60 \, \text{kg})} = 1.45 \, \text{m/s}^2$$

$$(3 \, \text{m/s})^2 - 0 = 2(1.45 \, \text{m/s}^2)(x - 0)$$

and $x = 3.1 \, \text{m}$.

(c) In situation (a) we have

$$\sum F_y = n - mg = 0$$

and

$$n = mg = (60 \, \text{kg})(9.8 \, \text{m/s}^2) = 588 \, \text{N}$$

In (b) we have

$$\sum F_y = n + F \sin \theta - mg = 0$$

and

$$n = (588 \, \text{N}) - (100 \, \text{N}) \sin 30° = 538 \, \text{N}$$

Example 3.3 A particle of mass 0.5 kg is moving along the curve given by $\mathbf{r} = (1/3t^4 \mathbf{i} - t^3 \mathbf{j})$ m where t is time. Determine the force acting on the particle.

Solution 3.3 The force acting on the particle is

$$\mathbf{F} = m \frac{d^2 \mathbf{r}}{dt^2} = m \frac{d}{dt} (4/3 t^3 \mathbf{i} - 3t^2 \mathbf{j}) = m(4t^2 \mathbf{i} - 6t \mathbf{j})$$

$$= (0.5 \, \text{kg})(4t^2 \mathbf{i} - 6t \mathbf{j}) = (2t^2 \mathbf{i} - 3t \mathbf{j}) \, \text{N}$$

Example 3.4 A particle of mass of 1 kg is moving under the influence of a force given by

$$\mathbf{F} = ((12t^2 - 5t)\mathbf{i} + (9t - 1)\mathbf{j} - 3t^2\mathbf{k})\,\text{N}$$

If at $t = 0$, $\mathbf{r}_0 = \mathbf{0}$ and $\mathbf{v}_0 = \mathbf{0}$, find the velocity and the position of the particle at any time.

Solution 3.4

$$\mathbf{F} = m\mathbf{a}$$

$$\mathbf{a} = \frac{\mathbf{F}}{m} = ((12t^2 - 5t)\mathbf{i} + (9t - 1)\mathbf{j} - 3t^2\mathbf{k})\,\text{m/s}^2$$

$$\mathbf{v} = \int \mathbf{a}\,dt = \left(4t^3 - \frac{5}{2}t^2\right)\mathbf{i} + \left(\frac{9}{2}t^2 - t\right)\mathbf{j} - t^3\mathbf{k} + \mathbf{c}_1$$

Since at $t = 0$, $\mathbf{v}_0 = \mathbf{0}$, then $\mathbf{c}_1 = \mathbf{0}$ and the velocity at any time is

$$\mathbf{v} = \left(\left(4t^3 - \frac{5}{2}t^2\right)\mathbf{i} + \left(\frac{9}{2}t^2 - t\right)\mathbf{j} - t^3\mathbf{k}\right)\,\text{m/s}$$

$$\mathbf{v} = \frac{d\mathbf{r}}{dt}$$

$$\mathbf{r} = \int \mathbf{v}\,dt = \int \left[\left(4t^3 - \frac{5}{2}t^2\right)\mathbf{i} + \left(\frac{9}{2}t^2 - t\right)\mathbf{j} - t^3\mathbf{k}\right]dt$$

$$\mathbf{r} = \left(\left(t^4 - \frac{5}{6}t^3\right)\mathbf{i} + \left(\frac{3}{2}t^3 - \frac{1}{2}t^2\right)\mathbf{j} - \frac{1}{4}t^4\mathbf{k}\right)\,\text{m}$$

Example 3.5 If a man weighs himself on an elevator that is accelerating upwards at a rate a relative to an observer outside the elevator (in an inertial frame) as shown in Fig. 3.3, what reading will he get for the normal force acting on him by the floor? what is the force if the elevator is accelerating downwards?

Solution 3.5 The normal force is n = m(a + g) for upward acceleration and n = m(g − a) for downward acceleration. Since a weighing scale measures the normal force and calculates the mass from it, the downward journey might be a more pleasant one!

3.2.5 Newton's Third Law

A force acting on an object is always due to another object in the surrounding environment. Newton's third law shows that: if body 1 exerts a force $\mathbf{F}_{21}$ on body 2 then body 2 will exert an equal and opposite force $\mathbf{F}_{12}$ on body 1. That is

$$\mathbf{F}_{12} = -\mathbf{F}_{21}$$

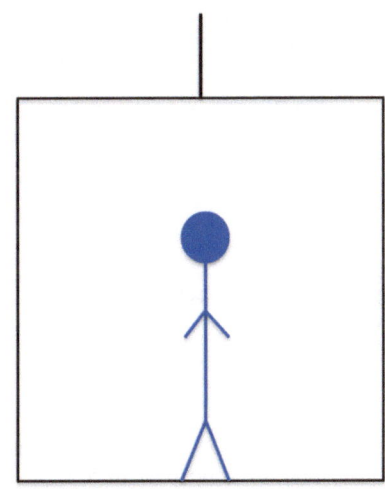

Fig. 3.3 A man weighing himself in an elevator

Any of these forces can be called an action force. When one of these forces is called an action force the other force is called a reaction (see Fig. 3.4). This law is sometimes stated *as "To every action there is an equal and opposite reaction."* Note that *the action and reaction forces always act on different objects*, i.e., they can't cancel each other out. This law also shows that forces come in pairs and that there is no such thing as a single isolated force.

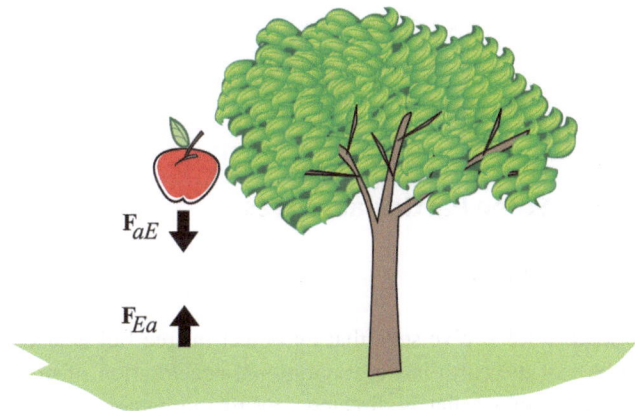

Fig. 3.4 The gravitational force exerted by the Earth on the apple and that exerted by the apple to the Earth form an action-reaction pair

Example 3.6 Three blocks of masses m_1, m_2, and m_3 are placed on a frictionless surface and pushed by a horizontal force F as in Fig. 3.5. Determine (a) the acceleration of the system; (b) the contact forces between m_1 and m_2 and between m_2 and m_3.

Solution 3.6 The free-body diagram of each block is shown in Fig. 3.5, where F_{21} is the force exerted on m_2 by m_1. (a) Applying Newton's second law for the ,

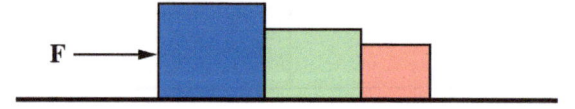

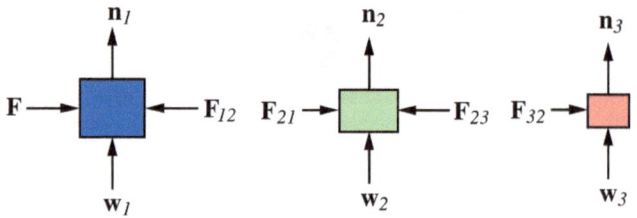

Fig. 3.5 Three blocks of masses m_1, m_2, and m_3 are placed on a frictionless surface and pushed by a horizontal force F

$$F = (m_1 + m_2 + m_3)a$$

$$a = \frac{F}{(m_1 + m_2 + m_3)}$$

(b) Applying Newton's second law for each block, we have $F - F_{12} = m_1a$, $F_{21} - F_{23} = m_2a$ and $F_{32} = m_3a$. From Newton's law of action and reaction we have $F_{12} = F_{21}$, and $F_{32} = F_{23}$, and therefore

$$F_{12} = m_2a + F_{23} = (m_2 + m_3)a = \frac{(m_2 + m_3)}{(m_1 + m_2 + m_3)}F$$

$$F_{32} = \frac{m_3 F}{(m_1 + m_2 + m_3)}$$

3.3   Some Particular Forces

3.3.1   Weight

In Sect. 2.4.2, we've seen that an object in free fall near the surface of the earth has a gravitational acceleration of magnitude $9.8 \, \text{m/s}^2$ that is directed toward the center of earth. Using Newton's second law, we can calculate the force that caused this acceleration. If an object has a mass m then the gravitational force is given by $m\mathbf{g}$, and is denoted by w, i.e., $\mathbf{w} = m\mathbf{g}$. w is known as the weight of an object and is defined as the gravitational force exerted on it by earth (or any other astronomical body, where g is different than that of earth). In Chap. 9, we will see that a gravitational force exists between any two bodies. When one of the bodies is an astronomical body, such as the earth or moon, and the other body is relatively smaller in size and mass the gravitational force is then called the weight of the body We will also see that the gravitational force varies with the distance between objects, and that the value of **g** becomes less at greater altitudes. Thus, weight

is not an intrinsic property of an object. In everyday life, it is common to use the word weight when measuring the mass of a body mass and weight represent different quantities but they are proportional for a given value of **g**. For two masses at the same location, the ratio of their weights is equal to the ratio of their masses.

3.3.2   The Normal Force

If an object is in contact with a surface, either at rest or moving on it, the surface exerts a supporting force **n** on the object that is always perpendicular to the surface of contact. This force is called the normal force.

3.3.3   Tension

The tension force **T** is the force that a cord, rope, cable, or any other similar object exerts on an object attached to it. This force is directed along the rope away from the object at the point where the rope is attached. In solving problems, ropes are usually assumed to be massless (referred to as light ropes) and unstretchable. For any light rope, the magnitude of the tension force T is the same at all points along the rope.

3.3.4   Friction

Imagine that everything around you is coated with an extremely good lubricant. Simple activities such as walking, sitting, driving a car, or holding objects would become extremely difficult or impossible. Therefore, friction plays a very important role in our everyday life. The frictional force is due to the interaction between the surface atoms of any two bodies in contact. The direction of this force is always parallel to the surface of contact, opposing the motion or the planned motion of one object relative to the other. Hence, the normal and frictional forces are both contact forces and they are always perpendicular to each other.

Consider a block resting on a table. If the block is pushed with a horizontal force **F** and remains stationary, it is because that the applied force is balanced by an equal and opposite force. This opposing force is known as the statistical frictional force $\mathbf{f}_s$ and it has the value $f_s = F$. The statistical frictional force increases with increasing **F** (see Fig. 3.6). The name statistical comes from the fact that the block remains stationary.

However, if F is increased to a certain maximum value, the block will eventually accelerate (see Fig. 3.7). This maximum value is equal to the maximum frictional force $f_{s \, \text{max}} = F$ and it represents the applied force when the block is at the verge of slipping, i.e., it is the minimum force necessary to initiate

Fig. 3.6 The opposing force is known as the statistical frictional force $\mathbf{f}_s$ and it has the value $f_s = F$. Its maximum value $f_{s\,max} = F$ represents the applied force when the block is at the verge of slipping i.e. it is the minimum force necessary to initiate motion. When the block moves, the retarding frictional force is then called the kinetic frictional force $\mathbf{f}_k$

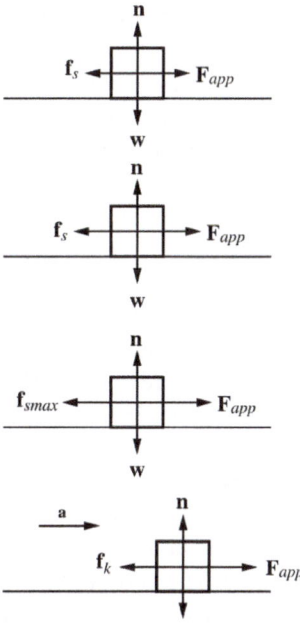

Table 3.1 Coefficients of Friction

Materials	μ_s (static)	μ_k (kinetic)
Steel on steel	0.74	0.57
Aluminum on steel	0.61	0.47
Copper on steel	0.53	0.36
Wood on wood	0.25–0.5	0.2
Rubber on concrete	1.0	0.8
Glass on glass	0.94	0.4
Copper on glass	0.68	0.53
Teflon on teflon	0.04	0.04

taneous velocity of the body relative to the surface. The dimensionless coefficients μ_s and μ_k depend on the nature of the surfaces in contact and are independent of the area of contact between these surfaces. μ_k is generally less than μ_s. Table 3.1 lists μ_s and μ_k for some materials. Note that μ_k may vary with speed but such variations are not included here. Friction is a very complex phenomenon. One reason behind this is that the actual area of contact viewed from a microscopic level is much less than the area of contact viewed from a macroscopic level, as in Fig. 3.8, even for very smooth surfaces. For our purposes here, the detailed friction mechanism will not be discussed.

motion. When the block moves, the retarding frictional force is then called the kinetic frictional force $\mathbf{f}_k$ and is usually less than $f_{s\,max}$. If $f_k = F$, the block will move with a constant speed. If $F < f_s$ or if $\mathbf{F}$ is removed, the block will decelerate and will eventually be brought to rest. Experiments show that $f_{s\,max}$ and f_k have the following properties:

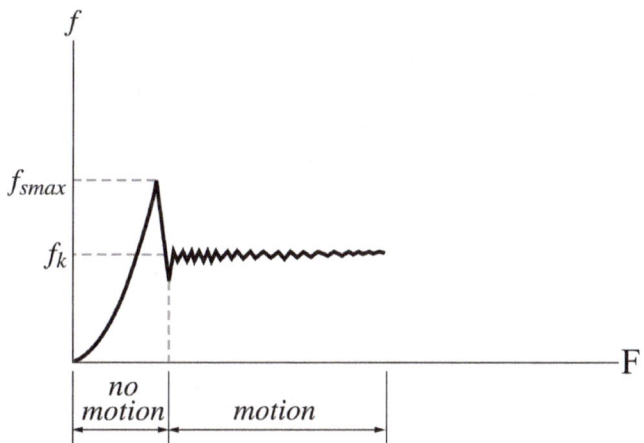

Fig. 3.7 A graph of the friction force versus the applied force

1. $f_{s\,max} = \mu_s n$, where μ_s is the coefficient of static friction and n is the normal force acting on the block. As long as the block is at rest $f_s = F$, where $f_s \le \mu_s n$.
2. $f_k = \mu_k n$ where μ_k is the coefficient of the kinetic friction.
3. The directions of $\mathbf{f}_s$ and $\mathbf{f}_k$ are always parallel to the surface. $\mathbf{f}_s$ is opposite to the component of the applied force that is parallel to the surface and $\mathbf{f}_k$ is opposite to the instan-

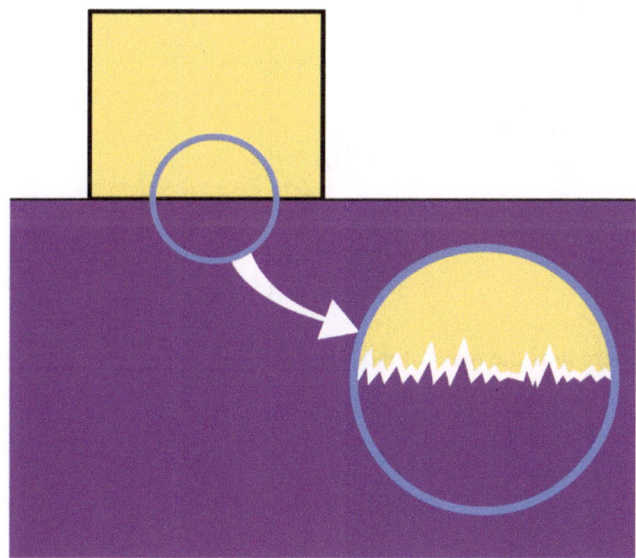

Fig. 3.8 The actual area of contact viewed from a microscopic level is much less than the area of contact viewed from a macroscopic level

3.3.5 The Drag Force

If a fluid (such as gas or liquid) and a body are in relative motion, the body will experience a resistive force opposing the relative motion called the drag force D. The direction of **D** is the direction in which the fluid is flowing relative to the body. The drag force depends on some factors, such as the speed of the object. We will consider the situation in which a relatively large object is moving through air at high speed (such as a skydiver or an airplane). In this case, the drag force D is proportional to the square of the body's speed and is given by

$$D = \frac{1}{2} c\rho A v^2$$

where ρ is the air density, A is the effective cross-sectional area of the body taken in a plane perpendicular to its velocity, and c is the drag coefficient. c is a dimensionless constant that has a value ranging from 0.4 to 2. The value of c may vary with v, but such variations will be ignored.

As an example, consider an object falling through air far from the earth's surface. The forces acting on the object are the drag force and the gravitational force as in Fig. 3.9. As the object falls, its speed increases and thus **D** increases from zero until it eventually becomes equal to the object's weight ($\mathbf{D} = -m\mathbf{g}$). The net force on the object when ($\mathbf{D} = -m\mathbf{g}$) will be equal to zero, and hence the object's acceleration will become zero ($\mathbf{a} = \mathbf{0}$). As a result the body will fall at a constant speed called the terminal speed v_t:

$$\Sigma \mathbf{F} = \mathbf{0}$$

$$\frac{1}{2} c\rho A v_t^2 - mg = 0$$

$$\frac{1}{2} c\rho A v_t^2 = mg$$

$$v_t = \sqrt{\frac{2mg}{c\rho A}}$$

where m is the mass of the body.

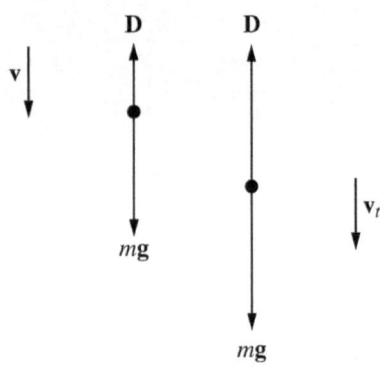

Fig. 3.9 An object falling through air experiences a drag force **D** and a gravitational force $m\mathbf{g}$

3.4   Applying Newton's Laws

It is necessary to follow some steps when solving problems using Newton's second law. These steps can be summarized in the following:

1. Draw a simple diagram of the objects in the system analyzed;
2. Draw a free-body diagram for each object in the system. In a free-body diagram, the body is represented by a dot, and all external forces (represented by vectors) acting on the body are shown. The forces exerted by the body on other bodies in the system are not included in its free-body diagram;
3. A coordinate system should be drawn in a free-body diagram with the body at its origin. Newton's second law is then applied along each axis using the components of each force. The coordinate system must be oriented in such a way that simplifies the analysis, i.e., some forces should be directed along the axes;
4. Solve obtained equations for the unknowns.

Note that from here until Chap. 5, any object is assumed to behave as a particle, i.e., all of its parts move in exactly the same way When applying Newton's second law, a particle is represented by a dot on the free-body diagram. Furthermore, the mass or friction of any rope or pulley is neglected.

Example 3.7 A 25 kg block is released from rest at the top of a rough 40° inclined surface. It then accelerates at a constant rate of 0.1 m/s². Find: (a) the coefficient of kinetic friction between the box and the surface; (b) the maximum angle the box would be at the verge of slipping if the angle of the incline is changeable.

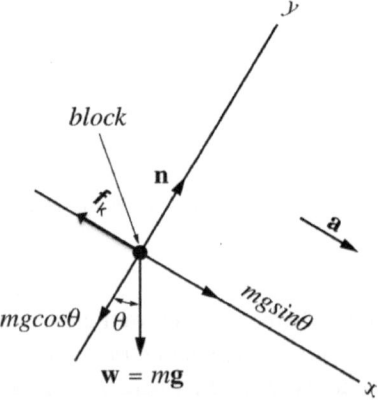

Fig. 3.10 The free-body diagram of a block accelerating down an incline

Solution 3.7 (a) The free-body diagram is shown in Fig. 3.10. Applying Newton's second law to the box gives

$$\sum F_x = mg \sin \theta - f_k = ma$$

$$\sum F_y = n - mg \cos \theta = 0$$

$$f_k = mg \sin \theta - ma = (25 \, \text{kg})(9.8 \, \text{m/s}^2) \sin 40° - (25 \, \text{kg})(0.1 \, \text{m/s}^2)$$

$$= 155 \, \text{N}$$

$$n = mg \cos \theta = (25 \, \text{kg})(9.8 \, \text{m/s}^2) \cos 40° = 187.7 \, \text{N}$$

The coefficient of kinetic friction is

$$\mu_k = \frac{f_k}{n}$$

$$\mu_k = \frac{(120 \, \text{N})}{(212.2 \, \text{N})} = 0.57$$

(c) At the verge of slipping the force of static friction is maximum:

$$f_{s \, \text{max}} = \mu_s n$$

Applying Newton's second law we get

$$\sum F_x = mg \sin \theta - f_{s \, \text{max}} = 0$$

$$f_{s \, \text{max}} = mg \sin \theta$$

Also we have $n = mg \cos \theta$, therefore

$$\mu_s = \frac{mg \sin \theta}{mg \cos \theta} = \tan \theta$$

Since $\mu_s = 0.74$ from Table 3.1, we have

$$\theta = \tan^{-1} 0.74 = 36.5°$$

Example 3.8 Two masses $m_1 = 2 \, \text{kg}$ and $m_2 = 5 \, \text{kg}$ are connected by a massless cord that passes over a massless and frictionless pulley (Atwood's machine) as shown in Fig. 3.11. Find the acceleration of the system and the tension in the cord.

Solution 3.8 The free-body diagram of each mass is shown in Fig. 3.11. Applying Newton's second law to each block (taking positive y to be upwards) gives

$$\sum F_{1y} = T - m_1 g = m_1 a$$

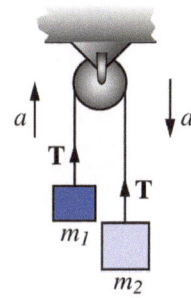

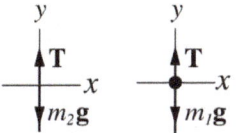

Fig. 3.11 The free-body diagram of an Atwoodõs machine

$$\sum F_{2y} = T - m_2 g = -m_2 a$$

adding the two equations we have

$$a = \left(\frac{m_2 - m_1}{m_1 + m_2} \right) g = \left(\frac{3 \, \text{kg}}{7 \, \text{kg}} \right)(9.8 \, \text{m/s}^2) = 4.2 \, \text{m/s}^2$$

Substituting a in any of the two equations gives

$$T = \left(\frac{2 m_1 m_2}{m_1 + m_2} \right) g = \left(\frac{2(10 \, \text{kg})}{(7 \, \text{kg})} \right)(9.8 \, \text{m/s}^2) = 28 \, \text{N}$$

Example 3.9 Two blocks of masses $m_1 = 1.5 \, \text{kg}$ and $m_2 = 3.2 \, \text{kg}$ are connected by a light string that passes over a massless frictionless pulley as shown in Fig. 3.12. If the surface is frictionless: (a) what is the distance that m_2 will drop during the first 0.6 s? (b) if a third block is attached to m_1 using strong glue, what must its mass be such that the system moves with constant speed?

Solution 3.9 (a) The acceleration value is the same for both masses since they are connected by a string. Figure 3.12 shows the free-body diagram for each mass. Applying Newton's law to m_1 and m_2 in the direction of motion we have for m_2

$$\sum F_{2y} = T - m_2 g = -m_2 a$$

and for m_1

$$\sum F_{1x} = m_1 g \sin \theta - T = -m_1 a$$

from this we have

$$T = m_2 (g - a)$$

and

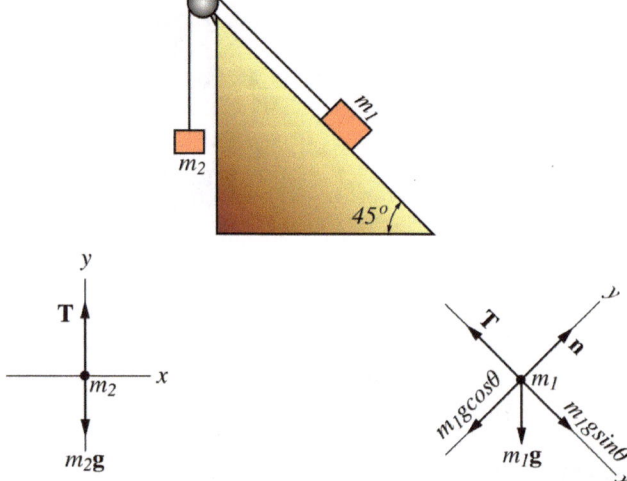

Fig. 3.12 The free-body diagram showing the forces on each block

$$m_1 g \sin\theta - m_2(g - a) = -m_1 a$$

and therefore

$$a = \frac{(m_2 - m_1 \sin\theta)g}{m_1 + m_2} = \frac{((3.2\,\text{kg}) - (1.5\,\text{kg})\sin 45°)(9.8\,\text{m/s}^2)}{4.7\,\text{kg}}$$
$$= 4.5\,\text{m/s}^2$$

After 0. 6s, the distance that m_2 falls is

$$h = \frac{1}{2}at^2 = \frac{1}{2}(-4.5\,\text{m/s}^2)(0.6\,\text{s})^2 = -0.81\,\text{m}$$

(b) If the system moves with constant speed, its acceleration is zero

$$\sum F_{1x} = (m_1 + m_3)g\sin\theta - T = 0$$

and

$$\sum F_{2y} = T - m_2 g = 0$$

that gives

$$T = m_2 g$$

and

$$(m_1 + m_3)g\sin\theta - m_2 g = 0$$

hence

$$m_3 = \frac{m_2 - m_1 \sin\theta}{\sin\theta} = \frac{((3.2\,\text{kg}) - (1.5\,\text{kg})\sin 45°)}{\sin 45°} = 3.03\,\text{kg}$$

Example 3.10 A 3 kg block is hanged from the ceiling as in Fig. 3.13. Find the magnitude of T_1, T_2, and T_3.

Solution 3.10 The free-body diagrams of the block and the knot are shown in Fig. 3.14. From Newton's second law T_3 is equal to the weight of the block, i.e.,

Fig. 3.13 A block hanged from the ceiling

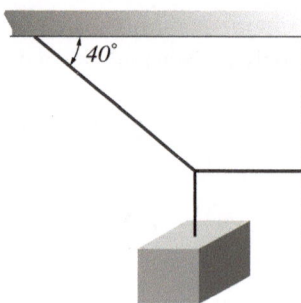

Fig. 3.14 The free-body diagram of the block

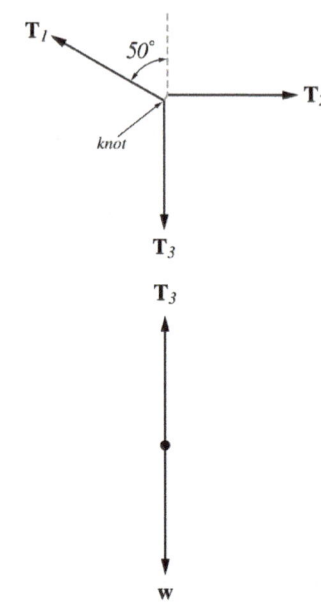

$$T_3 = w = mg = (3\,\text{kg})(9.8\,\text{m/s}^2) = 29.4\,\text{N}$$

For the knot, we have

$$\sum F_x = T_2 - T_1 \sin 50° = 0 \qquad (3.1)$$

$$\sum F_y = T_1 \cos 50° - T_3 = 0 \qquad (3.2)$$

Solving for T_1 from Eq. 3.2 gives

$$T_1 = \frac{T_3}{\cos 50°} = \frac{(29.4\,\text{N})}{(0.64)} = 45.7\,\text{N}$$

Substituting this result into Eq. 3.1 we get

$$T_2 = T_1 \sin 50° = (45.7\,\text{N})(0.76) = 35\,\text{N}$$

Example 3.11 Figure 3.15 shows a weight of 200 N that is lifted with a constant speed. Find the tension in each part of the rope and the force of lift.

Solution 3.11 Since the pulleys are massless and frictionless we have $2T_1 = T_2$ (T_1 is the tension in each rope)and $T_2 =$

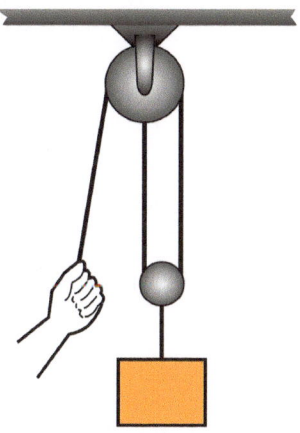

Fig. 3.15 Using two pulleys to reduce the force necessary to lift a weight

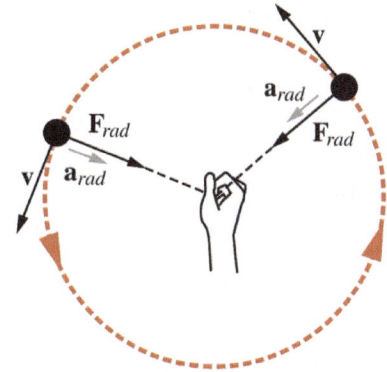

Fig. 3.16 An object attached to a string in uniform circular motion

$w = 200\,\text{N}$, thus

$$T_1 = \frac{T_2}{2} = \frac{(200\,\text{N})}{2} = 100\,\text{N}$$

we also have

$$T_1 = F = 100\,\text{N}$$

3.4.1 Uniform Circular Motion

In Sect. 2.4.5, it was found that a particle moving in a circle with a constant linear speed v (uniform circular motion) has a centripetal acceleration directed towards the center of the circle. Its magnitude is given by

$$a_{rad} = \frac{v^2}{r}$$

where r is the radius of the circle. Figure 3.16 shows an object attached to a string in uniform circular motion (the plane of motion is parallel to the Earth's surface). From Newton's second law, the centripetal acceleration is caused by a force or net force directed towards the center of the circle. Therefore, as a_{rad}, the centripetal (or radial) force F_{rad} has a constant magnitude but its direction changes continuously The magnitude of this centripetal force is given by

$$|\Sigma \mathbf{F}| = F_{rad} = ma_{rad} = m\frac{v^2}{r}$$

If at some instant the radial force becomes zero, the object would then move along a straight line path tangent to the circle. Hence, the centripetal force is necessary to keep the object in its circular path. The centripetal force may be any kind of force such as friction, gravity, or tension.

Example 3.12 A conical pendulum consists of a bob of mass m attached to a light string rotating in a horizontal circle as in

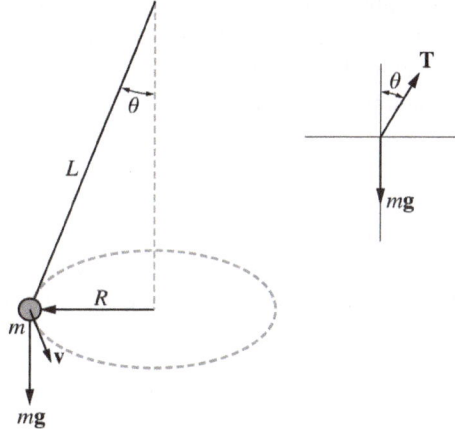

Fig. 3.17 A conical pendulum consisting of a bob of mass m attached to a light string rotating in a horizontal circle

Fig. 3.17. If the bob rotates with a constant speed and if θ and m are known, find: (a) the tension in the string; (b) the speed of the bob; (c) the period of motion.

Solution 3.12 The horizontal component of the tension force supplies the required centripetal force to keep the bob in its circular path while the vertical component balances the weight of the bob. (a) Applying Newton's second law in both the x- and y-directions we have

$$T\cos\theta - mg = 0$$

$$T = \frac{mg}{\cos\theta} \tag{3.3}$$

and

$$T\sin\theta = \frac{mv^2}{R} \tag{3.4}$$

(b) Dividing Eq. 3.4 by Eq. 3.3 gives

$$v = \sqrt{gR\tan\theta}$$

(c) The period of motion is given by

$$\tau = \frac{2\pi R}{v} = 2\pi \sqrt{\frac{R}{g \tan \theta}}$$

Since $R = L \sin \theta$, we have

$$\tau = 2\pi \sqrt{\frac{L \cos \theta}{g}}$$

Example 3.13 (a) A car needs to turn on a level road without skidding as in Fig. 3.18. Find the maximum speed for which the car can take the curved path of the level road safely (b) If the road is banked, i.e., the outer edge is raised relative to the inner edge as in Fig. 3.19, find the maximum speed for which the car can take the curved path of the level road safely without depending on friction.

Solution 3.13 (a) The centripetal force required for the car to remain in its circular path, is in this case, is the force of static friction. The maximum speed for which the car can take the curve without skidding is when the static frictional force is a maximum. That is,

$$f_{s\,max} = \mu mg = \frac{mv_{max}^2}{r}$$

Hence

$$v_{max} = \sqrt{\mu rg}$$

(b) If the road is banked the car can take the turn without depending on friction as the required centripetal force. In that case, the horizontal component of the normal force supplies

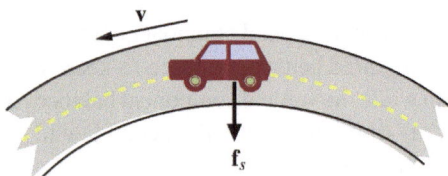

Fig. 3.18 A car turning without skidding

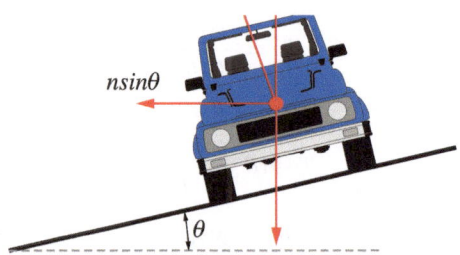

Fig. 3.19 A car turning on a banked road

the necessary centripetal force. Thus we have

$$n \sin \theta = \frac{mv^2}{r}$$

and

$$n \cos \theta = mg$$

where v is the speed of the car. Dividing these two equations gives

$$\tan \theta = \frac{v^2}{rg}$$

If the angle θ and the curvature r are known, then the safe speed limit can be found. If the car moves at a speed lower or higher than that speed then the frictional force must supply the additional centripetal force for the car to stay in its circular path.

3.4.2 Nonuniform Circular Motion

In Sect. 2.4.6, we saw that an object in nonuniform circular motion has both perpendicular (centripetal) and parallel components of acceleration given by

$$a_{rad} = \frac{v^2}{r}$$

and

$$a_t = \frac{d|\mathbf{v}|}{dt}$$

The total acceleration is

$$\mathbf{a} = \mathbf{a}_{rad} + \mathbf{a}_t$$

These radial and tangential accelerations are caused by radial and tangential forces respectively (see Fig. 3.20):

$$F_{rad} = ma_{rad} = m\frac{v^2}{r}$$

$$F_t = ma_t = m\frac{d|\mathbf{v}|}{dt}$$

The net force is

$$\mathbf{F} = \mathbf{F}_{rad} + \mathbf{F}_t$$

Example 3.14 An object attached to a light string is rotating in a vertical circle of radius r (see Fig. 3.21). Find: (a) the tension in the cord at the lowest and highest points; (b) the minimum speed at the highest point such that the object remains in its circular path.

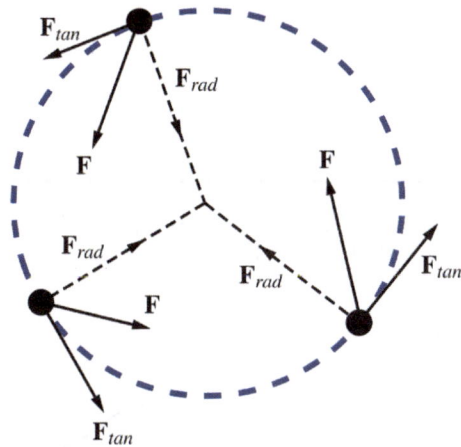

Fig. 3.20 The radial and tangential forces in nonuniform circular motion

and

$$T - mg \cos \theta = \frac{mv^2}{r}$$

At the bottom $\theta = 0$ and therefore $a_t = 0$ and

$$T_b = m \left(\frac{v_b^2}{r} + g \right)$$

At the top $\theta = 180^0$ and $a_t = 0$ and

$$T_t = m \left(\frac{v_t^2}{r} - g \right)$$

(b) For the object to remain in its circular path, the string must remain taut, i.e., T_t must be positive ($T_t > 0$). If $T_t = 0$ then $v_t = \sqrt{gr}$. Hence, the velocity must satisfy $v_t > \sqrt{gr}$.

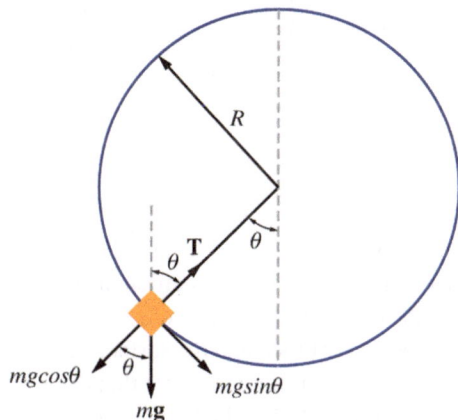

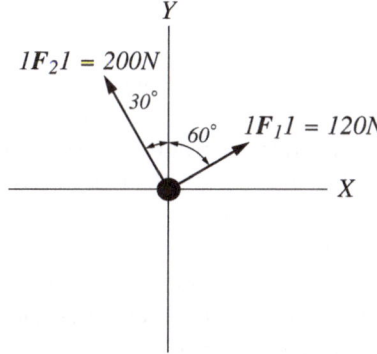

Fig. 3.22 An object subjected to two forces acting in different directions

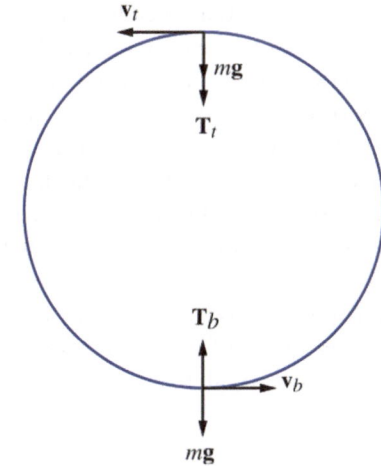

Fig. 3.21 An object attached to a string rotating in a vertical circle

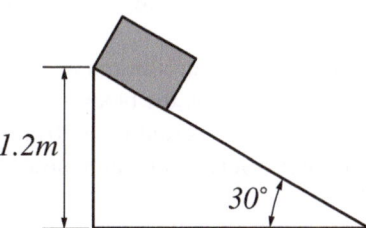

Fig. 3.23 A block released from the top of an incline

Solution 3.14 (a) Applying Newton's second law in both the tangential and radial directions gives

$$mg \sin \theta = ma_t$$

Fig. 3.24 Two masses connected by a light string over a frictionless pulley of negligible mass

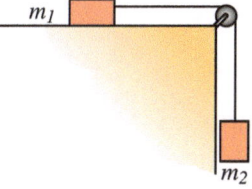

Fig. 3.25 A block hanged from
the ceiling

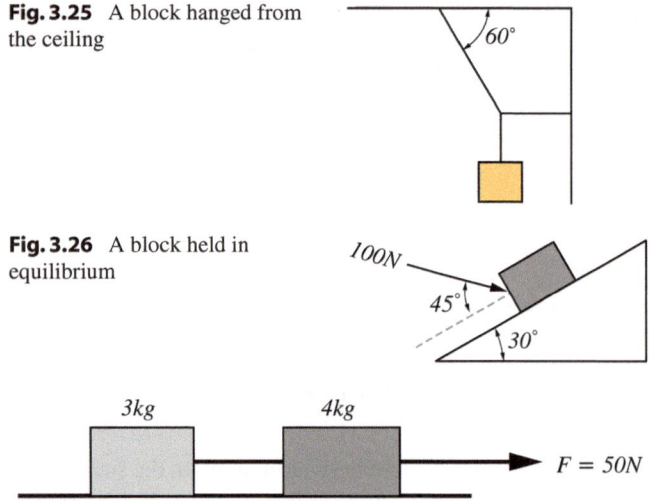

Fig. 3.26 A block held in
equilibrium

Fig. 3.27 Two blocks connected by a light rope and pulled by a force

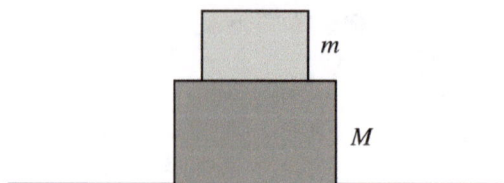

Fig. 3.28 Two blocks placed on top of each other, where a horizontal
force is applied to the lower block

Fig. 3.29 A car moving on a
curved path

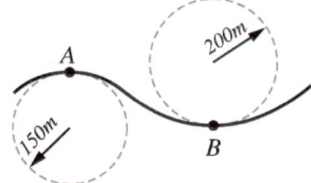

Fig. 3.30 A block of mass m on
a frictionless table is attached to
light string that passes through
the center of the table and is
connected to a larger block of
mass M

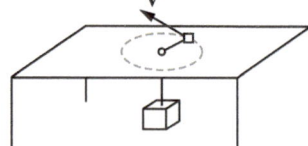

Problems

1. A 4 kg object is exposed to two forces (see Fig. 3.22).
   Find the magnitude and direction of the acceleration of
   the object.
2. A 0.2 kg block is released from the top of an inclined plane
   of angle 30° as in Fig. 3.23. Find the speed of the block
   just as it reaches the bottom.
3. Two masses are connected by a light string that is connected to a frictionless pulley of negligible mass as in
   Fig. 3.24. Find the magnitude of the acceleration of the
   masses and the tension in the string.
4. A block is pushed up along a smooth inclined plane of
   angle of 45° where it is given an initial velocity of 8 m/s.
   Determine the time it takes the block to return to its initial
   position.
5. Find the tension in each string in the system shown in
   Fig. 3.25.
6. A 5 kg block is held in equilibrium as in Fig. 3.26. Find
   the normal force acting on the block.
7. Find the normal force exerted on a 70 kg man standing
   inside an elevator that is accelerating upwards at a rate of
   2 m/s^2.
8. Find the acceleration of the system shown in Fig. 3.27
   and the tension in the string if $m_1 = 3$ kg and $m_2 = 4$ kg
   (assume massless string and frictionless surface).

9. Two blocks of masses 3 and 5 kg are placed on top of each
   other as in Fig. 3.28. If the coefficient of static friction
   between the blocks is 0.2 and assuming there is no friction
   between the lower block and the surface on which it rests,
   find the maximum horizontal force that can be applied to
   the lower block such that the blocks move together.
10. A 1000 kg car move along the track shown in Fig. 3.29.
    Find (a) the maximum speed the car can have at point A
    such that it does not leave the track (b) the normal force
    exerted on the car at B if its speed there is 15 m/s.
11. A block of mass m on a frictionless table is attached to
    light string that passes through the center of the table
    and is connected to a larger block of mass M (see
    Fig. 3.30). If m moves in uniform circular motion of radius
    r and speed v, find v such that M remains at rest when
    released.
12. A 1 kg particle moves in the force field given by $\mathbf{F} =
    2t\mathbf{i} + (5t - 1)\mathbf{j} - 6t^2\mathbf{k}$. Find the position of the particle
    at any time if at $t = 0$, $\mathbf{r}_0 = \mathbf{0}$, and $\mathbf{v}_0 = \mathbf{0}$.

Open Access This chapter is licensed under the terms of the Creative Commons Attribution 4.0 International License (http://creativecommons.org/licenses/by/4.0/), which permits use, sharing, adaptation, distribution and reproduction in any medium or format, as long as you give appropriate credit to the original author(s) and the source, provide a link to the Creative Commons license and indicate if changes were made.

The images or other third party material in this chapter are included in the chapter's Creative Commons license, unless indicated otherwise in a credit line to the material. If material is not included in the chapter's Creative Commons license and your intended use is not permitted by statutory regulation or exceeds the permitted use, you will need to obtain permission directly from the copyright holder.

Open Access This chapter is licensed under the terms of the Creative Commons Attribution 4.0 International License (http://creativecommons.org/licenses/by/4.0/), which permits use, sharing, adaptation, distribution and reproduction in any medium or format, as long as you give appropriate credit to the original author(s) and the source, provide a link to the Creative Commons license and indicate if changes were made.

The images or other third party material in this chapter are included in the chapter's Creative Commons license, unless indicated otherwise in a credit line to the material. If material is not included in the chapter's Creative Commons license and your intended use is not permitted by statutory regulation or exceeds the permitted use, you will need to obtain permission directly from the copyright holder.

4.1 Introduction

Energy is a very important concept that is heavily used in everyday life. Everything around us, including ourselves, needs energy to function. For example, electricity provides home appliances with the energy they require, food gives us energy to survive, and the sun provides earth with the energy needed for the existence of life!

Experiments show that energy is a scalar quantity related to the state of an object. Energy may exist in various forms: mechanical, chemical, gravitational, electromagnetic, nuclear, and thermal. Furthermore, energy cannot be created or destroyed; it can only be transformed from one form to another. In other words, if energy were to be exchanged between objects inside a system, then the total amount of energy (the sum of all forms of energy) in the system will remain constant.

A transformation of energy occurs due to the action of a force known as work or due to heat exchange between objects (or between an object and its environment). If energy is transferred due to work then it may be defined as the capacity of doing work. This book is concerned with mechanical energy which involves kinetic energy (associated with the object's motion) and potential energy (associated with the position of the object in space).

4.2 Work

Work may have many meanings. Sometimes, work is said to be done when a muscular activity is performed. Work may also refer to mental activity (mental work). In physics, the definition of work is different. Work is said to be done if a force is applied to an object while it is moving, i.e., if there is no resulting displacement, no work is done. Suppose that a person holds a heavy box for sometime and then starts to feel tired. The reason he/she feels tired is because chemical energy in his/her body is converted into internal microscopic motions of the muscles. Since the energy is not transferred to the box being carried (the box did not move), the work done on the box is equal to zero.

4.2.1 Work Done by a Constant Force

Consider an object exposed to a constant force F (see Fig. 4.1). If the object is displaced through a displacement s, then the work done on the object is a scalar quantity defined as

$$W = Fs \cos \theta = \mathbf{F} \cdot \mathbf{s}$$

where θ is the smaller angle between $\mathbf{F}$ and $\mathbf{s}$. The component of $\mathbf{F}$ in the direction of $\mathbf{s}$ ($F \cos \theta$) is the only effective component that produces motion. The work done represents energy transferred to or from the object via that force. If ($\theta = 0$), the work done on the object is positive, i.e. energy is transferred to the object. If ($\theta = 180^\circ$), the work done is negative, i.e., energy is transferred from the object. The SI unit of work is Newton meter (N.m) also named as the Joule.

$$1 \, \text{Joul} = 1 \, \text{J} = 1 \, \text{kg.m}^2/\text{s}^2$$

Note that energy and work have the same units.

4.2.2 Work Done by Several Forces

Consider an object exposed to several forces as in Fig. 4.2. The work done by all of these forces is the sum of the individual amounts of work done by each force:

$$W = \mathbf{F}_1 \cdot \mathbf{s} + \mathbf{F}_2 \cdot \mathbf{s} + \mathbf{F}_3 \cdot \mathbf{s} + \cdots$$

$$W = W_1 + W_2 + W_3 +$$

© The Author(s) 2019
S. Alrasheed, *Principles of Mechanics*, Advances in Science,
Technology & Innovation, https://doi.org/10.1007/978-3-030-15195-9_4

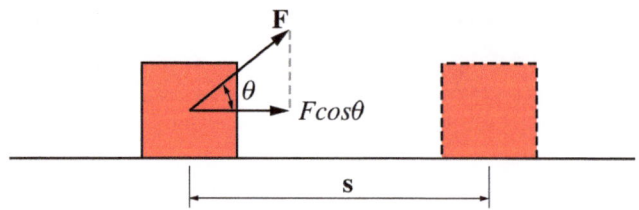

Fig. 4.1   An object exposed to a constant force **F** and undergoes a displacement of **s**

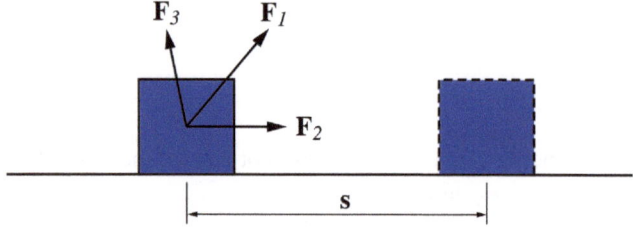

Fig. 4.2   An object exposed to several forces undergoes a displacement of **s**

Another method to find the work is by considering the resultant of these forces:

$$W = \left(\sum \mathbf{F}\right) \cdot \mathbf{s}$$

where

$$\Sigma \mathbf{F} = \mathbf{F}_1 + \mathbf{F}_2 + \mathbf{F}_3 +$$

Example 4.1   A lady pulls an 80 kg block horizontally on a rough surface by a constant force of 400 N that is at $20°$ to the horizontal. If the block is pulled a distance of 6 m and if the opposing force of friction has a magnitude of 118 N : (a) determine the work done on the block by each of the applied force, the frictional force, the normal force, and the force of gravity; (b) find the total work done on the block; (c) determine if it is easier for the lady to pull the block at an angle larger than $20°$.

Solution 4.1   (a) The work done by the applied force is

$$W_{app} = \mathbf{F} \cdot \mathbf{s} = Fs\cos\theta = (400\,\text{N})(6\,\text{m})\cos 20° = 2255.3\,\text{J}$$

The work done by the frictional force is

$$W_f = Fs\cos\theta = (118\,\text{N})(6\,\text{m})\cos 180° = -708\,\text{J}$$

The work done by the normal force and the force of gravity are both zero since each force is perpendicular to the displacement.

(b) The total work done is

$$W_{tot} = W_{app} + W_f = (2255.3\,\text{J}) - (708\,\text{J}) = 1547.3\,\text{J}$$

The total work done can also be found by computing the net force acting on the block and calculating its work.

(c) For ($0 \leq \theta \leq 90°$), If $\theta_2 > \theta_1$, then $\cos\theta_2 < \cos\theta_1$ and therefore $W_{app2} < W_{app1}$, i.e., it is easier for the man to pull at an angle larger than $20°$.

Example 4.2   A delivery man wants to push a crate up a ramp of length s: (a) find the minimum work the man must do to lift the crate to the top of the ramp; (b) determine if a ramp with a steeper incline would be more difficult for the man to push the crate.

Solution 4.2   (a) The minimum work that the delivery man must do is the work done against gravity The work done on the crate by the force of gravity is

$$W_g = -mgs\sin\theta$$

Hence the minimum work W_w that the delivery man must do is equal to $+mgs\sin\theta$.

(b) For angles between 0 and $90°$, if $\theta_2 > \theta_1$, then $\sin\theta_2 > \sin\theta_1$. Hence $W_{w2} \geq W_{w1}$, i.e., the more inclined the ramp is the more difficult it is to move the crate.

4.2.3   Work Done by a Varying Force

Previously, the work done in the special case of a force that is constant in both magnitude and direction was discussed. The object there moved along a straight line. In many situations, the force may vary in magnitude or in direction or in both, and the object may move along a curved path. To find the work done in this case, consider a particle moving along the curved path shown in Fig. 4.3. While it is moving, a force **F** that varies in both magnitude and direction with the position of the particle acts on it. Let us divide the path into a large number n of very small displacements where each is tangent to the path. For each displacement, the force can be approximated to be constant in both magnitude and direction. The total work done as the particle moves from P to Q is the sum of the individual amounts of work done along each displacement, that is

$$W = \mathbf{F}_1 \cdot \triangle \mathbf{r}_1 + \mathbf{F}_2 \cdot \triangle \mathbf{r}_2 + \mathbf{F}_3 \cdot \triangle \mathbf{r}_2 + \cdots \mathbf{F}_n \cdot \triangle \mathbf{r}_n$$

$$W = \sum_{i=1}^{n} \mathbf{F}_i \cdot \triangle \mathbf{r}_i$$

By dividing the path into more displacements we have

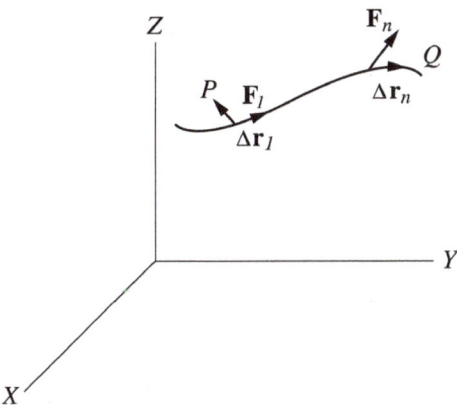

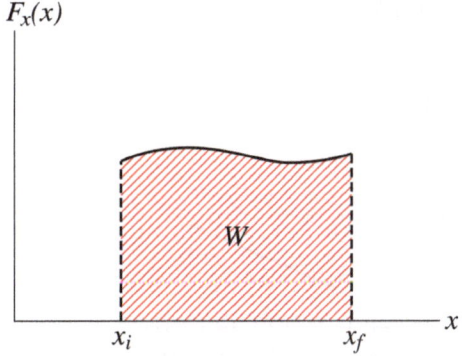

Fig. 4.4 The area under the curve represents the work

Fig. 4.3 A particle moving along a curved path. While itõs moving, a force **F** that varies in both magnitude and direction with the position of the particle acts on it

$$W = \lim_{\Delta \mathbf{r}_i \to 0} \sum_{i=1}^{n} \mathbf{F}_i \cdot \Delta \mathbf{r}_i$$

or

$$W = \int_C \mathbf{F} \cdot d\mathbf{r} = \int_P^Q \mathbf{F} \cdot d\mathbf{r}$$

As mentioned in Sect. 1.10.1, this integral is called the line integral. Each component of $\mathbf{F}(F_x, F_y$ or $F_z)$ may be a function of x, y, and z, and the curve can be determined by its equations that relates x, y, and z to each other. The component form of the above equation is

$$W = \int_{\mathbf{r}_i}^{\mathbf{r}_f} \mathbf{F} \cdot d\mathbf{r} = \int_{x_i}^{x_f} F_x dx + \int_{y_i}^{y_f} F_y dy + \int_{z_i}^{z_f} F_z dz \tag{4.1}$$

Now consider the case in which the particle moves along a straight line (for example the positive x-axis) and in which the force acting on the particle has a constant direction along the x-axis and a magnitude that varies with x. Equation 4.1 is then reduced to

$$W = \int_{x_i}^{x_f} F_x(x) dx \tag{4.2}$$

This equation represents the area under the curve in Fig. 4.4. If $F(x)$ is constant then we have

$$W = \int_{x_i}^{x_f} F_x(x) dx = F \int_{x_i}^{x_f} dx = F(x_f - x_i) = Fs$$

The work is then equal to the rectangular area shown in Fig. 4.5.

Example 4.3 In Example 3.3, find the work done by the force in moving the particle during the time interval from $t = 0$ to $t = 1$ s.

Solution 4.3 The work done from $t = 0$ to $t = 1$ s is

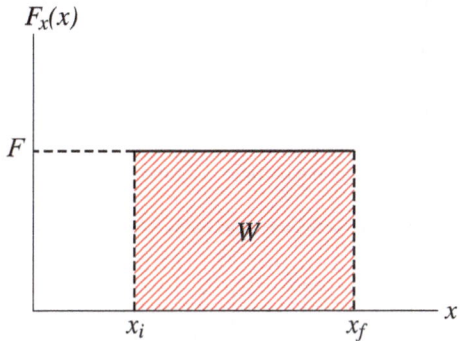

Fig. 4.5 The work is equal to the rectangular area

$$W = \int_{t=0}^{t=1} \mathbf{F} \cdot d\mathbf{r} = \int_{t=0}^{t=1} (2t^2 \mathbf{i} - 3t \mathbf{j}) \cdot (1/3t^4 \mathbf{i} - t^3 \mathbf{j}) dt$$

$$= \int_{t=0}^{t=1} (0.66t^6 + 3t^4) dt$$

$$= (0.1t^7 + 0.6t^5)|_{t=0}^{t=1} = 0.7 \, \text{J}$$

Example 4.4 A force acting on a particle is a function of position according to Fig. 4.6. Find the work done by this force as the particle moves from $x_i = 0$ to $x_f = 9$ m.

Solution 4.4 The work done is equal to the area of the triangle under the curve between $x_i = 0$ to $x_f = 9$ m, i.e.

$$W = \frac{1}{2}(9 \, \text{m})(4 \, \text{N}) = 18 \, \text{J}$$

Example 4.5 A ball that is suspended from a ceiling by a light rope is displaced a small distance to the position shown in Fig. 4.7. If it is released from rest at B, find the work done by the tension force and the force of gravity as the ball moves from B to A.

Solution 4.5 Because the tension force is always perpendicular to the displacement, the work done by it is zero at all times. The only component of the gravitational force that does work is its tangential component. Therefore,

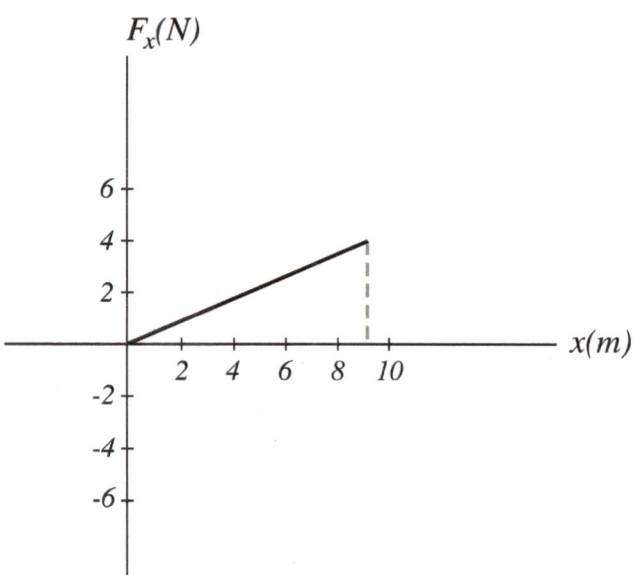

Fig. 4.6 A force acting on a particle is a function of position

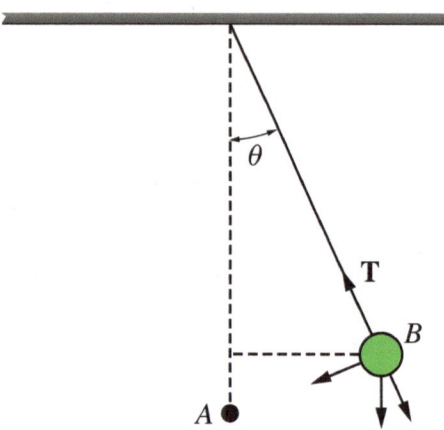

Fig. 4.7 A ball suspended by a light rope and displaced a small distance from the position of equilibrium

$$W = \int_0^{\theta_0} \mathbf{F} \cdot d\mathbf{s} = \int_0^{\theta_0} mg \sin\theta \cos(0) ds$$

Since $s = R\theta$, then $ds = Rd\theta$, and we have

$$W = mgR \int_0^{\theta_0} \sin\theta d\theta = -mgR \cos\theta|_0^{\theta_0} = mgR(1 - \cos\theta_0)$$

4.3 Kinetic Energy (KE) and the Work–Energy Theorem

Consider a particle that is exposed to a net field force and is moving along a curve in space. Suppose that the particle is at P at t_i and at Q at t_f and that its velocity at P and Q is $\mathbf{v}_i$ and $\mathbf{v}_f$, respectively. The net work done on the particle as it moves from P to Q is then given by

$$W = \int_P^Q \mathbf{F} \cdot d\mathbf{r} = \int_{t_i}^{t_f} \mathbf{F} \cdot \frac{d\mathbf{r}}{dt} dt = \int_{t_i}^{t_f} \mathbf{F} \cdot \mathbf{v} dt = \int_{t_i} m \frac{d\mathbf{v}}{dt} \cdot \mathbf{v} dt$$

$$= m \int_{\mathbf{v}_i}^{\mathbf{v}_f} \mathbf{v} \cdot d\mathbf{v} = \frac{1}{2} m \int_{\mathbf{v}_i}^{\mathbf{v}_f} d(\mathbf{v} \cdot \mathbf{v}) = \frac{1}{2} m (\mathbf{v} \cdot \mathbf{v})|_{\mathbf{v}_i}^{\mathbf{v}_f}$$

$$= \frac{1}{2} mv_f^2 - \frac{1}{2} mv_i^2$$

The quantity $\frac{1}{2}mv^2$ is the energy associated with the motion of the particle called the kinetic energy (KE). Thus, if a particle of constant mass m is moving with a speed v, its KE is a scalar quantity defined as

$$K = \frac{1}{2} mv^2$$

It also can be written as $K = \frac{1}{2} m(\mathbf{v} \cdot \mathbf{v})$. Hence, the total work done by the net force in displacing the particle is equal to the change in the KE of the particle

$$W_{net} = K_f - K_i = \triangle K$$

Similar to work, the SI unit of kinetic energy is the Joul. Note that the work–energy theorem is applied only if the object is treated as a particle (all of its parts move in exactly the same way). As an example of how the theorem is applied only for particle-like objects consider a man standing on a skateboard on a horizontal surface (see Fig. 4.8). If the man

Fig. 4.8 The center of mass of the system (man+skateboard) moves and the work-energy theorem can be applied to that point

pushes the bar then that would move him backwards along with his skateboard. This motion is due to the reaction force **F** exerted on him by the bar. The work done by **n** or **w** is equal to zero since each force is perpendicular to the displacement. Because the point of application of **F** did not move it follows that the work done by that force is zero. Thus, from the work–energy theorem the man should not move. The question is why did he move?

The fact here is that it is incorrect to treat the man as a particle, since different parts of his body move in different ways as he pushes the bar. Therefore, the work–energy theorem does not hold. The man must be treated as a system of particles. In Chap. 6, we will see that the motion of a system of particles can be represented by the motion of its center of mass. The center of mass behaves as if all of the mass of the object (or system) is concentrated there and as if the net external force is applied there. In the case of the skateboarder, the center of mass of the system (man + skateboard) moves and the work–energy theorem can be applied to that point.

The work–energy theorem is an alternative method for describing motion without using Newton's laws. It is especially useful in problems involving a varying force. Note that the work and the kinetic energy are not invariant quantities; they have different values when measured in different inertial frames of reference. However, from the principle of invariance, the equation $W_{net} = \triangle K$ still holds for any inertial frame.

Example 4.6 A 5 kg block resting on a surface is given an initial velocity of 5 m/s. If the coefficient of kinetic friction of the surface is $\mu_k = 0.2$, find the distance the block would move before it stops.

Solution 4.6 As we will see later in Sect. 4.3.1, the change in the kinetic energy of the block due to friction is $\triangle K = -f_k s$, where s is the displacement of the block.

$$W_f = \triangle K = -f_k d = -\mu_k mgd = \frac{1}{2}mv_f^2 - \frac{1}{2}mv_i^2$$

$$= -(0.2)(5\,\text{kg})(9.8\,\text{m/s}^2)d = 0 - \frac{1}{2}(5\,\text{kg})(5\,\text{m/s})^2$$

$$d = 6.4\,\text{m}$$

Example 4.7 A 10 kg block is pushed on a frictionless horizontal surface by a constant force of magnitude of 100 N and that is at $30°$ below the horizontal. If the block starts from rest, find its final speed after it has moved a distance of 3 m using work–energy theorem.

Solution 4.7

$$W = \mathbf{F} \cdot \mathbf{s} = Fs\cos\theta = (100\,\text{N})(3\,\text{m})\cos(-30°) = 259.8\,\text{J}$$

From the work–energy theorem, we have

$$W = \frac{1}{2}mv_f^2 - \frac{1}{2}mv_i^2$$

since $v_i = 0$ we get

$$v_f^2 = \frac{2W}{m} = \frac{2(259.8\,\text{J})}{(10\,\text{kg})} = 52\,\text{m}^2/\text{s}^2$$

$$v_f = 7.2\,\text{m/s}$$

4.3.1 Work Done by a Spring Force

Consider a block attached to a light spring fixed at the other end on a frictionless horizontal surface as in Fig. 4.9. Suppose an external force $\mathbf{F}_{ext}$ is applied to the block by either stretching or compressing it through a small displacement from its equilibrium (relaxed) position taken at $x = 0$. The spring will then exert a restoring force $\mathbf{F}_s$ on the block that opposes the applied force and restores the block to its equilibrium position. For many kinds of springs and in the case of small displacements, the spring force varies linearly with the displacement x of the block (or any other object) from its equilibrium position ($x = 0$). That is

$$F_s = -kx$$

where k is a constant called the force or spring constant. k measures the stiffness of the spring. The stiffer the spring the larger is k. This equation is known as Hook's law. The minus sign indicates that the spring force is always acting in a direction opposing the displacement. The work done by the spring force in moving the block from an initial position x_i to a final position x_f is:

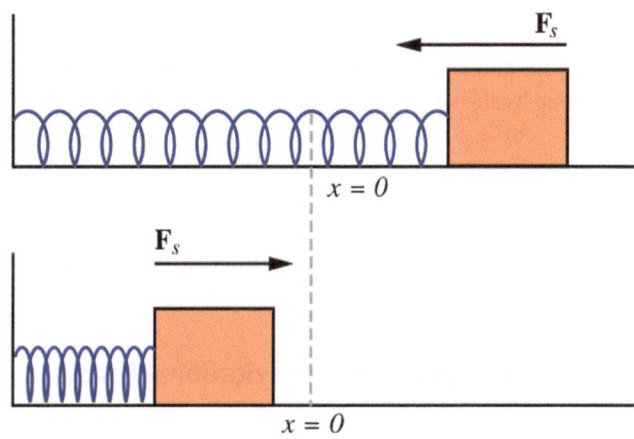

Fig. 4.9 A block attached to a light spring on a frictionless surface

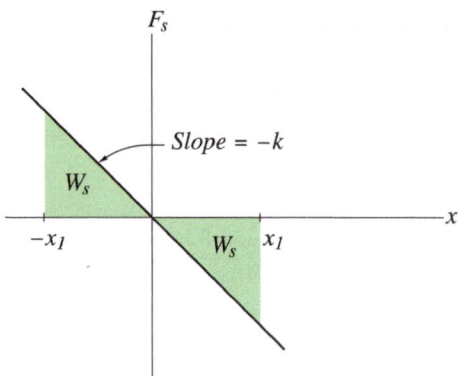

Fig. 4.10 A plot of F_s versus x for the mass-spring system

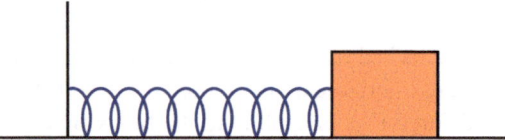

Fig. 4.11 A 2 kg block attached to a light spring of force constant 300 N/m on a horizontal smooth surface

$$W_s = \int_{x_i}^{x_f} F_x dx = \int_{x_i}^{x_f} (-kx)dx = -k \int_{x_i}^{x_f} x dx$$

$$W_s = \frac{1}{2}kx_i^2 - \frac{1}{2}kx_f^2$$

The work done on the block by the spring as it moves from an initial position $x_i = x$ to a final position $x_f = 0$ is

$$W_s = \frac{1}{2}kx^2$$

Figure 4.10 shows a plot of F_s versus x for the mass–spring system.

Example 4.8 A 2 kg block is attached to a light spring of force constant 300 N/m on a horizontal smooth surface as shown in Fig. 4.11. If the system is initially at rest at the position of equilibrium and is then stretched a distance of 3 cm, find the work done by the spring on the block as it moves from $x_i = 0$ to $x_f = 3$ cm.

Solution 4.8

$$W_s = \frac{1}{2}kx_i^2 - \frac{1}{2}kx_f^2 = 0 - \frac{1}{2}(300\,\text{N/m})(0.03\,\text{m})^2 = -0.135\,\text{J}$$

4.3.2 Work Done by the Gravitational Force (Weight)

If a particle-like object of mass m is moving vertically upward or downward near the surface of the earth where g is assumed

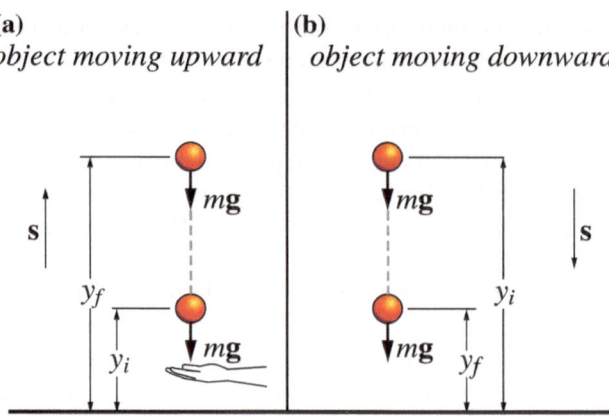

Fig. 4.12 By taking $y = 0$ at the hand level, in the work done by gravity **a** is $-mgy_f$ and in **b** is $+mgy_i$

to be constant (see Fig. 4.12), and if air resistance is neglected, then the only force that does work on the object is the gravitational force mg. By taking the y-axis along the line of motion (positive upwards) with $y = 0$ at the earth's surface, the work done by the gravitational force is

$$W_g = \int_{y_i}^{y_f} F_y dy = -mg \int_{y_i}^{y_f} dy$$

$$W_g = mgy_i - mgy_f$$

Note that unlike the spring force the reference point y_i may be chosen anywhere. If the object moved downwards from $y_i = y$ to $y_f = 0$, the work done by the gravitational force is

$$W_g = mgy$$

Now suppose the object moves along a curved path from P to Q as in Fig. 4.13. The work done by the gravitational force is

$$W = \int_P^Q m\mathbf{g} \cdot d\mathbf{s} = -\int_P^Q mg\mathbf{i} \cdot d(dx\mathbf{i} + dy\mathbf{j}) = -\int_{y_i}^{y_f} mgdy = mgy_i - mgy_f$$

This result is the same as if the object has followed a straight vertical path. Therefore, the work done by the gravitational force depends only on the initial and final positions of the object.

Example 4.9 A man lifts a 300 kg weight a distance of 2 m above the ground. Find the work done by the force of gravity on the weight.

Solution 4.9

$$W = mgy_i - mgy_f = 0 - (300\,\text{kg})(9.8\,\text{m/s}^2)(2\,\text{m}) = -5880\,\text{J}$$

(a)

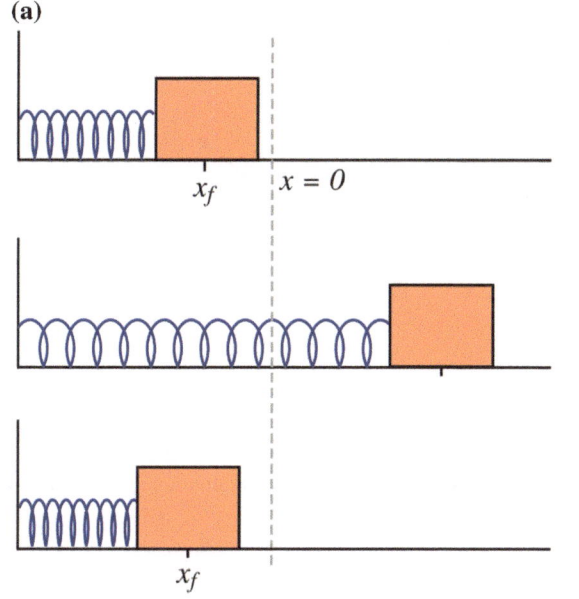

(b)

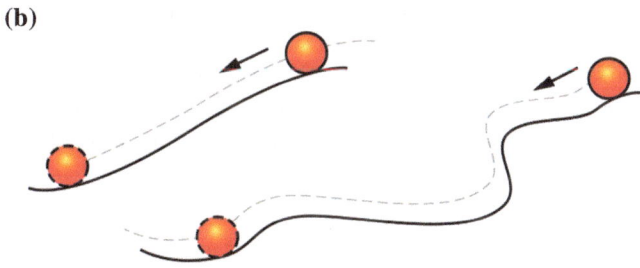

Fig. 4.13 a The total work done by the spring force on the block is zero since $x_i = x_f$. **b** Along any path the work done by the gravitational force is the same since the initial and final positions are the same

4.3.3 Power

Power is a quantity that defines how much work is done over a period of time, i.e., power is the time rate of doing work, or more generally, it is the time rate of energy transfer. If an external force F does work W on an object for a time interval $\triangle t$, then the average power during that time is

$$\overline{P} = \frac{W}{\triangle t}$$

The instantaneous power is

$$P = \lim_{\triangle t \to 0} \frac{W}{\triangle t} = \frac{dW}{dt}$$

Our concern in this book is the mechanical power since it involves mechanical work. If y is the velocity of the object, we have

$$P = \frac{dW}{dt} = \mathbf{F} \cdot \frac{d\mathbf{s}}{dt}$$

for a constant force, or

$$P = \mathbf{F} \cdot \mathbf{v}$$

The SI unit of power is joules per second (J/s) and is called the watt (W).

$$1\,\text{W} = 1\,\text{J/s} = 1\,\text{kg.m}^2/\text{s}^3$$

Another unit of power that is often used is the horsepower:

$$1\,\text{hp} = 746\,\text{W}$$

4.4 Conservative and Nonconservative Forces

In nature, there are two kinds of forces: conservative and non-conservative forces. A conservative force is a force that conserves the energy of a system when acting upon it. The action of this force results in changing the kinetic energy of any object in the system. This change will be stored in the system in the form of potential energy. For every conservative force, there is a certain potential energy that is associated with it. Such potential energy can be retransformed into kinetic energy Thus, the total energy of the system would not be dissipated, instead it would be conserved. A force that does not act in this way is said to be a nonconservative force. Properties of a conservative force are given as follows:

1. The net work done by a conservative force on a particle moving from one point to another is independent of the path taken by the particle;
2. The net work done by a conservative force in moving a particle through any closed path is equal to zero.

A force not meeting these conditions is a nonconservative force. As mentioned in Sect. 1.10.2, property 2 of a conservative force can be obtained from property 1 (if **A** is a vector field and the line integral of **A** between any two points is independent of path, then $\oint_C \mathbf{A} \cdot \mathbf{r} = 0$). That is, these two properties are equivalent. Examples of conservative forces in mechanics are the gravitational and spring forces. To show this let us go back to Sects. 4.3.1 and 4.3.2, where the work done by the gravitational force or the spring force was calculated. We have seen that the work done in each case depends only on the initial and final positions of the object. Therefore, the work done by any of these forces is independent of the path joining the initial and final positions. Furthermore, if $(x_i = x_f)$ in the case of the spring or $(y_i = y_f)$ in the case of the gravitational force the net work done is zero. Hence, these forces are conservative.

Fig. 4.14 The longer the path the more interaction between the block and the surface and the more the force of friction will act and do work on the block

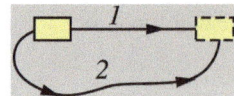

The force of friction is an example of a nonconservative force. To show that, consider a block sliding on a rough surface. Figure 4.14 shows two possible paths connecting two points. The longer the path the more interaction between the block and the surface and the more the force of friction will act and do work on the block. Thus, the work depends on the path taken between the two points and therefore the frictional force is a nonconservative force.

4.4.1   Potential Energy

For a system consisting of two or more objects, the potential energy U of the system is the energy associated with the configuration of the system. That is, the potential energy is the energy associated with the position of objects in the system relative to each other. If the configuration of the system is changed, then the potential energy of the system also changes. Such energy is defined only in terms of a conservative force because if such a force acts on a system then it can transform the kinetic energy of any object in the system into potential energy of the system and vice versa. The potential energy means that the system has potential to do work.

In Sect. 1.10.2 it has been proven that the line integral in Eq. (1.2) is independent of the path joining the points P and Q if and only if $\mathbf{A} = \nabla\phi$, or equivalently $\nabla \times \mathbf{A} = 0$. Where $\phi(x, y, z)$ is some scalar that has continuous partial derivatives. Therefore, for a conservative force field $\mathbf{F}(x, y, z)$, there always exist a scalar field $U = U(x, y, z)$ (called the potential energy) such that

$$\mathbf{F} = -\nabla U = -\left(\frac{\partial U}{\partial x}\mathbf{i} + \frac{\partial U}{\partial y}\mathbf{j} + \frac{\partial U}{\partial z}\mathbf{k}\right)$$

Furthermore

$$\nabla \times \mathbf{F} = 0$$

Thus, the total work done by a conservative force in moving a particle from P_i to P_f (see Fig. 4.15) is

$$W = \int_{P_i}^{P_f} \mathbf{F} \cdot d\mathbf{s} = \int_{P_i}^{P_f} -\nabla U \cdot d\mathbf{s} = \int_{P_i}^{P_f} -dU = U_i - U_f = -\Delta U$$

or

$$\Delta U = -\int_{P_i}^{P_f} \mathbf{F} \cdot d\mathbf{s}$$

where $U = U(x, y, z)$. Because only the change in the potential energy is significant, it does not matter where the

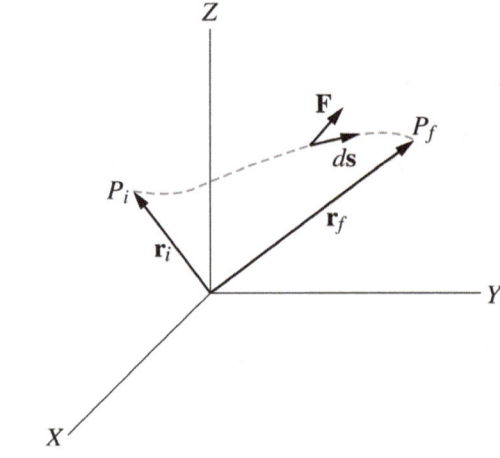

Fig. 4.15 The total work done by a conservative force in moving a particle from P_i to P_f

reference point (U) is chosen. This is because if U_i is changed U_f will be also changed but ΔU will remain constant.

Example 4.10 A force acting on a particle is given by $\mathbf{F} = -k\mathbf{r}$. Determine: (a) whether or not the force is conservative; (b) the potential energy associated with the force if it is conservative.

Solution 4.10   (a)

$$\nabla \times \mathbf{F} = \begin{vmatrix} \mathbf{i} & \mathbf{j} & \mathbf{k} \\ \frac{\partial}{\partial x} & \frac{\partial}{\partial y} & \frac{\partial}{\partial z} \\ -kx & -ky & -kz \end{vmatrix}$$

$$= \left[\frac{\partial}{\partial y}(-kz) - \frac{\partial}{\partial z}(-ky)\right]\mathbf{i} + \left[\frac{\partial}{\partial z}(-kx) - \frac{\partial}{\partial x}(-kz)\right]\mathbf{j} + \left[\frac{\partial}{\partial x}(-ky) - \frac{\partial}{\partial y}(-kx)\right]\mathbf{k} = \mathbf{0}$$

Therefore, the force is conservative.
(b)

$$U = -\int \mathbf{F} \cdot d\mathbf{r} = -\int -k\mathbf{r} \cdot d\mathbf{r} = \int kr\,dr = \frac{1}{2}kr^2 = \frac{1}{2}k(x^2 + y^2 + z^2)$$

Example 4.11 If a force acting on a particle is given by $\mathbf{F} = ay\mathbf{j}$, where a is a positive constant: (a) find the work done in moving the particle along the closed path shown in Fig. 4.16; (b) determine if the force is conservative.

Solution 4.11   (a) Along path 1 we have $y = 1$ and $dy = 0$ and along path 3 we have $y = 2$ and $dy = 0$.

$$W = \oint_c \mathbf{F} \cdot d\mathbf{r} = \int_1 \mathbf{F} \cdot d\mathbf{r} + \int_2 \mathbf{F} \cdot d\mathbf{r} + \int_3 \mathbf{F} \cdot d\mathbf{r} + \int_4 \mathbf{F} \cdot d\mathbf{r} = 0 + \int_{y=1}^{2} ay\,dy + 0 + \int_{y=2}^{1} ay\,dy = 0$$

(b) Since the total work done through the closed path is zero, the force is conservative.

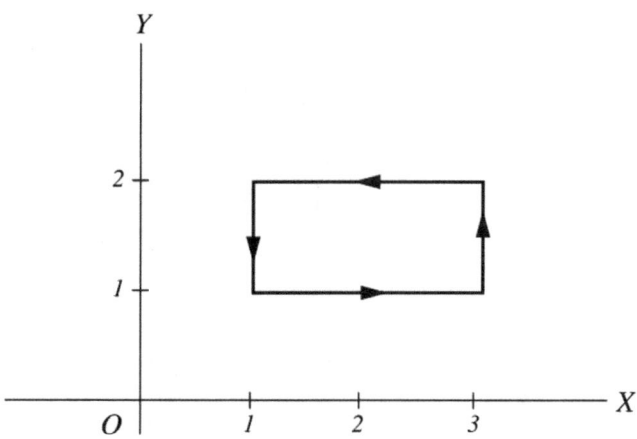

Fig. 4.16 The work done in moving the particle along a closed path

Example 4.12   Find the force acting on a particle if the potential energy associated with it is $U = 5y^2 - 3z$.

Solution 4.12

$$F_y = -\frac{\partial U}{\partial y} = -10y$$

$$F_z = -\frac{\partial U}{\partial z} = 3$$

and therefore $\mathbf{F} = -10y\mathbf{j} + 3\mathbf{k}$.

4.4.1.1 The Gravitational Potential Energy

As we have mentioned in Sect. (4.1.7), the work done by the gravitational force in displacing a particle near the earth's surface from y_i to y_f is

$$W_g = mgy_i - mgy_f$$

Since $W_g = -\triangle U_g = U_{gi} - U_{gf}$, we have

$$U_{gi} - U_{gf} = mgy_i - mgy_f$$

If $y_f = 0$ and $y_i = y$, the gravitational potential energy of the object–earth system may be written as

$$U_g = mgy$$

The force of gravity near the surface of the earth can be found from the gravitational potential energy In general we have $\mathbf{F} = -\nabla U$ here, since the motion is in one direction we have

$$\mathbf{F}_g = -\frac{dU}{dy}\mathbf{j} = -\frac{d}{dy}(mgy)\mathbf{j} = -mg\mathbf{j}$$

4.4.1.2 The Elastic Potential Energy

It was found in Sect. (4.1.6) that the work done by the spring force when moving a block from x_i to x_f (when it is stretched

or compressed) is

$$W_s = \frac{1}{2}kx_i^2 - \frac{1}{2}kx_f^2$$

Since $W_s = -\triangle U_s = U_{si} - U_{sf}$, we have

$$U_{si} - U_{sf} = \frac{1}{2}kx_i^2 - \frac{1}{2}kx_f^2$$

If $x_i = 0$ and $x_f = x$, the elastic potential energy of the block-spring system can be written as

$$U_s = \frac{1}{2}kx^2$$

The spring force can be found from the elastic potential energy

$$\mathbf{F} = -\frac{dU}{dx}\mathbf{i} = -\frac{d}{dx}\left(\frac{1}{2}kx^2\right)\mathbf{i} = -kx\mathbf{i}$$

4.5   Conservation of Mechanical Energy

The total mechanical energy of a system is defined as the sum of all of the kinetic energies of the objects within the system plus all of the potential energies of the system.

$$E_{tot} = K_{tot} + U_{tot}$$

Now, consider an isolated system in which there are no external forces acting on it, or the net external force is zero. The only forces acting on the system will be the internal forces within the system. These forces may be conservative or nonconservative. If only internal conservative forces exist, then the work done by any of these forces on an object in the system will transform its kinetic energy into potential energy (associated with that force), or vice versa. The internal conservative force can also transform one form of potential energy into another. The work done by such a force on an object in the system is

$$W = \triangle K$$

The change in potential energy due to this work is

$$W = -\triangle U$$

Thus,

$$\triangle K = -\triangle U$$

or

$$\triangle K + \triangle U = 0$$

or

$$K_i + U_i = K_f + U_f$$

If more than one conservative force acts, there will be a potential energy associated with each force. That is

$$K_i + \sum U_i = K_f + \sum U_f$$

Therefore we have

$$E_i = E_f$$

or

$$\triangle E = 0$$

From the previous discussion, we conclude that for an isolated system in which only conservative forces act, the total mechanical energy of the system remains constant (conserved). Figure 4.17 shows the changes of energy of a ball thrown upwards. Now suppose that the system is not isolated and that the external forces acting on the system are conservative. The change in the kinetic energy of the system is then equal to the work done on the system by an internal conservative force plus the amount of kinetic energy changed due to an external conservative force, that is,

$$\triangle K = W_{\text{intc}} + \triangle K_{ext}$$

or

$$\triangle K = -\triangle U - \triangle U_{ext}$$

Hence

$$\triangle K + \triangle U + \triangle U_{ext} = 0$$

Therefore, the total mechanical energy of the system remains constant under both external and internal conservative forces. If external nonconservative forces act on the system, or if there is heat transfer, or if internal nonconservative forces act, then the total mechanical energy may change and is no longer conserved.

4.5.1 Changes of the Mechanical Energy of a System due to External Nonconservative Forces

External nonconservative forces may act on a system if it is not isolated. Consider a system that is not isolated in which only internal conservative forces act. The change in the kinetic energy of the system is then equal to the work done on the system by an internal conservative force plus the amount of kinetic energy changed due to an external nonconservative force. This can be expressed as

$$\triangle K = W_{\text{intc}} + \triangle K_{ext}$$

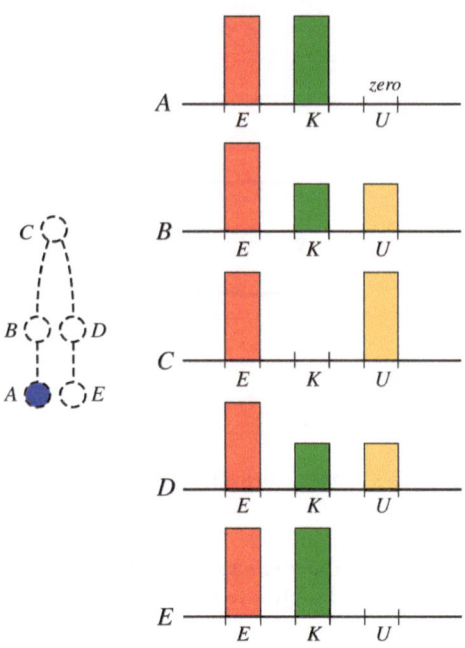

Fig. 4.17 Changes in the kinetic and potential energies of a ball thrown vertically upwards

or

$$\triangle K = -\triangle U + \triangle K_{ext}$$

Thus

$$\triangle E = \triangle K_{ext} \qquad (4.3)$$

This implies that the total mechanical energy has changed by an amount of $\triangle K_{ext}$. Not that the work done by a nonconservative force cannot be calculated generally but the change in the kinetic energy can be observed.

4.5.2 Friction

Friction is a nonconservative force as seen in Sect. 4.2. If this force is applied externally to a system in which only internal conservative forces act, it will decrease (dissipate) the kinetic energy of the system by transforming it into thermal energy The change in the mechanical energy of the system is

$$\triangle E = \triangle K_{ext}$$

The work done by friction or any other nonconservative force cannot be calculated. In other words, the work done by friction is not simply $-f_k s$, where s is the displacement of the object in the system. The reason behind not being able to calculate the work done by friction is that at a microscopic level the frictional force is not a single force that acts at one point. Rather, it is a combination of forces acting at different points in the object. However, the loss in kinetic energy of the object can be calculated as shown below: Consider a block sliding on a rough surface. Let's choose the block only to be our system. From the equation of motion, we have

$$v_f^2 - v_i^2 = 2as$$

Newton's second law gives

$$-f_k = ma$$

Thus

$$-fs = mas$$

or

$$-f_s s = \frac{1}{2}mv_f^2 - \frac{1}{2}mv_i^2$$

Therefore

$$\triangle K_{ext} = -f_k s$$

This quantity represents the magnitude of the loss in the kinetic energy of the block due to friction. This loss of energy appears as thermal energy of the block and of the surface on which it slides.

4.5.3 Changes in Mechanical Energy due to Internal Nonconservative Forces

In solving problems you are free to choose the system. If we considered the block plus the surface as our system, then friction will be an internal nonconservative force and we may write

$$\triangle E = -\triangle E_{int} = \triangle K_{in,nc} = -f_k s$$

where $\triangle K_{in,nc}$ is the change in the kinetic energy of the system due to an internal nonconservative force. Another example of a nonconservative force is the force that you exert on your body by your muscles. This force transfers the chemical energy of your body into kinetic energy In Sect. (4.1.5), we have seen that the motion of the skateboarder can be explained using the concept of the center of mass. Another way to explain the motion of the skateboarder is that the internal chemical energy of the man is transformed into kinetic energy, and we may write

$$\triangle E = \triangle K = -\triangle E_{int}$$

Since $\triangle U = 0$ in his case. An additional example of nonconservative forces is the forces that different parts in an object exert on each other when the object is deformed. These forces transform the kinetic energy of the object into internal energy.

In all cases, even though energy can transfer from one object to another or to the environment, the total amount of energy in the universe is constant. That is, energy gained by a system is lost by another system. In other words, *energy cannot be created or destroyed it can only be transformed from one form to another and the total energy of an isolated system is conserved (constant)*. This statement is known as the law of conservation of energy The law of conservation of energy is also valid in relativity and quantum mechanics.

4.5.4 Changes in Mechanical Energy due to All Forces

Consider a system in which there are both internal and external conservative and nonconservative forces acting on it. In this case, the change in the total mechanical energy of the system can be written as

$$\triangle E = \triangle K + \triangle U + \triangle U_{ext} = \triangle K_{ext} - \triangle E_{int}$$

Example 4.13 A 0.2 kg apple falls from a tree at a distance of 3 m above the ground. Find: (a) the velocity of the apple at an altitude of 2 m and at the instance just before it hits the ground; (b) the altitude of the apple when its velocity is 4 m/s.

Solution 4.13 (a) Consider the system to be the earth + the apple. By neglecting air resistance (the apple is in free-fall), the only internal force that acts within the earth–apple system is the force of gravity. Because the gravitational force is a conservative force, the total mechanical energy of the system is conserved. Therefore as the apple falls its gravitational potential energy is converted into kinetic energy such that at any instant the total mechanical energy of the system is constant. Applying the law of conservation of energy to the system and by taking $y = 0$ at the earth's surface and the gravitational potential energy to be zero at $y = 0$, we have

$$K_f + U_f = K_i + U_i$$

$$\frac{1}{2}mv_f^2 + mgy = 0 + mgh$$

where h is its initial altitude. That gives

$$v_f = \sqrt{2g(h - y)}$$

At $y = 2$ m,

$$v_f = \sqrt{2(9.8 \, \text{m/s}^2)(1 \, \text{m})} = 4.43 \, \text{m/s}$$

At $y = 0$

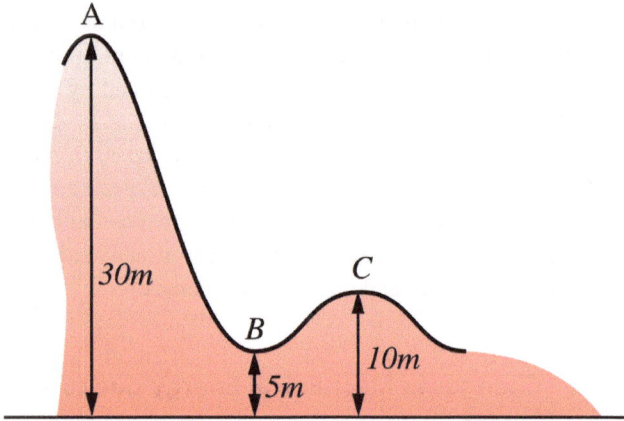

Fig. 4.18 By ignoring friction, the total energy of the roller coaster can be considered to be conserved

$$v_f = \sqrt{2(9.8\,\text{m/s}^2)(3\,\text{m})} = 7.7\,\text{m/s}$$

(b)

$$y = h - \frac{v_f^2}{2g} = (3\,\text{m}) - \frac{(4\,\text{m/s})^2}{2(9.8\,\text{m/s}^2)} = 2.2\,\text{m}$$

Example 4.14 A roller coaster of mass 500 kg starts from rest at point A, and rolls down the track as shown in Fig. 4.18. Ignoring friction, determine: (a) the roller coaster speed at B and C; (b) the work done by gravity as the rollercoaster moves from A to B.

Solution 4.14 (a) Consider the system to consist of the roller-coaster + the track + the earth. Taking the gravitational potential energy to be zero at the earth's surface and from the conservation of energy we have

$$K_f + U_f = K_i + U_i$$

$$\frac{1}{2}mv_B^2 + mgh_B = 0 + mgh_A$$

Therefore,

$$v_B = \sqrt{2g(h_A - h_B)} = \sqrt{2(9.8\,\text{m/s}^2)(25\,\text{m})} = 22.13\,\text{m/s}$$

Similarly, the velocity at C is

$$v_C = \sqrt{2g(h_A - h_C)} = \sqrt{2(9.8\,\text{m/s}^2)(20\,\text{m})} = 19.8\,\text{m/s}$$

You may also calculate the velocity at C by taking B as the initial point.

(b) As the car moves from A to B the work done by gravity is

$$W_g = -\Delta U$$

Fig. 4.19 A block released from rest on top of an incline

$$W_g = -(mgh_b - mgh_a) = 1.22 \times 10^5\,\text{J}$$

Example 4.15 A block of mass 5 kg is released from rest at the top of a 45° incline that is 0.5 m long as shown in Fig. 4.19. It then slides on a horizontal surface that is 0.7 m long and goes up again on a second ramp that is at 30° to the horizontal. If the coefficient of kinetic friction between the block and all three surfaces is 0.2, find the maximum distance that the block would move up the second ramp?

Solution 4.15 First, we divide the path into three parts. Let us consider the system as the block only Along the first part the change in the total mechanical energy of the system is equal to the energy dissipated by friction. Thus,

$$\Delta E = \Delta K_{ext}$$

$$K_f + U_f = K_i + U_i + \Delta K_{ext}$$

$$\frac{1}{2}mv_{f1}^2 + 0 = 0 + mgh - f_{k1}s_1$$

the force of kinetic friction is

$$f_{k1} = \mu_k n = \mu_k mg\cos\theta_1 = (0.2)(5\,\text{kg})(9.8\,\text{m/s}^2)\cos 45^o = 6.93\,\text{N}$$

That gives

$$\frac{1}{2}mv_{f1}^2 = mgs_1\sin\theta_1 - f_{k1}s_1 = (5\,\text{kg})(9.8\,\text{m/s}^2)(0.5\,\text{m})\sin 45^\circ$$
$$- (6.93\,\text{N})(0.5\,\text{m}) = 13.9\,\text{J}$$

$v_{f1} = 2.35\,\text{m/s}$. Along the second path we have again

$$K_f + U_f = K_i + U_i + \Delta K_{ext}$$

$$\frac{1}{2}mv_{f2}^2 + 0 = \frac{1}{2}mv_{i2}^2 + 0 - f_{k2}s_2$$

The force of kinetic friction is given by

$$f_{k2} = \mu_k mg = (0.2)(5\,\text{kg})(9.8\,\text{m/s}^2) = 9.8\,\text{N}$$

and therefore

$$\frac{1}{2}mv_{f2}^2 = \frac{1}{2}(5\,\text{kg})(2.35)^2 - (9.8\,\text{N})(0.7\,\text{m}) = 6.95\,\text{J}$$

$$v_{f2} = \sqrt{2\frac{(6.94\,\text{J})}{(5\,\text{kg})}} = 1.7\,\text{m/s}$$

Finally, along the third path, we also have

$$K_f + U_f = K_i + U_i + \triangle K_{ext}$$

and

$$0 + mgs_3 \sin 30^o = \frac{1}{2}mv_{i3}^2 + 0 - f_{k3}s_3$$

but we have

$$f_{k3} = \mu_k mg \cos 30^o = (0.2)(5\text{kg})(9.8\,\text{m/s}^2)(0.866) = 8.5\,\text{N}$$

and thus

$$(5\,\text{kg})(9.8\,\text{m/s}^2)s_3(0.5) = \frac{1}{2}(5\,\text{kg})(1.7\,\text{m/s})^2 - (8.5\,\text{N})s_3$$

That gives $s_3 = 0.2\,\text{m}$.

Example 4.16 Two masses $m_1 = 5$ kg and $m_2 = 9$ kg are connected by a light rope that passes over a massless frictionless pulley as in Fig. 4.20. If the system is released from rest when m_2 is at 0. 5 m above the ground, use the principle of conservation of energy to determine the speed with which m_2 will hit the ground.

Solution 4.16 If air resistance is neglected, the only force acting in the masses-earth system is the gravitational force between them and hence the total mechanical energy of the system is conserved, i.e.,

$$K_f + U_f = K_i + U_i$$

Fig. 4.20 Two masses connected by a light rope that passes over a massless frictionless pulley

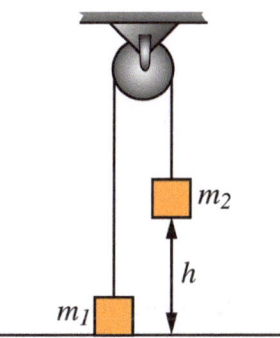

Because the two masses are connected by a rope, they have the same speed at any instant. If m_2 descends a distance h, m_1 will rise through the same distance and we have

$$\frac{1}{2}m_1v^2 + \frac{1}{2}m_2v^2 + m_1gh = m_2gh$$

$$\frac{1}{2}(m_1 + m_2)v^2 = g(m_2 - m_1)h$$

and therefore

$$v = \sqrt{\frac{2gh(m_2 - m_1)}{(m_1 + m_2)}} = \sqrt{\frac{2(0.5\,\text{m})(9.8\,\text{m/s}^2)(4\,\text{kg})}{(14\,\text{kg})}} = 1.7\,\text{m/s}$$

Example 4.17 A 0.25 kg ball is attached to alight string of length $L = 0.5$ m as in Fig. 4.21. Find (a) the tension in the string at B($\theta = 10°$) if the ball is given an initial velocity $v_a = 0.5$ m/s at its lowest position; (b) the velocity of the ball at A if the ball is released from rest at B.

Solution 4.17 (a) At point B some of the kinetic energy of the ball is converted into potential energy By taking the origin of the x-y coordinates at the lowest point A, we have

$$K_f + U_f = K_i + U_i$$

$$\frac{1}{2}mv_b^2 + mgL(1 - \cos\theta) = \frac{1}{2}mv_a^2 + 0$$

and therefore we get

$$v_b^2 = v_a^2 - 2gL(1 - \cos\theta) \qquad (4.4)$$

Applying Newton's second law along the radial direction to the ball at B we have

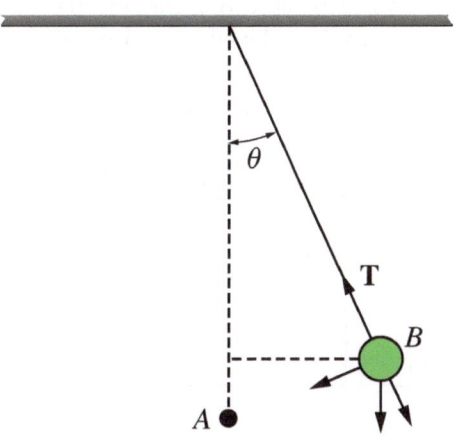

Fig. 4.21 A ball attached to a light string

$$T - mg \cos \theta = \frac{mv_b^2}{L} \qquad (4.5)$$

Substituting Eq. 4.4 into Eq. 4.5 gives

$$T = mg \cos \theta + \frac{m}{L}(v_a^2 - 2gL(1 - \cos \theta))$$

thus

$$T = m\left(g \cos \theta + \frac{v_a^2}{L} - 2g + 2g \cos \theta\right)$$

and hence

$$T = m\left(3g \cos \theta - 2g + \frac{v_a^2}{L}\right)$$

Substituting the values of θ and v_a gives $T = 2.5$ N.

(b) If $v_b = 0$, we have from $K_f + U_f = K_i + U_i$

$$\frac{1}{2}mv_a^2 + 0 = 0 + mgL(1 - \cos \theta)$$

hence

$$v_a = \sqrt{2gL(1 - \cos \theta)} = \sqrt{2(9.8 \, \text{m/s}^2)(0.5 \, \text{m})(1 - \cos 10^\circ)} = 0.4 \, \text{m/s}$$

Example 4.18 A 3 kg block compresses a spring of negligible mass a distance of 0.1 m from its equilibrium position as in Fig. 4.22. If the surface is frictionless and the force constant of the spring is 200 N/m, and the block is free to move, find: (a) the speed of the block just as it leaves the spring; (b) the maximum height that the block will reach; (c) suppose that a part of the horizontal track is rough with a length of 0.05 m, find the coefficient of kinetic friction if the block reaches a maximum height of 0.014 m.

Solution 4.18 (a) The only force acting inside the spring–mass–earth system is the spring force that acts on the block. This force is conservative and therefore the total mechanical energy of the system is conserved. The potential energy of the spring is transformed into kinetic energy of the block,

$$K_f + U_f = K_i + U_i$$

$$\frac{1}{2}mv_f^2 + 0 = 0 + \frac{1}{2}kx^2$$

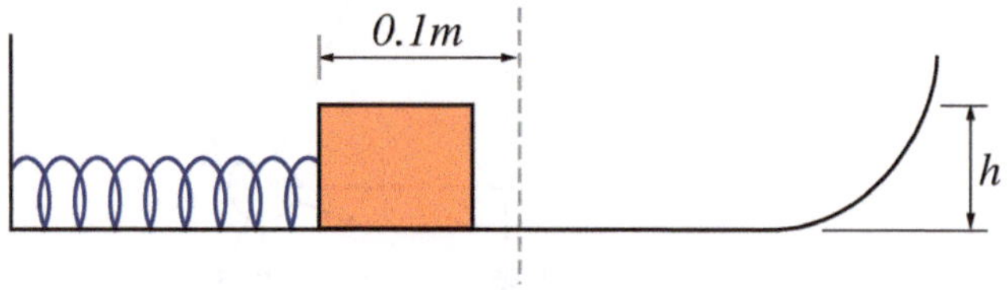

and therefore

$$v_f^2 = \frac{k}{m}x^2 = \frac{(200 \, \text{N/m})}{(3 \, \text{kg})}(-0.1 \, \text{m})^2$$

this gives $v_f = 0.82$ m/s.

(b)

$$K_f + U_f = K_i + U_i$$

$$0 + mgh = \frac{1}{2}mv_i^2 + 0$$

and hence

$$h = \frac{v_i^2}{2g} = \frac{(0.8.2 \, \text{m/s})^2}{2(9.8 \, \text{m/s}^2)} = 0.034 \, \text{m}$$

We can also take the initial position before the block is released.

(c)

$$K_f + U_f = K_i + U_i + \Delta K_{in,nc}$$

$$0 + mgh = 0 + \frac{1}{2}kx^2 - f_k d$$

along the rough surface $f_k = \mu_k mgd$, and therefore

$$\mu_k mgd = \frac{1}{2}kx^2 - mgh$$

thus

$$\mu_k(3 \, \text{kg})(9.8 \, \text{m/s}^2)(0.05 \, \text{m}) = \frac{1}{2}(200 \, \text{N/m})(0.1)^2 - (3 \, \text{kg})(9.8 \, \text{m/s}^2)(0.014)$$

That gives $\mu_k = 0.2$

Example 4.19 A small stone of mass 0.1 kg is released from rest inside a large hemispherical bowl of radius $R = 0.2$ m. It then slides along the surface as in Fig. 4.23. (a) Find the speed of the stone at point B and C; (b) If the surface of the bowl is not frictionless, how much energy is dissipated by friction as the stone moves from A to B if the speed at B is 1.7 m/s?

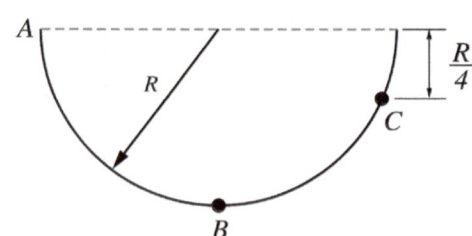

Fig. 4.23 A small stone of mass 0.1 kg is released from rest inside a large hemispherical bowl of radius $R = 0.2$m

Fig. 4.22 A 3 kg block compresses a spring of negligible mass a distance of 0.1 m from its equilibrium position

Solution 4.19   (a)

$$K_f + U_f = K_i + U_i$$

$$\frac{1}{2}mv_B^2 + 0 = 0 + mgR$$

thus

$$v_B = \sqrt{2gR} = \sqrt{2(9.8\,\text{m/s}^2)(0.2\,\text{m})} = 2\,\text{m/s}$$

At point C some of the kinetic energy at B is converted into potential energy and we have

$$\frac{1}{2}mv_C^2 + mg\left(R - \frac{R}{4}\right) = \frac{1}{2}mv_B^2 + 0$$

$$v_C^2 = v_B^2 - \frac{3}{2}gR = (2\,\text{m/s})^2 - \frac{3}{2}(9.8\,\text{m/s}^2)(0.2\,\text{m})$$

and therefore $v_C = 1\,\text{m/s}$.

(b) If a force of kinetic friction exists between the stone and the bowl, the total mechanical energy at point B is given by

$$E_f = E_i + \Delta K_{ext}$$

where the stone is considered as the system, therefore

$$K_f + U_f = K_i + U_i + \Delta K_{ext}$$

$$\frac{1}{2}mv_b^2 + 0 = 0 + mgR + \Delta K_{ext}$$

hence the energy dissipated by friction is

$$\Delta K_{ext} = \frac{1}{2}mv_b^2 - mgR = (0.1\,\text{kg})\left(\frac{1}{2}(1.7\,\text{m/s})^2 - (9.8\,\text{m/s}^2)(0.2\,\text{m})\right) = -0.05\,\text{J}$$

Example 4.20   A skier starts at the top of a frictionless incline as in Fig. 4.24. Find the velocity with which he will leave the second incline.

Solution 4.20   From the conservation of energy the velocity when he leaves the track is

$$K_f + U_f = K_i + U_i$$

$$\frac{1}{2}mv^2 + mgh_2 = mgh_1$$

$$v = \sqrt{2g(h_1 - h_2)} = \sqrt{2(98\,\text{m/s}^2)((20\,\text{m}) - (10\,\text{m}))}$$

That gives $v = 14\,\text{m/s}$.

Fig. 4.24   A skier slides from rest on top of an incline

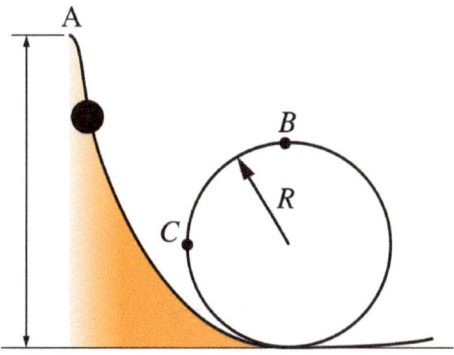

Fig. 4.25   A 0.4 kg stone is released from rest at point A where $h_A = 2\text{m}$

Example 4.21   A 0.4 kg stone is released from rest at point A where $h_A = 2\,\text{m}$ (see Fig. 4.25). It then slides without friction along the track shown where $R = 0.5\,\text{m}$. Determine: (a) the speed of the stone at B; (b) the normal force exerted on the stone at B; (c) the magnitude of the total acceleration of the stone at C; (d) the minimum height in which the stone must be released such that it does not fall off the track.

Solution 4.21   (a) From the conservation of energy, we have

$$mgh_A = \frac{1}{2}mv_B^2 + 2mgR$$

$$v_B = \sqrt{2g(h_A - 2R)} = \sqrt{2(9.8\,\text{m/s}^2)((2\,\text{m}) - 2(0.5\,\text{m}))} = 4.43\,\text{m/s}$$

(b) From Newton's second law, we have

$$n + mg = m\frac{v_B^2}{R}$$

$$n = m\frac{v_B^2}{R} - mg = (0.4\,\text{kg})\left[\frac{(4.43\,\text{m/s})^2}{(0.5\,\text{m})} - (9.8\,\text{m/s}^2)\right] = 11.78\,\text{N}$$

(c) The velocity of the stone at C is

$$v_C = \sqrt{2g(h_A - R)} = \sqrt{2(9.8\,\text{m/s}^2)((2\,\text{m}) - (0.5\,\text{m}))} = 5.42\,\text{m/s}$$

Therefore, the radial acceleration at C is

$$a_r = \frac{v_C^2}{R} = \frac{(5.42\,\text{m/s})^2}{(0.5\,\text{m})} = 58.8\,\text{m/s}^2$$

The tangential force exerted on the stone at C is its weight $F_t = -mg$, hence the tangential acceleration of the stone at

C is $a_t = -g$ and the magnitude of the total acceleration is

$$a = \sqrt{a_r^2 + a_t^2} = \sqrt{(58.8\,\text{m/s}^2)^2 + (-9.8\,\text{m/s}^2)^2} = 59.6\,\text{m/s}^2$$

(d) When the stone is at the verge of falling at B, then the only force acting on it is the force of gravity and we have $mg = mv_B^2/R$, $v_B^2 = gR$. From conservation of energy

$$v_B = \sqrt{2g(h_{A\,\text{min}} - 2R)}$$

or

$$2g(h_{A\,\text{min}} - 2R) = gR$$

and

$$h_{A\,\text{min}} = \frac{R}{2} + 2R = (0.25\,\text{m}) + (1\,\text{m}) = 1.25\,\text{m}$$

4.5.5 Power

Expanding on the definition of power, power is the rate of energy transfer due to a force. If ΔE is the amount of energy transferred in an amount of time Δt, the average power is

$$\overline{P} = \frac{\Delta E}{\Delta t}$$

The instantaneous power is then

$$P = \lim_{\Delta t \to 0} \frac{\Delta E}{\Delta t} = \frac{dE}{dt}$$

4.5.6 Energy Diagrams

Consider a particle that is a part of an isolated system where only internal conservative forces act. Suppose this particle is moving along the x-axis while a conservative force that depends only on the position of the particle acts on it. For simplicity, we will assume that is the only force acting on the system and that it does work only on that particle. The potential energy of the system as a function of the particle's position (x) is shown in Fig. 4.26. At any point $F(x)$ is given by

$$F(x) = -\frac{dU(x)}{dx}$$

That is, it is the negative of the slope of the curve at that point. Because this force is conservative it follows that the total mechanical energy of the system is conserved. Therefore the kinetic energy of the particle as a function of position is given by

$$K(x) = E - U(x)$$

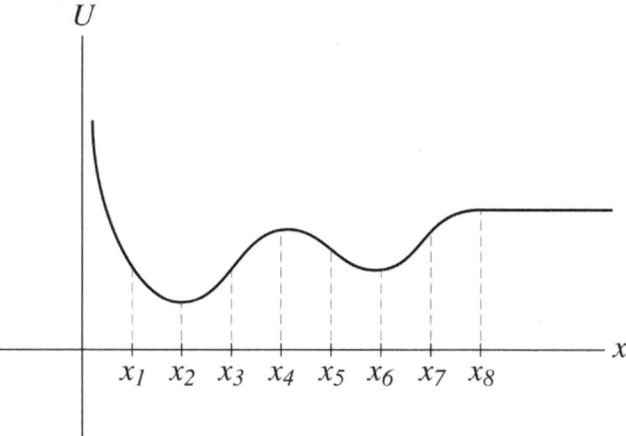

Fig. 4.26 The potential energy of the system as a function of the particle's position (x)

On the U versus x curve, the kinetic energy at any point can be found by subtracting the value of U (at that certain point) from E.

4.5.7 Turning Points

A turning point is a point in which the particle changes its direction of motion. The points x_1, x_3, x_5 and x_7 are all turning points.

4.5.8 Equilibrium Points

Equilibrium points occur in general when $\nabla U = 0$. In the case of one dimensional motion it occurs when $dU(x)/dx = 0$, i.e. when $F(x) = 0$.

4.5.9 Positions of Stable Equilibrium

If at an equilibrium point $d^2U(x)/dx^2 > 0$, then $U(x)$ is a minimum at that point. The point is then said to be a position of stable equilibrium, i.e., any minimum on the $U(x)$ curve is a position of stable equilibrium. Another method to find the position of stable equilibrium is to find the sign of $F(x)$ at each side of the point. As an example, consider the point x_2.

This point is a position of stable equilibrium since if the particle is displaced slightly to the right of x_2 then $dU(x)/dx$ is positive which leads to $F(x)$ being negative and the particle will accelerate back towards x_2. On the other hand, if the particle is displaced slightly to the left of x_2, then $dU(x)/dx$ is negative and thus $F(x)$ is positive and the particle will also accelerates back to x_2. Therefore, because $F(x)$ tends to restore the particle back to that position when the particle is

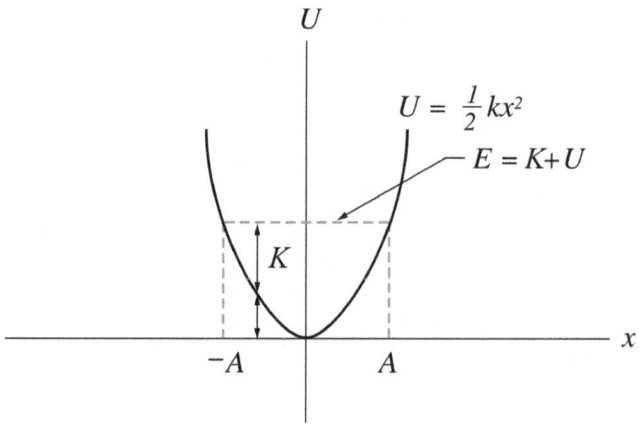

Fig. 4.27 The potential energy of a mass-spring system as a function of x

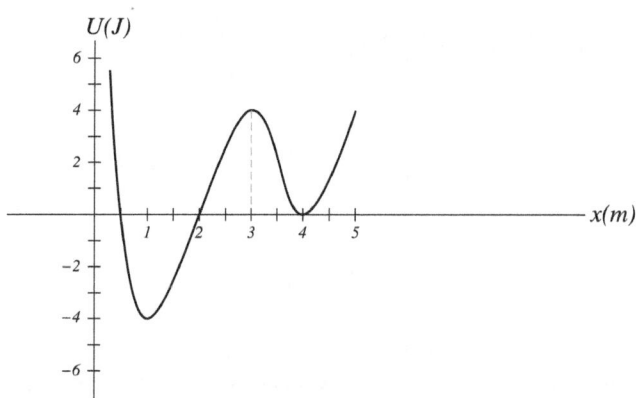

Fig. 4.28 The potential energy of a particle as a function of its displacement

displaced in either direction, it is called a position of stable equilibrium. x_6 is also a position of stable equilibrium.

4.5.10 Positions of Unstable Equilibrium

If at an equilibrium point $d^2U(x)/dx^2 < 0$, then $U(x)$ is maximum at that point, and the point is called a position of unstable equilibrium. In Fig. 4.26, x_4 is a position of unstable equilibrium since if the particle is slightly displaced to the right of x_4, $F(x)$ is positive and the particle will accelerate away from x_4. If the particle is displaced to the left of x_4, $F(x)$ is negative and the particle will also accelerate away from that position. Therefore, because $F(x)$ tends to repel the particle away from that position, it is called a position of unstable equilibrium. In general this force tends to move the particle towards the minimum value of $U(x)$. Figure 4.27 shows the potential energy of a mass–spring system as a function of x.

4.5.11 Positions of Neutral Equilibrium

Any point in a region where $U(x)$ is constant and $F(x) = 0$ is called a position of neutral equilibrium. x_8 is a position of neutral equilibrium. If the particle is slightly displaced to the right or left of x_8, no restoring or repelling forces will act on the particle and it will remain stationary The position of the particle as a function of time can be obtained from

$$U(x) + K(x) = E$$

$$U(x) + \frac{1}{2}mv^2 = E$$

$$v = \pm\sqrt{\frac{2}{m}[E - U(x)]}$$

or

$$\frac{dx}{dt} = \pm\sqrt{\frac{2}{m}[E - U(x)]}$$

hence

$$t = \int_{x_i}^{x} \frac{dx}{\pm\sqrt{\frac{2}{m}[E - U(x)]}}$$

By evaluating this integral, we would obtain the time as a function of the position, then by solving for x we get the position as a function of time.

Example 4.22 Figure 4.28 shows the potential energy of a particle as a function of its displacement. Find: (a) the values of x where the particle is in stable or unstable equilibrium; (b) the direction of the force acting on the particle at 0.5 m.

Solution 4.22 (a) We have $x = 1$ m and $x = 4$ m are positions of stable equilibrium, $x = 3$ m is a position of unstable equilibrium.

(b) At 0.5 m, $dU(x)/dx$ is negative and hence $F(x)$ is positive which means that the particle will accelerate in the positive x-direction.

Example 4.23 Consider a block attached to a light spring and released from rest at $x = A$. Find the position of the block as a function of time using energy methods.

Solution 4.23

$$t = \int_{x_i=A}^{x_f=x} \frac{dx}{\pm\sqrt{\frac{2}{m}[E - U(x)]}} = \int_{x_i=A}^{x_f=x} \frac{dx}{\pm\sqrt{\frac{2}{m}[(1/2)kA^2 - (1/2)kx^2]}}$$

$$= \int_{x_i=A}^{x_f=x} \frac{dx}{\pm\sqrt{\frac{k}{m}[A^2 - x^2]}} = \pm\sqrt{\frac{m}{k}} \int_{x_i=A}^{x_f=x} \frac{dx}{\sqrt{[A^2 - x^2]}}$$

$$= \pm\sqrt{\frac{m}{k}}\left[-\cos^{-1}\left(\frac{x}{A}\right)\right]_{x_i=A}^{x_f=x} = \pm\sqrt{\frac{m}{k}}\left[-\cos^{-1}(\frac{x}{A})\right]$$

$$-\cos\left(\pm\sqrt{\frac{k}{m}}t\right) = \cos\left(\cos^{-1}\left(\frac{x}{A}\right)\right)$$

or

$$x = A\cos\sqrt{\frac{k}{m}}t$$

Since $\cos(\pm\theta) = \theta$ and $-\cos\theta = \cos\theta$. In Chap. 10, we will see that this equation represents the equation of a simple harmonic motion.

Problems

1. A force acting on a particle varies with position as in Fig. 4.29. Find the work done by the force as the particle moves from $x = 0$ to $x = 8$ m.
2. A force $\mathbf{F} = (3\mathbf{i} + \mathbf{j} - 5\mathbf{k})$ N acts on a particle that undergoes a displacement $\mathbf{r} = (-2\mathbf{i} + 3\mathbf{j} - \mathbf{k})$ m. Find the work done by the force on the particle.
3. A 5 kg block is pulled from rest on a rough surface by a constant force of 10 N that is at 30° to the horizontal. If the coefficient of kinetic friction between the block and the surface is 0.15, find the final speed of the block as it moves through a displacement of 2 m using the work–energy theorem.
4. Calculate the work done against gravity in moving a 30 kg box through a height of 6 m.
5. A 1600 kg car accelerates from rest at a rate of $1\,\text{m/s}^2$. Find the average power delivered to the car during the first 5 s.
6. Determine whether or not the force $\mathbf{F} = -m\omega^2(x\mathbf{i} + y\mathbf{j})$ is conservative, where ω is constant and m is the mass of the particle. If the force is conservative determine the potential energy associated with it.

7. A 5 kg block slides down an inclined plane of angle 50° (see Fig. 4.30). Using energy methods, find the speed of the block just as it reaches the bottom if the coefficient of kinetic friction is $\mu_k = 0.2$.
8. A block of mass of 2 kg is pressed against a light spring of force constant $400\,\text{N/m}$ (see Fig. 4.31). If the compression of the spring is 10 cm, find the maximum height the block will reach when it is released.
9. A force acting on a particle is given by $\mathbf{F} = -\beta y^2\mathbf{j}$. Find the work done in moving the particle along the path shown in Fig. 4.32.

Fig. 4.30 A block slides down an inclined plane

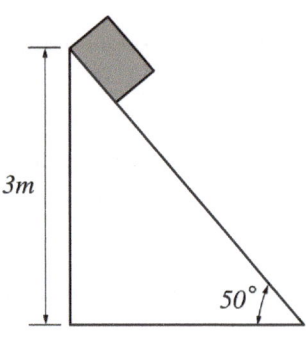

Fig. 4.31 A block pressed against a light spring and released

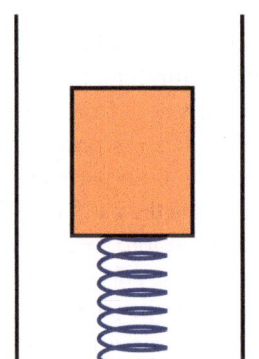

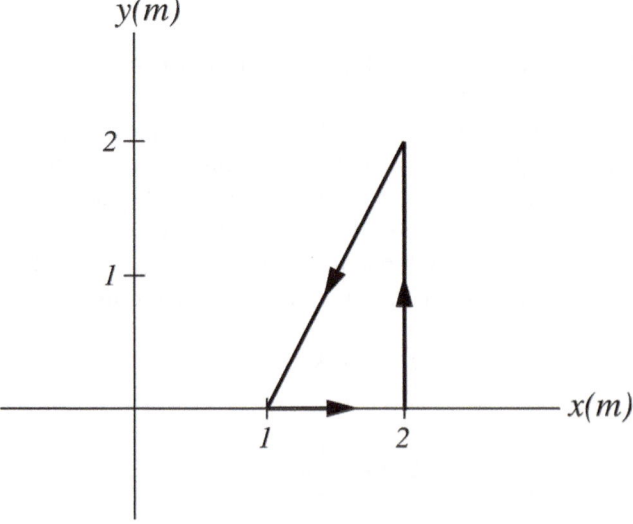

Fig. 4.32 The work done in moving the particle along a closed path

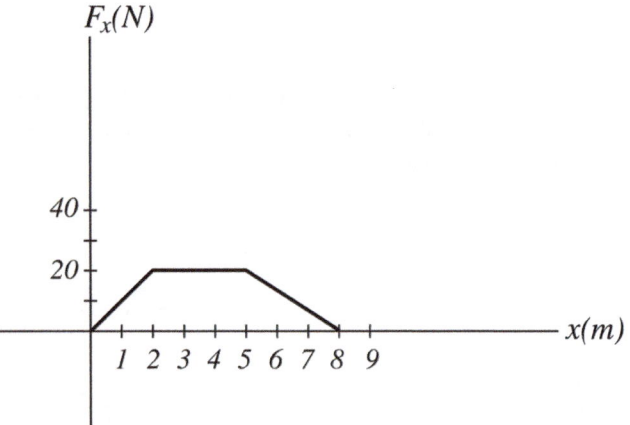

Fig. 4.29 A force acting on a particle varies with position

Fig. 4.33 Two blocks connected by a light rope that passes over a massless frictionless pulley

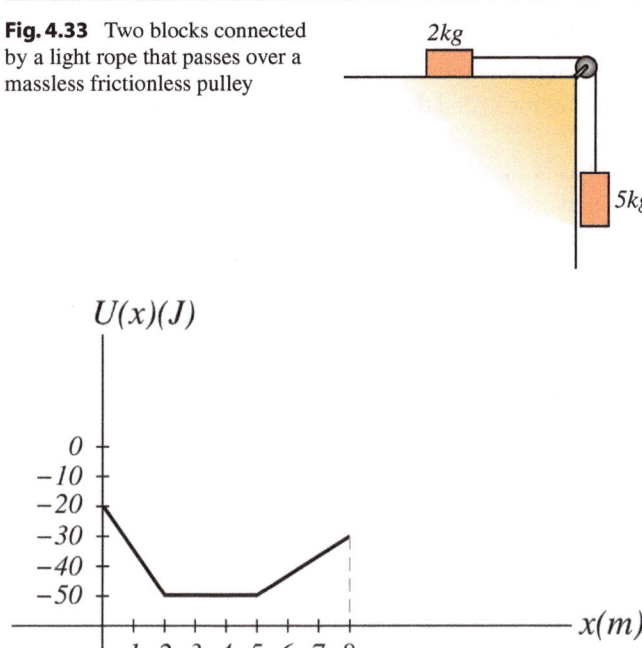

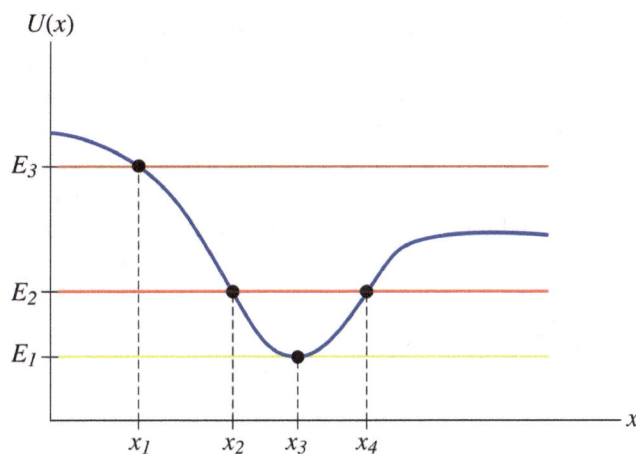

Fig. 4.37 The potential energy versus position of a particle

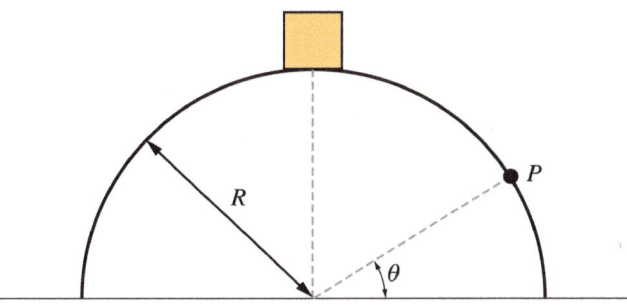

Fig. 4.34 The potential energy versus displacement of a particle

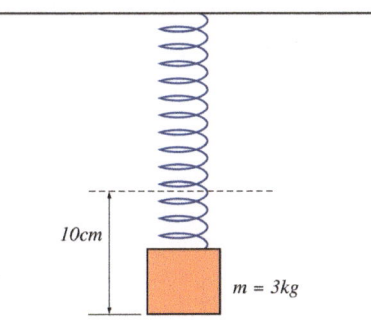

Fig. 4.35 A block of mass m resting on a hemispherical mound of ice

Fig. 4.36 A block hanging from a spring

10. Two blocks are connected by a light rope that passes over a massless frictionless pulley (see Fig. 4.33). If the system is released from rest, find the total kinetic energy of the blocks when the 5 kg block descends a distance of 0.5 m assuming that the surface is frictionless.

11. A particle of mass 1.5 kg moves along the x-axis where its potential energy varies as in Fig. 4.34. Plot the force $F_x(x)$ versus x from $x = 0$ to $x = 8$ m.

12. A block of mass m rests on a hemispherical mound of ice as shown in Fig. 4.35. If it is given a very small push and start sliding, find the height of the point in which the block will lose contact with the mound.

13. A 3 kg block hangs from a spring as in Fig. 4.36. If the spring stretches a distance of 10 cm, find (a) the force constant of the spring (b) the work done in expanding the spring a distance of 5 cm without accelerating it.

14. In Fig. 4.37, determine the Turning points and the positions of stable and unstable equilibrium.

Open Access This chapter is licensed under the terms of the Creative Commons Attribution 4.0 International License (http://creativecommons.org/licenses/by/4.0/), which permits use, sharing, adaptation, distribution and reproduction in any medium or format, as long as you give appropriate credit to the original author(s) and the source, provide a link to the Creative Commons license and indicate if changes were made.

The images or other third party material in this chapter are included in the chapter's Creative Commons license, unless indicated otherwise in a credit line to the material. If material is not included in the chapter's Creative Commons license and your intended use is not permitted by statutory regulation or exceeds the permitted use, you will need to obtain permission directly from the copyright holder.

Impulse, Momentum, and Collisions

5.1   Linear Momentum and Collisions

When two billiard balls collide, in which direction would they travel after the collision? If a meteorite hits the earth, why does the earth remain in its orbit? When two cars collide with each other, why is one of the cars more damaged than the other? We will find that to answer such questions, new concepts must be introduced.

Consider the situation where two bodies collide with each other. During the collision, each body exerts a force on the other. This force is called an impulsive force, because it acts for a short period of time compared to the whole motion of the objects, and its value is usually large. To solve collision problems by using Newton's second law, it is required to know the exact form of the impulsive forces. Because these forces are complex functions of the collision time, it is difficult to find their exact form and would make it difficult to use Newton's second law to solve such problems. Thus, new concepts known as momentum and impulse were introduced. These concepts enable us to analyze problems that involve collisions, as well as many other problems.

The law of conservation of momentum is especially used in analyzing collisions and is applied immediately before and immediately after the collision. Therefore, it is not necessary to know the exact form of the impulsive forces, which makes the problem easy to analyze. Next, we will discuss and verify the concepts of momentum and impulse, and the law of conservation of momentum. The linear momentum (or quantity of motion as was called by Newton) of a particle of mass m is a vector quantity defined as

$$\mathbf{p} = m\mathbf{v}$$

where y is the velocity of the particle. A fast moving car has more momentum than a slow moving car of the same mass. Another example is that a bowling ball has more momentum than a basketball moving at the same speed. The SI unit of

linear momentum is kg.m/s. In terms of components, we may write $p_x = mv_x$, $p_y = mv_y$, and $p_z = mv_z$. Newton's second law can be expressed in terms of momentum for a particle-like object of constant mass as

$$\Sigma \mathbf{F} = m\mathbf{a} = m\frac{d\mathbf{v}}{dt} = \frac{d(m\mathbf{v})}{dt}$$

or

$$\Sigma \mathbf{F} = \frac{d\mathbf{p}}{dt}$$

That is, the rate of change of the linear momentum of an object is equal to the resultant force acting on the object and is in the same direction as that force.

5.2   Conservation of Linear Momentum

The law of conservation of linear momentum states that if the net external force acting on a system equals zero (isolated) and if there is no mass exchange with the surroundings of the system (closed), then the total linear momentum of the system remains constant. To show that, consider an isolated system consisting of two particles where the only forces that act in the system are internal forces (see Fig. 5.1). The total linear momentum of the system at any particular time is given by

$$\mathbf{p}_{tot} = \mathbf{p}_1 + \mathbf{p}_2 \tag{5.1}$$

If the net force exerted on particle 2 by particle 1 is $\mathbf{F}_{21}$, then from Newton's third law, the net force exerted on particle 1 by particle 2 is $\mathbf{F}_{12}$, That is

$$\mathbf{F}_{12} = -\mathbf{F}_{21}$$

Differentiating Eq. 5.1 with respect to time and by using Newton's second law, we have

© The Author(s) 2019
S. Alrasheed, *Principles of Mechanics*, Advances in Science,
Technology & Innovation, https://doi.org/10.1007/978-3-030-15195-9_5

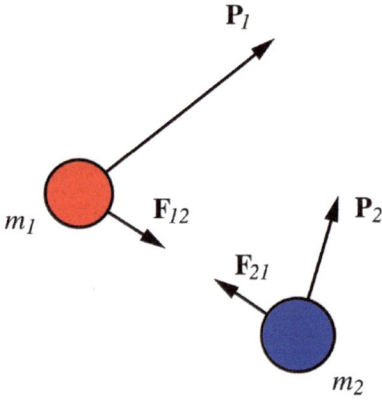

Fig. 5.1 An isolated system consisting of two particles where the only forces that act in the system are internal forces

$$\frac{d\mathbf{p}_{tot}}{dt} = \frac{d\mathbf{p}_1}{dt} + \frac{d\mathbf{p}_2}{dt} = \mathbf{F}_{12} + \mathbf{F}_{21} = \mathbf{F}_{12} - \mathbf{F}_{12} = 0$$

That is,

$$\mathbf{p}_{tot} = \text{constant}$$

or

$$\mathbf{p}_i = \mathbf{p}_f$$

That is, the linear momentum of each particle may change, but the total linear momentum of the system is the same at all times. This statement is known as the law of conservation of linear momentum: *If the net external force on a system is zero, the total linear momentum of the system remains unchanged (constant).* In terms of components, we have $p_{ix} = p_{fx}$, $p_{iy} = p_{fy}$, and $p_{iz} = p_{fz}$. In solving problems involving collisions, $\mathbf{p}_i$ and $\mathbf{p}_f$ refers to the total momentum of the system immediately before and immediately after the collision, respectively. For a two-particle system, we have

$$\mathbf{p}_{1i} + \mathbf{p}_{2i} = \mathbf{p}_{1f} + \mathbf{p}_{2f}$$

From the principle of invariance, the law of conservation of momentum is valid with respect to any inertial frame of reference. Furthermore, as the law of conservation of energy, the law of conservation of momentum is valid in relativity and quantum mechanics.

5.3 Impulse and Momentum

Impulse is a quantity that defines how a certain force acting on a particle changes the linear momentum of that particle. Now, consider a time-dependent force acting on a particle. From Newton's second law ($\mathbf{F} = d\mathbf{p}/dt$), we have

$$d\mathbf{p} = \mathbf{F}dt$$

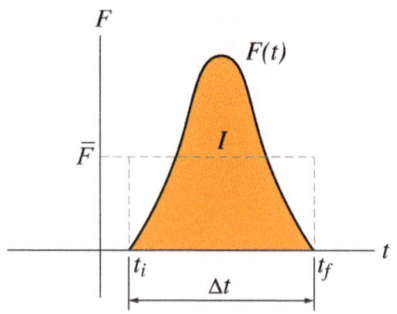

Fig. 5.2 One example of the variation of F over time

$$\int_{p_i}^{p_f} d\mathbf{p} = \int_{t_i} \mathbf{F}dt$$

$$\mathbf{p}_f - \mathbf{p}_i = \triangle\mathbf{p} = \int_{t_i}^{t_f} \mathbf{F}dt$$

The right side of the equation is a vector quantity known as the impulse I

$$\mathbf{I} = \int_{t_i}^{t_f} \mathbf{F}dt$$

Hence,

$$\mathbf{I} = \triangle\mathbf{P}$$

Which is known as the impulse–momentum theorem. In component form, we have $I_x = \triangle p_x$, $I_y = \triangle p_y$, and $I_z = \triangle p_z$. That is, *the impulse of a force that acts on a particle during a time interval is equal to the change in the momentum of the particle during that interval.* The direction of the impulse is in the same direction as the change of momentum. If F has a constant direction, the variation of its magnitude with time may be of the form as shown in Fig. 5.2. The average of F is given by

$$\overline{\mathbf{F}} = \frac{1}{\triangle t} \int_{t_i}^{t_f} \mathbf{F}dt$$

And thus, I can be written as

$$\mathbf{I} = \triangle\mathbf{p} = \overline{\mathbf{F}}\triangle t$$

That is, $\overline{\mathbf{F}}$ is a constant force that gives the same impulse as F. In the case of a collision between two bodies, the variation of the impulsive force that each body exerts on the other during the collision time takes the form as shown in Fig. 5.2.

5.4 Collisions

As discussed previously, when two bodies collide, they exert large forces on one another (during the time of the collision) called impulsive forces. These forces are very large such that any other forces (e.g., friction or gravity) present during the short time of the collision can be neglected. This approximation is known as the impulse approximation. For example, if a golf ball was hit by a golf club, the change in the momentum of the ball can be assumed to be only due to the impulsive force exerted on it by the club. The change in its momentum due to any other force present during the collision can be neglected. That is, the force in the expression $I = \Delta p = \overline{F}\Delta t$ can be assumed to be the impulsive force only The neglected forces present during the collision time are external to the two-body system, whereas the impulsive forces are internal. The two-body system can therefore be considered to be isolated during the short time of the collision (which is in the order of a few milliseconds). Hence, the total linear momentum of the system is conserved during the collision, which enables us to apply the law of conservation of momentum immediately before and immediately after the collision. In general, for any type of collision, the total linear momentum is conserved during the time of the collision. That is, $p_i = p_f$., where p_i and p_f are the momenta immediately before and after the collision. In the next sections, we will define various types of two- body collisions, depending on whether or not the kinetic energy of the system is conserved.

Example 5.1 A 50 g golf ball initially at rest is struck by a golf club. The golf club exerts a force on the ball that varies during a very short time interval from zero before impact, to a maximum value and back to zero when the ball is no longer in contact with the club. If the ball is given a speed of 25 m/s, and if the club is in contact with the ball for 7×10^{-4} s, find the average force exerted by the club on the ball.

Solution 5.1 The impulse of the force is

$$I = \Delta p = mv_f - 0 = (0.05\,\text{kg})(25\,\text{m/s}) = 1.25\,\text{kg} \cdot \text{m/s}$$

the average force exerted on the ball by the club is then

$$\overline{F} = \frac{I}{\Delta t} = \frac{(1.25\,\text{kgm/s})}{(7 \times 10^{-4}\,\text{s})} = 1785.7\,\text{N}$$

Example 5.2 A canon placed on a carriage fires a 250 kg ball to the horizontal with a speed of 50 m/s. If the mass of the canon and the carriage is 4000 kg, find the recoil speed of the canon.

Solution 5.2 Because there are no external horizontal forces acting on the cannon-carriage-ball system, then the total momentum of the system is constant (conserved) in the x-direction

$$p_{fx} = p_{ix}$$

$$m_1 v_{1f} + m_2 v_{2f} = 0$$

therefore,

$$v_{2f} = \frac{-m_1}{m_2}v_{1f} = -\frac{(250\,\text{kg})}{(4000\,\text{kg})}(50\,\text{m/s}) = -3.1\,\text{m/s}$$

i.e., the cannon recoils in the negative x-direction.

Example 5.3 A hockey puck of mass 0.16 kg traveling on a smooth ice surface collides with the court's edge. If its initial and final velocities are $v_i = -2\,\textbf{i}$m/s and $v_f = 1\,\textbf{i}$m/s and if the hockey puck is in contact with the wall for 2 ms, find the impulse delivered to the puck and the average force exerted on it by the wall.

Solution 5.3

$$I = \Delta p = p_f - p_i = mv_f - mv_i = (0.16\,\text{kg})((1\,\text{m/s}) - (-2\,\text{m/s}))\textbf{i} = 0.48\textbf{i}\,\text{kg} \cdot \text{m/s}$$

$$\overline{F} = \frac{I}{\Delta t} = \frac{(0.48\,\textbf{i}\,\text{kg} \cdot \text{m/s})}{(0.002\,\text{s})} = 240\,\textbf{i}\,\text{N}$$

Example 5.4 A 0.5 kg hockey puck is initially moving in the negative y-direction as shown in Fig. 5.3, with a speed of 7 m/s. If a hockey player hits the puck giving it a velocity of magnitude 12 m/s in a direction of 60° to the vertical, and if the collision lasts for 0.008 s, find the impulse due to the collision and the average force exerted on the puck.

Solution 5.4 Along the x-direction, we have

$$p_{ix} = mv_{ix} = 0$$

and

$$p_{fx} = mv_{fx} = (0.5\,\text{kg})(12\text{m/s})\cos 30^{\circ} = 5.2\,\text{kg} \cdot \text{m/s}$$

along the y-direction, we have

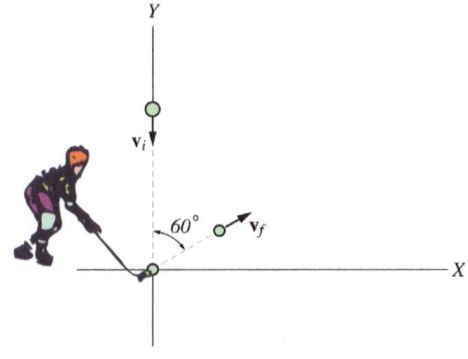

Fig. 5.3 A hockey player changing the momentum of the puck

$$p_{iy} = mv_{iy} = (0.5 \, \text{kg})(-7 \, \text{m/s}) = -3.5 \, \text{kg} \cdot \text{m/s}$$

and

$$p_{fy} = mv_{fy} = (0.5 \, \text{kg})(12 \, \text{m/s}) \sin 30° = 3 \, \text{kg} \cdot \text{m/s}$$

Therefore, the impulse of the force in each direction is

$$I_x = p_{fx} - p_{ix} = (5.2 \, \text{kg} \cdot \text{m/s}) - 0 = 5.2 \, \text{kg} \cdot \text{m/s}$$

and

$$I_y = p_{fy} - p_{iy} = (3 \, \text{kg·m/s}) - (-3.5 \, \text{kg·m/s}) = 6.5 \, \text{kg·m/s}$$

$$\mathbf{I} = (5.2 \, \mathbf{i} + 6.5 \, \mathbf{j}) \, \text{kg} \cdot \text{m/s}$$

$$I = \sqrt{(5.2 \, \text{kgm/s})^2 + (6.5 \, \text{kgm/s})^2} = 8.3 \, \text{kg} \cdot \text{m/s}$$

The direction of the impulse is

$$\tan \theta = \frac{I_y}{I_x} = \frac{(6.5 \, \text{kg} \cdot \text{m/s})}{(5.2 \, \text{kg} \cdot \text{m/s})} = 1.25$$

$$\theta = 51.3°$$

where θ is measured from the positive x-axis. The average force acting on the puck is

$$\overline{F} = \frac{I}{\Delta t} = \frac{(8.3 \, \text{kg} \cdot \text{m/s})}{(0.008 \, \text{s})} = 1037.5 \, \text{N}$$

Example 5.5 Two ice skaters of masses $m_1 = 50 \, \text{kg}$ and $m_1 = 62 \, \text{kg}$ standing face to face push each other on a frictionless horizontal surface. If skater (1) recoils with a speed of 5 m/s, find the recoil speed of the other skater.

Solution 5.5   For the two-skater system, the sum of the vertical forces are zero (weight and normal forces) and the forces exerted by one skater on the other is internal to the system. That is, there are no external forces acting on the system and the total momentum is conserved. Because the motion takes place in a straight line, we have

$$p_{1i} + p_{2i} = p_{1f} + p_{2f}$$

$$0 = m_1 v_{1f} + m_2 v_{2f}$$

and hence,

$$v_{2f} = \frac{-m_1}{m_2} v_{1f} = \frac{-(50 \, \text{kg})}{(62 \, \text{kg})} (5 \, \text{m/s}) = -4.03 \, \text{m/s}$$

Example 5.6 A particle is moving in space under the influence of a force. If its momentum as a function of time is

$$\mathbf{p} = ((4t^2 + t)\mathbf{i} - (3t - 1)\mathbf{j} + (5t^3 + 2t)\mathbf{k}) \, \text{kg. m/s}$$

(a) Find the force acting on the particle at any time; (b) Find the impulse of the force from $t = 0$ to $t = 1 \, \text{s}$.

Solution 5.6   (a)

$$\mathbf{F} = \frac{d\mathbf{p}}{dt} = ((8t + 1)\mathbf{i} - 3\mathbf{j} + (15t^2 + 2)\mathbf{k}) \, \text{N}$$

(b)

$$\mathbf{I} = \triangle \mathbf{p} = (5\mathbf{i} - 2\mathbf{j} + 7\mathbf{k}) - \mathbf{j} = (5\mathbf{i} - 3\mathbf{j} + 7\mathbf{k}) \, \text{kg.m/s}$$

5.4.1   Elastic Collisions

An elastic collision is one in which the total kinetic energy, as well as momentum, of the two-colliding-body system is conserved. These collisions exist when the impulsive force exerted by one body on the other is conservative. Such force converts the kinetic energy of the body into elastic potential energy when the two bodies are in contact. It then reconverts the elastic potential energy into kinetic energy when there is no more contact. After collision, each body may have a different velocity and therefore a different kinetic energy. However, the total energy as well as the total momentum of the system is constant during the time of the collision. An example of such collisions is those between billiard balls.

5.4.2   Inelastic Collisions

An inelastic collision is one in which the total kinetic energy of the two-colliding-body system is not conserved, although momentum is conserved. In such a collision, some of the kinetic energy of the system is lost due to deformation and appear as internal or thermal energy. In other words, the (internal) impulsive forces are not conservative. Therefore, the kinetic energy of the system before the collision is less than that after the collision. If the two colliding objects stick together, the collision is said to be perfectly inelastic. There are some types of collisions in which the total kinetic energy after the collision occurs is greater than that before it occurs. This type of collision is called an explosive collision.

Fig. 5.4 Two particles of masses m_1 and m_2 experiencing an elastic head-on collision

5.4.3 Elastic Collision in One Dimension

When a collision takes place in one dimension, it is referred to as a head-on collision. Consider two particles of masses m_1 and m_2 experiencing an elastic head-on collision as in Fig. 5.4. Applying the law of conservation of energy and the law of conservation of linear momentum gives

$$m_1\mathbf{v}_{1i} + m_2\mathbf{v}_{2i} = m_1\mathbf{v}_{1f} + m_2\mathbf{v}_{2f}$$

$$\frac{1}{2}m_1v_{1i}^2 + \frac{1}{2}m_2v_{2i}^2 = \frac{1}{2}m_1v_{1f}^2 + \frac{1}{2}m_2v_{2f}^2$$

Solving these equations for v_{1f} and v_{2f}, we get

$$v_{1f} = \left(\frac{m_1 - m_2}{m_1 + m_2}\right)v_{1i} + \left(\frac{2m_2}{m_1 + m_2}\right)v_{2i} \qquad (5.2)$$

$$v_{2f} = \left(\frac{2m_1}{m_1 + m_2}\right)v_{1i} + \left(\frac{m_2 - m_1}{m_1 + m_2}\right)v_{2i} \qquad (5.3)$$

5.4.3.1 Special Cases
1. If $m_1 = m_2$, it follows from Eqs. 5.2 and 5.3 that $v_{1f} = v_{2i}$ and $v_{2f} = v_{1i}$. In other words, if the particles have equal masses they exchange velocities.

2. If m_2 is stationary ($v_{2i} = 0$) , then from Eqs. 5.2 and 5.3, we have

$$v_{1f} = \left(\frac{m_1 - m_2}{m_1 + m_2}\right)v_{1i} \qquad (5.4)$$

$$v_{2f} = \left(\frac{2m_1}{m_1 + m_2}\right)v_{1i} \qquad (5.5)$$

In that case m_2 is called the target and m_1 is called the projectile. Furthermore, if $m_1 \gg m_2$, then from Eqs. 5.4 and 5.5, we find that $v_{1f} \approx v_{1i}$ and $v_{2f} \approx 2v_{1i}$. While if $m_2 \gg m_1$, then from Eqs. 5.4 and 5.5, we see that $v_{1f} \approx -v_{1i}$, and $v_{2f} \approx v_{2i} = 0$.

5.4.4 Inelastic Collision in One Dimension

Figure 5.5 shows a one-dimensional (head-on) perfectly inelastic collision between two particles of mass m_1 and m_2. Here, the kinetic energy of the system is not conserved, but the law of conservation of linear momentum still holds

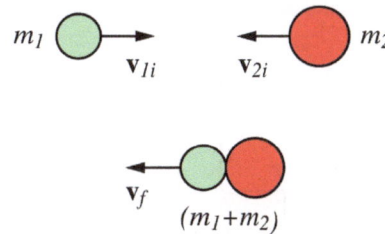

Fig. 5.5 A one dimensional (head-on) perfectly inelastic collision between two particles of mass m_1 and m_2

$$m_1\mathbf{v}_{1i} + m_2\mathbf{v}_{2i} = (m_1 + m_2)\mathbf{v}_f$$

$$\mathbf{v}_f = \frac{m_1\mathbf{v}_{1i} + m_2\mathbf{v}_{2i}}{m_1 + m_2}$$

5.4.5 Coefficient of Restitution

For any collision between two bodies in one dimension, the coefficient of restitution is defined as

$$e = \frac{v_{2f} - v_{1f}}{v_{1i} - v_{2i}}$$

where v_{1i} and v_{2i} are velocities before the collision. v_{1f} and v_{2f} are velocities after the collision. $|v_{1i} - v_{2i}|$ is called the relative speed of approach and $|v_{2f} - v_{1f}|$ is the relative speed of recession.

- If $e = 1$ the collision is perfectly elastic.
- If $e < 1$ the collision is inelastic.
- If $e = 0$ the collision is perfectly inelastic (the two bodies stick together).

Example 5.7 Two marble balls of masses $m_1 = 7\,\text{kg}$ and $m_2 = 3\,\text{kg}$ are sliding toward each other on a straight frictionless track. If they experience a head-on elastic collision and if the initial velocities of m_1 and m_2 are $0.5\,\text{m/s}$ to the right and $2\,\text{m/s}$ to the left, respectively, find the final velocities of m_1 and m_2.

Solution 5.7 For an elastic head-on collision, we have

$$v_{1f} = \left(\frac{m_1 - m_2}{m_1 + m_2}\right)v_{1i} + \left(\frac{2m_2}{m_1 + m_2}\right)v_{2i} = (0.4)(0.5\,\text{m/s}) + (0.6)(-2\,\text{m/s}) = -1\,\text{m/s}$$

$$v_{2f} = \left(\frac{2m_1}{m_1 + m_2}\right)v_{1i} + \left(\frac{m_2 - m_1}{m_1 + m_2}\right)v_{2i} = (1.4)(0.5\,\text{m/s}) + (-0.4)(-2\,\text{m/s}) = 1.5\,\text{m/s}$$

Example 5.8 The ballistic pendulum consists of a large wooden block suspended by a light wire (see Fig. 5.6). The system is used to measure the speed of a bullet where the bullet is fired horizontally into the block. The collision is perfectly inelastic and the system (bullet+block) swings up a height h.

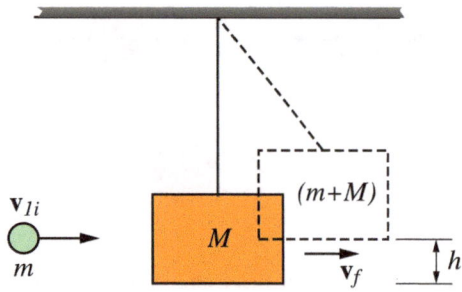

Fig. 5.6 The ballistic pendulum consists of a large wooden block suspended by a light wire

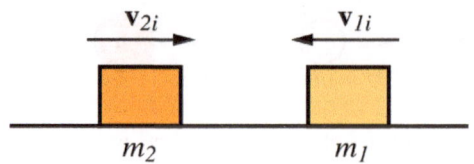

Fig. 5.7 Two blocks colliding head-on on a frictionless surface

If $M = 3\,\text{kg}, m = 5\,\text{g}$ and $h = 5\,\text{cm}$, find (a) the initial speed of the bullet; (b) the mechanical energy lost due to the collision.

Solution 5.8 (a) Using the impulse approximation, the law of conservation of momentum gives the velocities just before and after the collision when the string is still nearly vertical. For a perfectly inelastic collision, the total momentum is conserved but the total kinetic energy is not conserved during the collision. Thus, we have

$$mv_{1i} = (m + M)v_f$$

$$v_{1i} = \frac{(m + M)}{m}v_f$$

After the collision, the energy of the (bullet +block+earth) system is conserved since the gravitational force is the only force acting in the system.

$$E_i = E_f$$

$$\frac{1}{2}(m + M)v_f^2 = (m + M)gh$$

$$v_f = \sqrt{2gh}$$

That gives

$$v_{1i} = \frac{(m + M)}{m}\sqrt{2gh} = \frac{(3.005\,\text{kg})}{(0.005\,\text{kg})}\sqrt{2(9.8\,\text{m/s}^2)(0.05\,\text{m})} = 595\,\text{m/s}$$

(b) The kinetic energy of the bullet before collision is

$$\frac{1}{2}mv_{1i}^2 = \frac{1}{2}0.005\,\text{kg})(595\,\text{m/s})^2 = 885\,\text{J}$$

After collision, the kinetic energy of the (bullet+block) is

$$\frac{1}{2}(m + M)v_f^2 = (m + M)\,(gh) = (3.005\,\text{kg})(9.8\,\text{m/s}^2)(0.05\,\text{m}) = 1.5\,\text{J}$$

therefore,

$$\Delta E = (885\,\text{J}) - (1.5\,\text{J}) = 883.5\,\text{J}$$

That is, nearly, all the mechanical energy is dissipated and converted into internal (thermal) energy of the (block+bullet) system.

Example 5.9 Two masses $m_1 = 0.8\,\text{kg}$ and $m_2 = 0.5\,\text{kg}$ are heading toward each other with speeds of $0.25\,\text{m/s}$ and $-0.5\,\text{m/s}$, respectively. If they have a perfectly inelastic collision, find the final velocity of the system just after the collision.

Solution 5.9

$$v_f = \frac{m_1v_{1i} + m_2v_{2i}}{(m_1 + m_2)} = \frac{(0.8\,\text{kg})(0.25\,\text{m/s}) - (0.5\,\text{kg})(0.5\,\text{m/s})}{(1.3\,\text{kg})} = -0.04\,\text{m/s}$$

Example 5.10 Two blocks $m_1 = 2\,\text{kg}$ and $m_2 = 1\,\text{kg}$ collide head-on with each other on a frictionless surface (see Fig. 5.7. If $v_{1i} = -10\,\text{m/s}$ and $v_{2i} = 15\,\text{m/s}$ and the coefficient of restitution is $e = 1/4$, determine the final velocities of the masses just after the collision.

Solution 5.10

$$e = \frac{v_{2f} - v_{1f}}{v_{1i} - v_{2i}}$$

$$\frac{1}{4} = \frac{v_{2f} - v_{1f}}{(-25\text{m/s})}$$

$$v_{2f} - v_{1f} = -6.25\,\text{m/s} \qquad (5.6)$$

From the conservation of momentum, we have

$$m_1v_{1i} + m_2v_{2i} = m_1v_{1f} + m_2v_{2f}$$

$$(2\,\text{kg})(-10\,\text{m/s}) + (1\,\text{kg})(15\,\text{m/s}) = (2\,\text{kg})v_{1f} + (1\,\text{kg})v_{2f}$$

That gives

$$v_{2f} + (2\,\text{kg})v_{1f} = -5\,\text{m/s} \qquad (5.7)$$

Solving Eqs. 5.6 and 5.7 gives $v_{1f} = 0.42\,\text{m/s}$ and $v_{2f} = -5.83$ m/s.

Example 5.11 A $m_1 = 5\,\text{g}$ bullet is fired horizontally at the center of a wooden block with a mass of $m_2 = 2\,\text{kg}$. The bullet embeds itself in the block and the two slides a distance of 0.5 m on a rough surface ($\mu_k = 0.2$) before coming to rest. Find the initial speed of the bullet.

Solution 5.11 Applying the law of conservation of momentum immediately before and after the collision gives

$$p_{ix} = p_{fx}$$

$$m_1 v_{1i} + 0 = (m_1 + m_2)v_f$$

$$v_{1i} = \frac{(2.005\,\text{kg})}{(0.005\,\text{kg})} v_f = (401)v_f$$

by taking the (block+bullet) as the system after the collision until it comes to rest, we have

$$K_f + U_f = K_i + U_i + \Delta K_{ext}$$

that gives

$$0 = \frac{1}{2}(m_1 + m_2)v_f^2 - \mu_k(m_1 + m_2)gd$$

$$v_f = \sqrt{2\mu_k g d} = \sqrt{2(0.2)(9.8\,\text{m/s}^2)(0.5\,\text{m})} = 1.4\,\text{m/s}$$

Hence,

$$v_{1i} = (401)(1.4\,\text{m/s}) = 561.4\,\text{m/s}$$

5.4.6 Collision in Two Dimension

When a collision takes place in space, the total linear momentum is conserved along each of the $x-$, y-, and z-directions. That is, $p_{ix} = p_{fx}$, $p_{iy} = p_{fy}$, and $p_{iz} = p_{fz}$. Here, we will analyze a two-dimensional elastic collision between two particles where one particle is moving and the other is at rest as shown in Fig. 5.8. This type of collision is known as a glancing collision. Since the collision is elastic, it follows that the total linear momentum as well as the kinetic energy of the system are conserved. Applying these laws immediately before and immediately after the collision, we have $p_{ix} = p_{fx}$ and $p_{iy} = p_{fy}$ or

$$m_1 v_{1ix} + m_2 v_{2ix} = m_1 v_{1fx} + m_2 v_{2fx}$$

and

$$m_1 v_{1iy} + m_2 v_{2iy} = m_1 v_{1fy} + m_2 v_{2fy}$$

From Fig. 5.8, we have

$$m_1 v_{1i} = m_1 v_{1f} \cos\alpha_1 + m_2 v_{2f} \cos\alpha_2$$

and

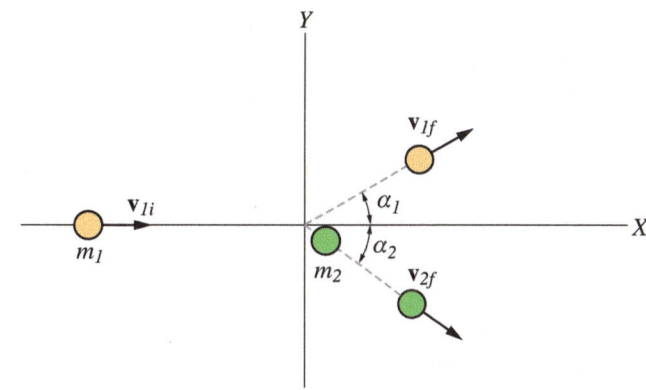

Fig. 5.8 A two dimensional elastic collision between two particles where one particle is moving and the other is at rest

$$0 = m_1 v_{1f} \sin\alpha_1 + m_2 v_{2f} \sin\alpha_2$$

Furthermore,

$$\frac{1}{2}m_1 v_{1i}^2 = \frac{1}{2}m_1 v_{1f}^2 + \frac{1}{2}m_2 v_{2f}^2$$

Therefore, we have three equations and seven unknown quantities. By knowing any four of these quantities, the three equations for the three variables can be solved.

Example 5.12 A ball of mass of 2 kg is sliding along a horizontal frictionless surface at a speed of 3 m/s. It then collides with a second ball of mass of 5 kg that is initially at rest. After the collision, the second ball is deflected with a speed of 1 m/s at an angle of 30° below the horizontal as shown in Fig. 5.9. (a) Find the final velocity of the first ball; (b) show that the collision is inelastic; (c) suppose that the two balls have equal masses and the collision is perfectly elastic, show that $\theta_1 + \theta_2 = 90°$.

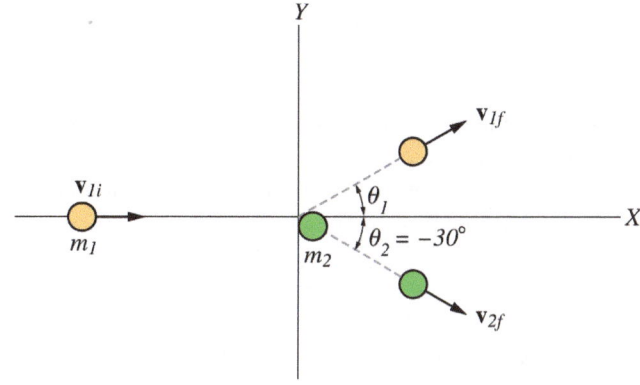

Fig. 5.9 A ball sliding along a horizontal frictionless surface collides with a second ball that is initially at rest

Solution 5.12 Applying the law of conservation of momentum immediately before and after the collision in each direction gives $p_{ix} = p_{fx}$ and $p_{iy} = p_{fy}$. Thus,

$$m_1 v_{1ix} + m_2 v_{2ix} = m_1 v_{1fx} + m_2 v_{2fx}$$

$$v_{1fx} = \frac{m_1 v_{1ix} + m_2 v_{2ix} - m_2 v_{2fx}}{m_1} = \frac{(2\,\text{kg})(3\,\text{m/s}) + 0 - ((5\,\text{kg})(1\,\text{m/s})\cos(-30))}{(2\,\text{kg})}$$

$$v_{1fx} = 0.84\,\text{m/s}$$

Along the y-direction, we have

$$m_1 v_{1iy} + m_2 v_{2iy} = m_1 v_{1fy} + m_2 v_{2fy}$$

$$v_{1fy} = \frac{m_1 v_{1iy} + m_2 v_{2iy} - m_2 v_{2fy}}{m_1} = \frac{0 - ((5\,\text{kg})(1\,\text{m/s})\sin(-30^\circ))}{(2\,\text{kg})}$$

$$v_{1fy} = 1.25\,\text{m/s}$$

Thus, the final velocity of the first ball is

$$v_{1f} = \sqrt{v_{1fx}^2 + v_{1fy}^2} = \sqrt{(0.84\,\text{m/s})^2 + (1.25\,\text{m/s})^2} = 1.5\,\text{m/s}$$

The direction of the velocity is

$$\tan\theta_1 = \frac{v_{1fy}}{v_{1fx}} = \frac{(1.25\,\text{m/s})}{(0.84\,\text{m/s})} = 1.5$$

$$\theta_1 = 56^\circ$$

(b) The total kinetic energy before the collision is

$$K_i = \frac{1}{2}m_1 v_{1i}^2 = \frac{1}{2}(2\,\text{kg})(3\,\text{m/s})^2 = 9\,\text{J}$$

The total kinetic energy after the collision is

$$K_f = \frac{1}{2}m_1 v_{1f}^2 + \frac{1}{2}m_2 v_{2f}^2 = \frac{1}{2}(2\,\text{kg})(1.5\,\text{m/s})^2 + \frac{1}{2}(5\,\text{kg})(1\,\text{m/s})^2 = 4.75\,\text{J}$$

That is, some of the energy of the system is lost and thus the collision is inelastic.

(c) In a perfectly elastic collision, both the total momentum and the total mechanical energy of the system are conserved. That is

$$p_{ix} = p_{fx}$$

$$m_1 v_{1ix} + m_2 v_{2ix} = m_1 v_{1fx} + m_2 v_{2fx}$$

$$v_{1i} = v_{1f}\cos\theta_1 + v_{2f}\cos\theta_2 \tag{5.8}$$

$$p_{iy} = p_{fy}$$

$$0 = v_{1f}\sin\theta_1 - v_{2f}\sin\theta_2$$

$$v_{1f}\sin\theta_1 = v_{2f}\sin\theta_2 \tag{5.9}$$

From the conservation of kinetic energy, we have

$$\frac{1}{2}m_1 v_{1i}^2 = \frac{1}{2}m_1 v_{1f}^2 + \frac{1}{2}m_2 v_{2f}^2$$

or

$$v_{1i}^2 = v_{1f}^2 + v_{2f}^2 \tag{5.10}$$

Substituting Eq. 5.8 into Eq. 5.9 gives

$$v_{1i} = v_{2f}\frac{\sin\theta_2}{\sin\theta_1}\cos\theta_1 + v_{2f}\cos\theta_2$$

or

$$v_{1i} = \frac{v_{2f}\sin(\theta_1 + \theta_2)}{\sin\theta_1} \tag{5.11}$$

Substituting Eq. 5.11 into Eq. 5.10 gives

$$\frac{v_{2f}^2\sin^2(\theta_1 + \theta_2)}{\sin^2\theta_1} = \frac{v_{2f}^2\sin^2\theta_2}{\sin^2\theta_1} + v_{2f}^2$$

Therefore,

$$\sin^2(\theta_1 + \theta_2) = \sin^2\theta_1 + \sin^2\theta_2$$

This is satisfied only if $\theta_1 + \theta_2 = 90^\circ$.

Example 5.13 A 1200 kg car traveling east at a speed of 18 m/s collides with another car of mass of 2500 kg that is traveling north at a speed of 23 m/s as shown in Fig. 5.10. If the collision is perfectly inelastic, how much mechanical energy is lost due to the collision?

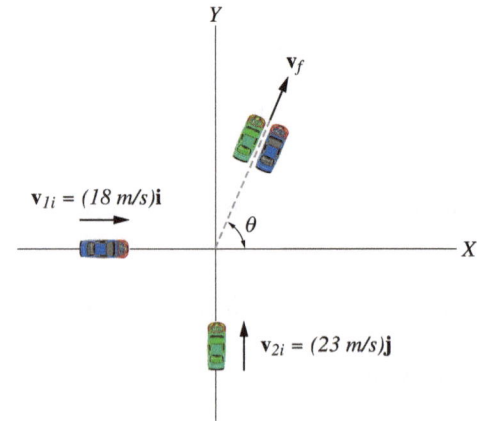

Fig. 5.10 A 1200 kg car traveling east at a speed of 18 m/s collides with another car of mass of 2500 kg that is traveling north at a speed of 23 m/s

Solution 5.13

$$p_{ix} = p_{fx}$$

$$m_1 v_{1ix} = (m_1 + m_2)v_{fx}$$

$$v_{fx} = \frac{m_1 v_{1ix}}{(m_1 + m_2)} = \frac{(1200\,\text{kg})(18\,\text{m/s})}{(3700\,\text{kg})} = 5.8\,\text{m/s}$$

$$p_{iy} = p_{fy}$$

$$m_2 v_{2iy} = (m_1 + m_2)v_{fy}$$

$$v_{fy} = \frac{m_2 v_{2iy}}{(m_1 + m_2)} = \frac{(2500\,\text{kg})(23\,\text{m/s})}{(3700\,\text{kg})} = 15.5\,\text{m/s}$$

$$v_f = \sqrt{v_{fx}^2 + v_{fy}^2} = \sqrt{(5.8\,\text{m/s})^2 + (15.5\,\text{m/s})^2} = 16.5\,\text{m/s}$$

The direction of v_f is

$$\theta = \tan^{-1}\frac{v_{fy}}{v_{fx}} = \tan^{-1}\frac{(15.5\,\text{m/s})}{(5.8\,\text{m/s})} = 69.5^\circ$$

from the positive x-axis. The change in the kinetic energy of the system is

$$\Delta K = K_f - K_i = \frac{1}{2}(m_1 + m_2)v_f^2 - \left(\frac{1}{2}m_1 v_{1i}^2 + \frac{1}{2}m_2 v_{2i}^2\right)$$

$$= \frac{1}{2}(3700\,\text{kg})(16.5\,\text{m/s})^2 - \left(\frac{1}{2}(1200\,\text{kg})(18\,\text{m/s})^2 + \frac{1}{2}(2500\,\text{kg})(23\,\text{m/s})^2\right)$$

$$\Delta K = -3.5 \times 10^5\,\text{J}$$

5.5 Torque

Consider a force F acting on a particle that has a position vector r with respect to some origin O that is in an inertial frame. The torque is a vector quantity that measures the tendency of that force to rotate the particle about O and is defined as

$$\boldsymbol{\tau} = \mathbf{r} \times \mathbf{F}$$

The direction of $\boldsymbol{\tau}$ is perpendicular to the plane formed by r and F and its sense is given by the right-hand rule or of advance of a right-handed screw rotating from r to F. From the vector product definition, this quantity has a magnitude given by

$$\tau = rF \sin\phi$$

where ϕ is the smaller angle between r and F, τ is positive if the force tends to rotate the particle counterclockwise and negative if it tends to rotate it clockwise. If $\phi = 0$ or 180°,

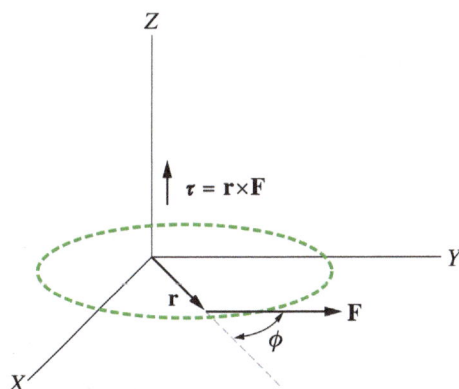

Fig. 5.11 A particle in the x-y plane exposed to a force that lies in that plane. The resulting torque is then perpendicular to the x-y plane parallel to the z-axis

the force is radial and thus it has no rotating tendency. In component form, we may write

$$\boldsymbol{\tau} = \mathbf{r} \times \mathbf{F} = (x\mathbf{i} + y\mathbf{j} + z\mathbf{k}) \times (F_x\mathbf{i} + F_y\mathbf{j} + F_z\mathbf{k})$$

$$= (yF_z - zF_y)\mathbf{i} + (zF_x - xF_z)\mathbf{j} + (xF_y - yF_x)\mathbf{k}$$

Let us consider a particle in the x–y plane exposed to a force that lies in that plane (see Fig. 5.11). The resulting torque is then perpendicular to the x–y plane parallel to the z-axis. τ can also be written as

$$\tau = Fd$$

where $d = r\sin\phi$ is called the moment arm of F where it represents the perpendicular distance from the axis of rotation to the line of action of F as shown in Fig. 5.12. Note that because τ depends on r, it follows that τ depends on the choice of the origin O. The force F can be resolved into two components $F_t = F\sin\phi$ and $F_r = F\cos\phi$. Since the line of action of F_r passes through O, it has no rotating effect. Hence, F_t is the only component of F that causes rotation. The SI unit of torque is the Newton-metre (N m). This unit is the same unit of work, but they are different quantities and the torque should never be expressed in joules.

Example 5.14 A force $\mathbf{F} = (-2t\mathbf{i} - (t^2 - 3)\mathbf{j} + 4t^5\mathbf{k})$ N acts on a particle that has a position vector $\mathbf{r} = \left(-6\mathbf{i} + 5t\mathbf{j} + (\frac{t}{2} - 1)\mathbf{k}\right)$ m find the torque of the particle about the origin at $t = 1$ s.

Solution 5.14

$$\boldsymbol{\tau} = \mathbf{r} \times \mathbf{F} = \begin{vmatrix} \mathbf{i} & \mathbf{j} & \mathbf{k} \\ -6 & 5t & (\frac{t}{2} - 1) \\ -2t & -(t^2 - 3) & 4t^5 \end{vmatrix}$$

Evaluating this at $t = 1$ s gives

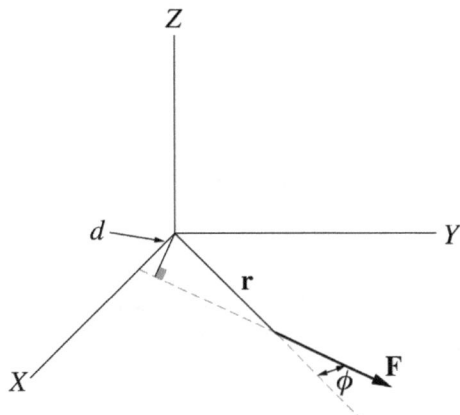

Fig. 5.12 $d = r \sin \phi$ is called the moment arm of F and it represents the perpendicular distance from the axis of rotation to the line of action of F

$$\boldsymbol{\tau} = (21\mathbf{i} + 25\mathbf{j} - 2\mathbf{k}) \, \text{N/m}$$

5.6   Angular Momentum

The angular momentum **L** of a particle of mass m and linear momentum **p** is a vector quantity defined as

$$\mathbf{L} = \mathbf{r} \times \mathbf{p}$$

where r is the position vector of the particle relative to an origin O that is in an inertial frame. Therefore, as $\boldsymbol{\tau}$, **L** also depends on the choice of the origin. Suppose the particle moves in the x–y plane (see Fig. 5.13). The direction of L is then perpendicular to the plane containing r and p and its sense is found by the right-hand rule. The magnitude of **L** is given by

$$L = mvr \sin \phi$$

where ϕ is the smaller angle between **r** and **p**. This quantity is the rotational analog of linear momentum in translational motion. If $\phi = 0$ or $180°$ the particle will move along a line passing through O and its angular momentum is zero. The SI unit of angular momentum is $\text{kg.m}^2/\text{s}$. In terms of rectangular components, we have

$$\mathbf{L} = \mathbf{r} \times \mathbf{p} = (x\mathbf{i} + y\mathbf{j} + z\mathbf{k}) \times (p_x\mathbf{i} + p_y\mathbf{j} + p_z\mathbf{k})$$

$$= (yp_z - zp_y)\mathbf{i} + (zp_x - xp_z)\mathbf{j} + (xp_y - yp_x)\mathbf{k}$$

5.6.1   Newton's Second Law in Angular Form

From the definition of torque, we have

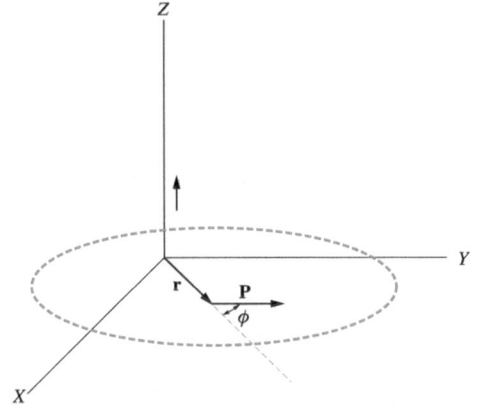

Fig. 5.13 If the particle is moving in the x-y plane, then the direction of L is perpendicular to the plane containing r and p and is found by the right-hand rule

$$\boldsymbol{\tau} = \mathbf{r} \times \mathbf{F} = \mathbf{r} \times \frac{d(m\mathbf{v})}{dt}$$

$$\frac{d\mathbf{L}}{dt} = \frac{d(\mathbf{r} \times m\mathbf{v})}{dt} = \frac{d\mathbf{r}}{dt} \times (m\mathbf{v}) + \mathbf{r} \times \frac{d(m\mathbf{v})}{dt}$$

$$= \mathbf{v} \times (m\mathbf{v}) + \mathbf{r} \times \frac{d(m\mathbf{v})}{dt} = \mathbf{0} + \mathbf{r} \times \mathbf{F} = \boldsymbol{\tau}$$

$$\boldsymbol{\tau} = \frac{d\mathbf{L}}{dt} \qquad (5.12)$$

This implies that the torque acting on a particle is equal to the time rate of change of the angular momentum for that particle. This equation is valid only if $\boldsymbol{\tau}$ and **L** are evaluated with respect to the same origin or any other fixed point in an inertial frame. If several forces act on the particle, Eq. 5.12 can be written as

$$\Sigma\boldsymbol{\tau} = \frac{d\mathbf{L}}{dt}$$

where $\Sigma\boldsymbol{\tau}$ is the net torque on the particle. This is the rotational analog of Newton's second law in linear form, which states that the net force acting on a particle is equal to the time rate of change of its linear momentum. In component form, we have $\Sigma\tau_x = dL_x/dt$, $\Sigma\tau_y = dL_y/dt$ and $\Sigma\tau_z = dL_z/dt$.

5.6.2   Conservation of Angular Momentum

The total angular momentum of a particle is constant if the net external torque acting on it is zero:

$$\Sigma\boldsymbol{\tau}_{ext} = \frac{d\mathbf{L}}{dt} = \mathbf{0}$$

$$\mathbf{L} = \text{constant}$$

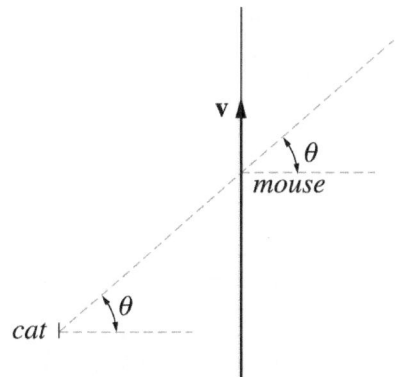

Fig. 5.14 A cat watching a mouse run by

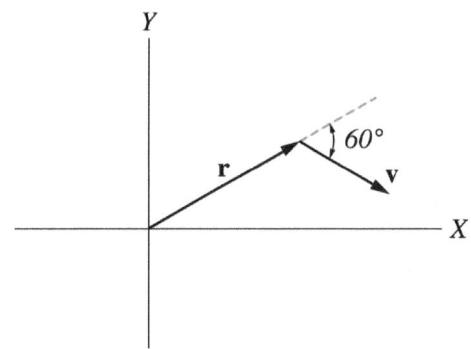

Fig. 5.15 A particle moving in the x-y plane

$$m(\mathbf{r} \times \mathbf{v}) = \text{contant}$$

or

$$\mathbf{L}_i = \mathbf{L}_f$$

The law of conservation of angular momentum is a fundamental law of physics and it holds in relativity and quantum mechanics. Thus, for an isolated system, the linear momentum and angular momentum are conserved.

Example 5.15 A cat watches a mouse of mass m run by, as shown in Fig. 5.14. Determine the mouse's angular momentum relative to the cat as a function of time if the mouse has a constant acceleration a and if it starts from rest.

Solution 5.15 Suppose the plane is the x–y plane. Since $v = at$, we have

$$\mathbf{L} = m(\mathbf{r} \times \mathbf{v}) = mrat \cos \theta \mathbf{k}$$

Example 5.16 A 0.2 kg particle is moving in the x–y plane. If at a certain instant $r = 3$ m and $v = 10$ m/s (see Fig. 5.15), find the magnitude and direction of the angular momentum of the particle at that instant relative to the origin.

Solution 5.16

$$\mathbf{L} = m(\mathbf{r} \times \mathbf{v}) = -(mvr \sin \phi)\mathbf{k} = -(0.2 \, \text{kg})(10 \, \text{m/s})(3 \, \text{m}) \sin 60° \mathbf{k} = (-5.2 \, \mathbf{k})\text{kg.m}^2/\text{s}$$

Example 5.17 A particle is moving under the influence of a force given by $\mathbf{F} = -k\mathbf{r}$. Prove that the angular momentum of the particle is conserved.

Solution 5.17

$$\boldsymbol{\tau} = \mathbf{r} \times \mathbf{F} = -k(\mathbf{r} \times \mathbf{r}) = \mathbf{0}$$

Since $\boldsymbol{\tau} = d\mathbf{L}/dt$, it follows that the total angular momentum of the particle is conserved. That is,

$$\mathbf{L} = \text{constant}$$

Example 5.18 A particle is moving in a circle where its position as a function of time is given by the expression $\mathbf{r} = a(\cos \omega t \mathbf{i} + \sin \omega t \mathbf{j})$, where ω is a constant. Show that the total angular momentum of the particle is constant.

Solution 5.18

$$\mathbf{v} = \frac{d\mathbf{r}}{dt} = a(-\omega \sin \omega t \mathbf{i} + \omega \cos \omega t \mathbf{j})$$

$$\mathbf{L} = m(\mathbf{r} \times \mathbf{v}) = ma^2[(\cos \omega t \mathbf{i} + \sin \omega t \mathbf{j}) \times (-\omega \sin \omega t \mathbf{i} + \omega \cos \omega t \mathbf{j})]$$

$$= ma^2(\omega \cos^2 \omega t \mathbf{k} + \omega \sin^2 \omega t \mathbf{k})$$

$$= m\omega a^2 \mathbf{k} = \text{constant}$$

Problems

1. A tennis ball of mass of 0.06 kg is initially traveling at an angle of 47° to the horizontal at a speed of 45 m/s. It then was shot by the tennis player and return horizontally at a speed of 35 m/s. Find the impulse delivered to the ball.
2. A force on a 0.5 kg particle varies with time according to Fig. 5.16. Find (a) The impulse delivered to the particle, (b) the average force exerted on the particle from $t = 0$ to $t = 6$ s(c). The final velocity of the particle if its initial velocity is 2 m/s.
3. A 1 kg particle moves in a force field given by $\mathbf{F} = (2t^2\mathbf{i} + (5t - 3)\mathbf{j} - 6t\mathbf{k})$ N. Find the impulse delivered to the particle during the time interval from $t = 1$ s to $t = 3$ s.
4. A boy of mass 45 kg runs and jump with a horizontal speed of 4.5 m/s into a 70 kg cart that is initially at rest (see Fig. 5.17). Find the final velocity of the boy and the cart.

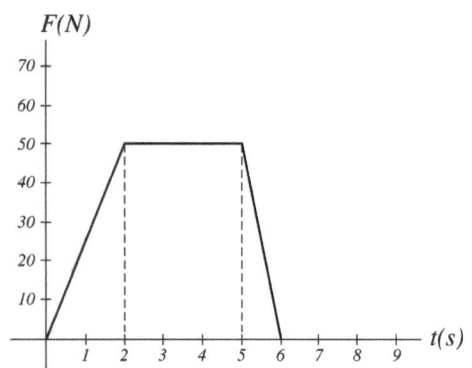

Fig. 5.16 A force acting on a particle varies with time

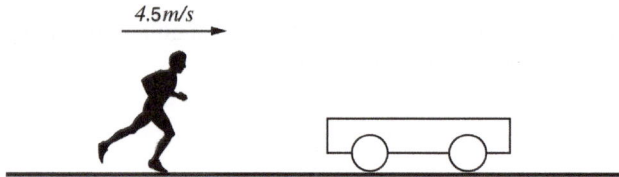

Fig. 5.17 A boy jumps on a cart that is initially at rest

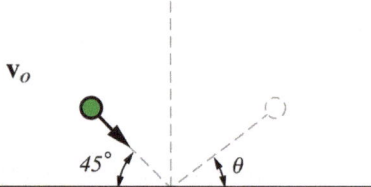

Fig. 5.18 A ball bouncing off a smooth surface

5. A rubber ball of mass of 0.2 kg is dropped from a height of 2.2 m. It re- bounds to a height of 1.1 m. Find (a) the coefficient of restitution, (b) the energy lost due to impact.

6. A 1200 kg car initially traveling at 12 m/s due east collides with another car of mass of 1600 kg that is initially at rest. If the cars become entangled after the collision, find the common final speed of the cars.

7. Figure 5.18 shows a ball that strikes a smooth surface with a velocity of 20 m/s at an angle of 45° with the horizontal. If the coefficient of restitution for the impact between the ball and the surface is $e = 0.85$, find the magnitude and direction of the velocity in which the ball rebounds from the surface. (Hint: use the velocity components in the direction perpendicular to the surface for the coefficient of restitution).

8. Two gliders moving on a frictionless linear air track experience a perfectly elastic collision (see Fig. 5.19). Find the velocity of each glider after the collision.

9. A bullet of mass of m is fired with a horizontal velocity v into a block of mass M. The block is initially at rest on a frictionless surface and is connected to a spring of force

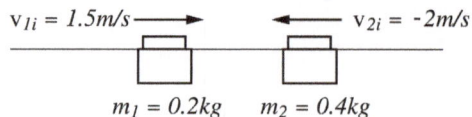

Fig. 5.19 Two gliders moving on a frictionless linear air track experience a perfectly elastic collision

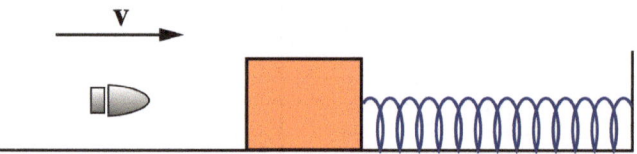

Fig. 5.20 A bullet of mass of m is fired with a horizontal velocity v into a block of mass M

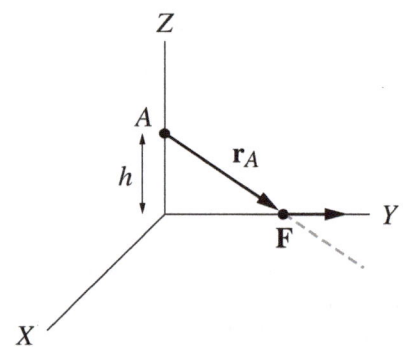

Fig. 5.21 A block moving along the y-axis subject to a force

Fig. 5.22 A conical pendulum of mass m and length L is in uniform circular motion with a velocity v

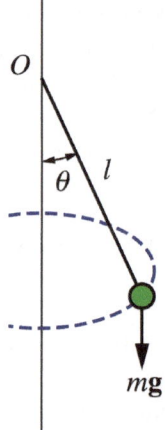

constant of k (see Fig. 5.20). If the bullet embeds itself in the block causing the spring to compress to a maximum distance d, find the initial speed of the bullet.

10. A block moves along the y-axis due to a force given by $\mathbf{F} = a\mathbf{i}$ (see Fig. 5.21). Find the torque on the block about (a) the origin (b) point A.

11. A conical pendulum of mass m and length L is in uniform circular motion with a velocity v (see Fig. 5.22). Find the angular momentum and torque on the mass about O.

Open Access This chapter is licensed under the terms of the Creative Commons Attribution 4.0 International License (http://creativecommons.org/licenses/by/4.0/), which permits use, sharing, adaptation, distribution and reproduction in any medium or format, as long as you give appropriate credit to the original author(s) and the source, provide a link to the Creative Commons license and indicate if changes were made.

The images or other third party material in this chapter are included in the chapter's Creative Commons license, unless indicated otherwise in a credit line to the material. If material is not included in the chapter's Creative Commons license and your intended use is not permitted by statutory regulation or exceeds the permitted use, you will need to obtain permission directly from the copyright holder.

System of Particles

6.1 System of Particles

In the previous chapters, objects that can be treated as particles were only considered. We have seen that this is possible only if all parts of the object move in exactly the same way. An object that does not meet this condition must be treated as a system of particles. Next, we will see that the complex motion of this object or system of particles can be represented by the motion of a point located at the center of mass of the system. The center of mass moves as if all of the mass of the object is concentrated there and as if the net external force acting on the system is applied there (at the center of mass). As well as representing an object by a particle, the concept of the center of mass is used to analyze the motion of many systems such as a system of two colliding blocks (particle-like objects) and the system of two colliding subatomic particles such as the neutron with the nucleus.

6.2 Discrete and Continuous System of Particles

6.2.1 Discrete System of Particles

A discrete system of particles is a system in which particles are separated from each other.

6.2.2 Continuous System of Particles

A continuous system of particles is a system where the separation of particles is very small such that it approaches zero. An extended object is a continuous system of particles. Now, consider the skateboarder example mentioned in Sect. 4.3. It has been shown that the system (man+skateboard) cannot be treated as a particle since different parts of the system move in different ways. By representing the skateboarder as a system of particles its motion can be represented by the motion of

its center of mass, hence, the work–energy theorem can be applied to that point. The work done by the force, exerted on the skateboarder by the bar, is not zero because the point of application of that force (which is at the center of mass) has moved.

6.3 The Center of Mass of a System of Particles

For a system of particles of total mass M the acceleration of its center of mass is given by

$$\mathbf{a} = \frac{\mathbf{F}}{M}$$

6.3.1 Two Particle System

Consider two particles of masses m_1 and m_2 moving in space. Suppose that their position vectors at a particular instant of time are given by $\mathbf{r}_1$ and $\mathbf{r}_2$ as shown in Fig. 6.1. The center of mass of the system lies somewhere along the line joining the two particles and its position vector is given by

$$\mathbf{r}_{cm} = \frac{m_1\mathbf{r}_1 + m_2\mathbf{r}_2}{m_1 + m_2}$$

The x, y and z components of the center of mass is

$$x_{cm} = \frac{m_1 x_1 + m_2 x_2}{m_1 + m_2}$$

$$y_{cm} = \frac{m_1 y_1 + m_2 y_2}{m_1 + m_2}$$

and

$$z_{cm} = \frac{m_1 z_1 + m_2 z_2}{m_1 + m_2}$$

© The Author(s) 2019
S. Alrasheed, *Principles of Mechanics*, Advances in Science, Technology & Innovation, https://doi.org/10.1007/978-3-030-15195-9_6

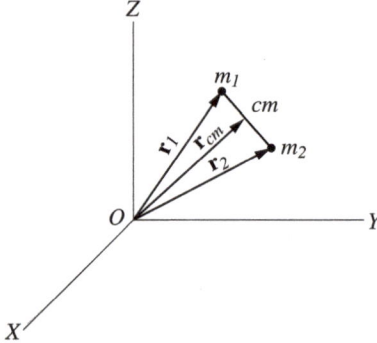

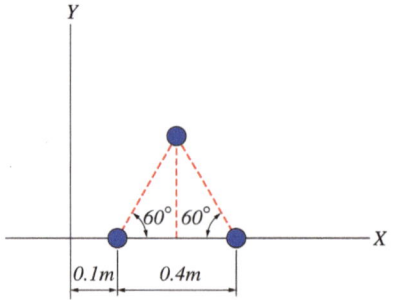

Fig. 6.3 The center of mass of a system in the x-y plane

Fig. 6.1 Two particles of masses m_1 and m_2 moving in space. Their position vectors at a particular instant of time are given by $\mathbf{r}_1$ and $\mathbf{r}_2$

Example 6.1 Find the center of mass of the system shown in Fig. 6.3 where the three particles have an equal mass of $m = 1$ kg.

Solution 6.1

$$x_{cm} = \frac{(1\,\text{kg})((0.1\,\text{m}) + (0.5\,\text{m}) + (0.3\,\text{m}))}{(3\,\text{kg})} = 0.3\,\text{m}$$

$$y_{cm} = \frac{0 + 0 + (1\,\text{kg})(0.2\,\text{m})\tan(60°)}{(3\,\text{kg})} = 0.12\,\text{m}$$

$$\mathbf{r}_{cm} = x_{cm}\mathbf{i} + y_{cm}\mathbf{j} = (0.3\,\text{m})\,\mathbf{i} + (0.12\,\text{m})\,\mathbf{j}$$

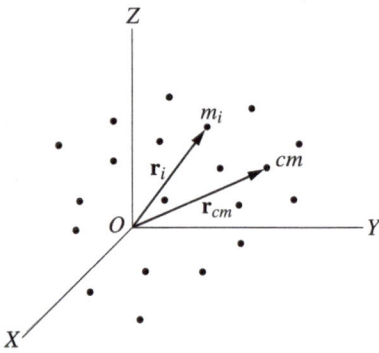

Fig. 6.2 A discrete system of particles consisting of n particles

6.3.2 Discrete System of Particles

Consider a discrete system of particles consisting of n particles (see Fig. 6.2). The position vector of the center of mass at a particular instant is given by

$$\mathbf{r}_{cm} = \frac{m_1\mathbf{r}_1 + m_2\mathbf{r}_2 + m_3\mathbf{r}_3. + \cdots\cdots\cdots \cdot m_n\mathbf{r}_n}{m_1 + m_2 + m_3 + \cdots + m_n} = \frac{\Sigma_{i=1}^n m_i\mathbf{r}_i}{M}$$

where $\mathbf{r}_i$ is the position vector of the ith particle and $M = \sum_{i=1}^{n} m_i$ is the total mass of the system. In component form, r_i can be written as

$$\mathbf{r}_i = x_i\mathbf{i} + y_i\mathbf{j} + z_i\mathbf{k}$$

The x, y and z components of the center of mass vector are

$$x_{cm} = \frac{\sum_{i=1}^n m_i x_i}{M}$$

$$y_{cm} = \frac{\sum_{i=1}^n m_i y_i}{M}$$

and

$$z_{cm} = \frac{\sum_{i=1}^n m_i z_i}{M}$$

Example 6.2 A system of particles consists of three masses $m_A = 0.5$ kg, $m_B = 2$ kg and $m_C = 5$ kg located at $P_A(-3, 1, 2)$, $P_B(0, 1, 2)$ and $P_C(-1, 3, 0)$, respectively. Find the position vector of the center of mass of the system.

Solution 6.2 The position vector of each particle is

$$\mathbf{r}_A = (-3\mathbf{i} + \mathbf{j} + 2\mathbf{k})\,\text{m}$$

$$\mathbf{r}_B = (\mathbf{j} + 2\mathbf{k})\,\text{m}$$

and

$$\mathbf{r}_C = (-\mathbf{i} + 3\mathbf{j})\,\text{m}$$

The center of mass of the system is

$$\mathbf{r}_{cm} = \frac{\sum_{i=1}^n m_i\mathbf{r}_i}{\sum_{i=1}^n m_i} = \frac{(0.5\,\text{kg})((-3\mathbf{i} + \mathbf{j} + 2\mathbf{k})\,\text{m}) + (2\,\text{kg})((\mathbf{j} + 2\mathbf{k})\,\text{m}) + (5\,\text{kg})((-\mathbf{i} + 3\mathbf{j})\,\text{m})}{(7.5\,\text{kg})}$$

That gives

$$\mathbf{r}_{cm} = (-0.87\mathbf{i} + 2.3\mathbf{j} + 0.7\mathbf{k})\,\text{m}.$$

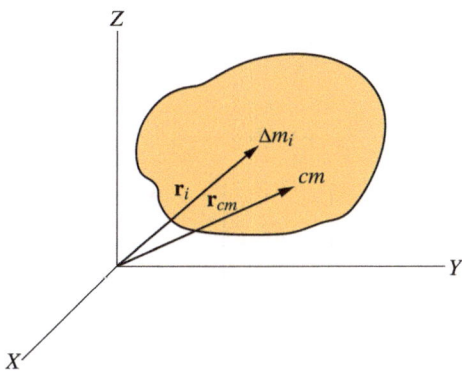

Fig. 6.4 An extended object of mass M divided into small volume elements each of mass $\triangle m_i$ and a vector position $\mathbf{r}_I$

6.3.3 Continuous System of Particles (Extended Object)

A continuous system of particles is a system consisting of a large number of particles separated by very small distances. Consider an extended object of mass M divided into small volume elements each of mass $\triangle m_i$ and a vector position $\mathbf{r}_i$ (see Fig. 6.4). The position vector of the center of mass at a particular instant is then approximately given by

$$\mathbf{r}_{cm} \approx \frac{\sum_{i=1}^{n} \mathbf{r}_i \triangle m_i}{M}$$

For a very large number of particles where $n \to \infty$ we have $\triangle m_i \to 0$, that gives

$$\mathbf{r}_{cm} = \lim_{\triangle m_i} \frac{\sum_{i=1}^{n} \mathbf{r}_i \triangle m_i}{M} = \frac{1}{M} \int \mathbf{r} dm$$

Since $\mathbf{r} = x\mathbf{i} + y\mathbf{j} + z\mathbf{k}$, the x, y and z components of the center of mass are given by

$$x_{cm} = \frac{1}{M} \int x dm$$

$$y_{cm} = \frac{1}{M} \int y dm$$

and

$$z_{cm} = \frac{1}{M} \int z dm$$

6.3.4 Elastic and Rigid Bodies

A body is called an elastic (deformable) body if the separation between its particles changes when a force is applied to it. This change or deformation is sometimes so small that it can

be neglected. A body that behaves in this way is called a rigid body. A rigid body can be defined as a body in which the separation between its particles remain constant with time despite the applied force, i.e., the body has a constant size and shape. Therefore, the center of mass of a rigid object remains fixed at the same location at all times. In this book, only rigid bodies are discussed. In solving problems, it is common to use the volume density ρ defined as the mass per unit volume given by

$$\rho = \frac{dm}{dV}$$

Therefore, the total mass of a rigid object is

$$M = \int \rho dV$$

The center of mass of a rigid object can thus be written as

$$\mathbf{r}_{cm} = \frac{1}{M} \int \mathbf{r} dm = \frac{\int \rho \mathbf{r} dV}{\int \rho dV}$$

ρ may be a function of position, i.e., it can vary from point to point in the body If the body has a uniform density (homogeneous body), then ρ can be written as

$$\rho = \frac{dm}{dV} = \frac{\mathrm{Total\,Mass}}{\mathrm{Total\,Volume}} = \mathrm{constant}$$

If the continuous distribution of particles occupies a surface, then the surface density σ is used and is given by

$$\sigma = \frac{dm}{dA} \ (\mathrm{mass\,per\,unit\,area})$$

$$\sigma = \frac{\mathrm{Total\,Mass}}{\mathrm{Total\,Area}} = \mathrm{constant} \ (\mathrm{homogeneous\,body})$$

If the particles occupy a curve or a line, the linear density λ is used given by

$$\lambda = \frac{dm}{dl} \ (\mathrm{mass\,per\,unit\,length})$$

$$\lambda = \frac{\mathrm{Total\,Mass}}{\mathrm{Total\,Length}} = \mathrm{constant} \ (\mathrm{homogeneous\,body})$$

The center of mass of any homogeneous symmetric object is at its geometrical center and it is not necessarily located within the object.

Example 6.3 A thin rod of length $L = 2\,\mathrm{m}$ has a linear density that increases with x according to the expression

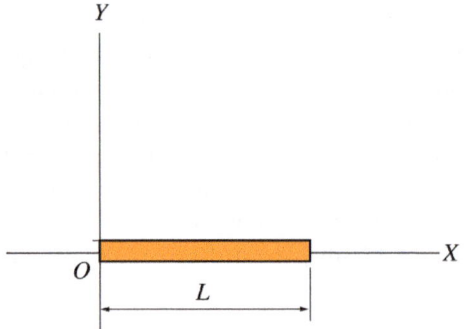

Fig. 6.5 A thin rod of length $L = 2$ m has a linear density that increases with x

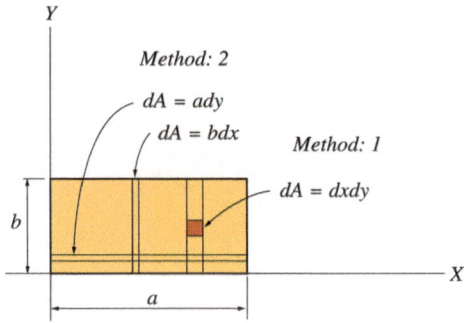

Fig. 6.7 The center of mass of a rectangular plate

$\lambda(x) = (2x - 1)$ kg/m (see Fig. 6.5). Locate the center of mass of the rod relative to O.

Solution 6.3

$$x_{cm} = \frac{1}{M} \int x\,dm = \frac{\int_0^L x\lambda(x)\,dx}{\int_0^L \lambda(x)\,dx} = \frac{\int_0^L (2x^2 - x)\,dx}{\int_0^L (2x - 1)\,dx}$$

$$= \frac{((2/3)x^3 - x^2/2)|_{x=0}^{L}}{(x^2 - x)|_{x=0}^{L}} = \frac{L((2/3)L - 1/2)}{(L - 1)}$$

Substituting $L = 2$ m gives $x_{cm} = 1.7$ m.

Example 6.4 A uniform square sheet is suspended by a uniform rod where they both lie in the same plane as shown in Fig. 6.6. Find the center of mass of the system.

Solution 6.4 Because the sheet and the rod are homogeneous, the center of mass of each is at its geometric center. Since the center of the sheet is at the origin we have

$$x_{cm} = \frac{\sum_i m_i x_i}{\sum_i m_i} = \frac{0 + (M_2 L/2)}{M_1 + M_2} = \frac{LM_2}{2(M_1 + M_2)}$$

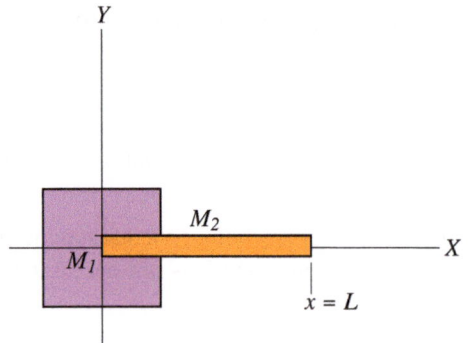

Fig. 6.6 A uniform square sheet suspended by a uniform rod where they both lie in the same plane

Example 6.5 Find the center of mass of the rectangular plate shown in Fig. 6.7. The plate has a uniform surface density σ.

Solution 6.5 • Method 1:

$$x_{cm} = \frac{\int x\,dm}{M} = \frac{\int x\sigma\,dA}{\int \sigma\,dA} = \frac{\int_{y=0}^{b}\int_{x=0}^{a} x\,dx\,dy}{\int_{y=0}^{b}\int_{x=0}^{a} dx\,dy} = \frac{ba^2}{2ab} = \frac{a}{2}$$

$$y_{cm} = \frac{\int y\,dm}{M} = \frac{\int x\sigma\,dA}{\int \sigma\,dA} = \frac{\int_{x=0}^{a}\int_{y=0}^{b} y\,dx\,dy}{\int_{x=0}^{a}\int_{y=0}^{b} dx\,dy} = \frac{ab^2}{2ab} = \frac{b}{2}$$

Hence

$$\mathbf{r}_{cm} = \frac{a}{2}\mathbf{i} + \frac{b}{2}\mathbf{j}$$

• Method 2:

Dividing the plate into very thin rods each of mass $\sigma b\,dx$ gives

$$x_{cm} = \frac{\int x\,dm}{M} = \frac{1}{M}\int x\sigma\,dA = \frac{1}{M}\left(\frac{M}{ab}\right)\int_{x=0}^{a} xb\,dx = \frac{1}{a}\left[\frac{x^2}{2}\right]_{x=0}^{a} = \frac{a}{2}$$

Similarly by dividing the plate into thin horizontal rods each of mass $\sigma a\,dy$ gives

$$y_{cm} = \frac{\int y\,dm}{M} = \frac{1}{M}\int y\sigma\,dA = \frac{1}{M}\left(\frac{M}{ab}\right)\int_{y=0}^{b} ay\,dy = \frac{1}{b}\left[\frac{y^2}{2}\right]_{y=0}^{b} = \frac{b}{2}$$

and

$$\mathbf{r}_{cm} = \frac{a}{2}\mathbf{i} + \frac{b}{2}\mathbf{j}$$

Example 6.6 An object of uniform surface density σ and mass M has the shape shown in Fig. 6.8 (half of an ellipse). Find the center of mass of the object.

Solution 6.6 The equation of an ellipse is

$$\frac{x^2}{a^2} + \frac{y^2}{b^2} = 1$$

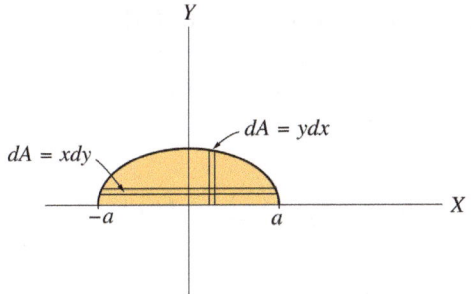

Fig. 6.8 The center of mass of half an ellipse

therefore

$$\frac{2xdx}{a^2} + \frac{2ydy}{b^2} = 0$$

or

$$xdx = \frac{-a^2}{b^2}ydy$$

By dividing the area into very thin rectangles each of mass $\sigma y dx$ gives

$$x_{cm} = \frac{\int x dm}{M} = \frac{1}{M}\int x\sigma dA = \frac{1}{M}\int_{x=-a}^{a} x\left(\frac{2M}{\pi ab}\right)ydx$$

$$= \frac{2}{\pi ab}\int_{y=0}^{0}\left(\frac{-a^2}{b^2}\right)y^2 dy = \frac{-2a}{\pi b^3}\left[\frac{y^3}{3}\right]_{y=0}^{0} = 0$$

To obtain the y coordinate of the center of mass we divide the area into very thin rectangles each of mass $\sigma x dy$ as in Fig. 6.8. That gives

$$y_{cm} = \frac{1}{M}\int y dm = \frac{1}{M}\int y\sigma dA = \frac{2}{\pi ab}\int_{y=0}^{b} yxdy$$

$$= \frac{2}{\pi ab}\int_{x=a}^{-a}\left(\frac{-b^2}{a^2}\right)x^2 dx = \frac{-2b}{\pi a^3}\int_{x=a}^{-a}x^2 dx = \frac{-2b}{\pi a^3}\left[\frac{x^3}{3}\right]_{x=a}^{-a}$$

$$\frac{-2b}{\pi a^3}\left[\frac{x^3}{3}\right]_{x=a}^{-a} = \frac{-2b}{\pi a^3}\left(\frac{-a^3}{3}-\frac{a^3}{3}\right) = \frac{4b}{3\pi}$$

Example 6.7 Determine the center of mass of the cylindrical shell shown in Fig. 6.9. The shell has a uniform surface density σ.

Solution 6.7 From symmetry, the center of mass lies on the z-axis. By dividing the shell into very thin rings each of mass $\sigma 2\pi R dz$ we have

$$z_{cm} = \frac{\int z dm}{M} = \frac{\int z\sigma dA}{M} = \frac{1}{M}\int_{z=0}^{h} z\sigma 2\pi R dz = \frac{1}{M}\left(\frac{M}{2\pi Rh}\right)\int_{z=0}^{h} 2\pi Rz dz$$

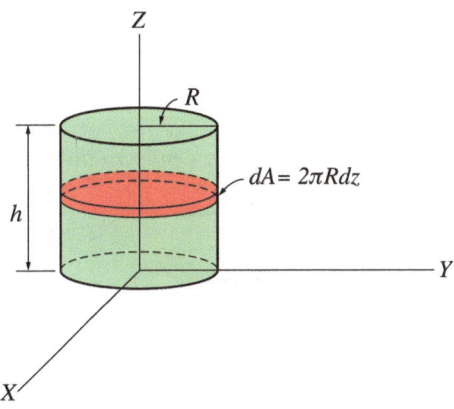

Fig. 6.9 The center of mass of a cylindrical shell

$$= \frac{1}{h}\left[\frac{z^2}{2}\right]_{z=0}^{h} = \frac{h}{2}$$

Example 6.8 A boy standing on a smooth ice surface wants to fetch a container that is at a distance of 10 m away from him. To do that, he throws a rope around the container and start to pull. Because the surface is smooth, both the boy and the container will move until they meet. If the masses of the boy and of the container are 40 kg and 70 kg respectively, how far will the container move when the boy has moved a distance of 2 m?

Solution 6.8 By taking the midpoint between the boy and the container as the origin (see Fig. 6.10) and by neglecting the mass of the rope, the center of mass of the system is

$$x_{cm} = \frac{\Sigma_i m_i x_i}{\Sigma_i m_i} = \frac{(70\,\text{kg})(5\,\text{m}) + (40\,\text{kg})(-5\,\text{m})}{(110\,\text{kg})} = 1.36\,\text{m}$$

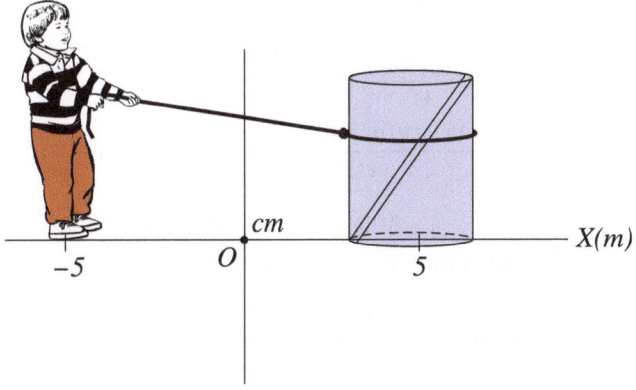

Fig. 6.10 A boy pulling a container on a smooth surface

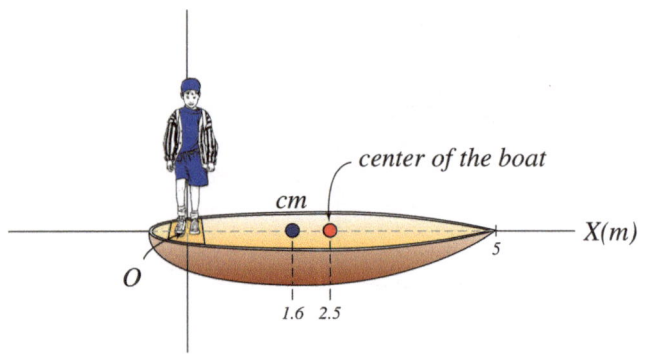

Fig. 6.11 A boy walking on a small boat

Because the surface may be assumed to be frictionless, the resultant external force on the system is zero and therefore the center of mass must remain stationary at all times. Hence, if the boy has moved a distance of 2 m, he will be at a distance of −3 m from the origin. Thus, we have

$$(1.36\,\text{m}) = \frac{(70\,\text{kg})x_c + (40\,\text{kg})(-3\,\text{m})}{(110\,\text{kg})}$$

That gives $x_c = 3.86$ m, therefore the distance moved by the container towards the center of mass is $(5\,\text{m}) - (3.86\,\text{m}) = 1.14$ m.

Example 6.9 A boy is standing at the rear of a boat as shown in Fig. 6.11. The masses of the boy and of the boat are 45 kg and 80 kg respectively Find the distance that the boat would move relative to the origin if the boy moves a distance of lm from the rear of the boat (the length of the boat is 5 m).

Solution 6.9 By neglecting air and water resistance, the net external force on the (boy+ boat) system is zero. Therefore the center of mass of the system must remain at rest. Suppose that the boat is a symmetrical homogeneous object where its center of mass is at its geometrical center. The center of mass of the boat is therefore at a distance of 2.5 m from the origin. Thus, the center of mass of the system is

$$x_{cm} = \frac{\sum_{i=1}^{n} m_i x_i}{M} = \frac{m_1 x_1 + m_2 x_2}{m_1 + m_2}$$
$$= \frac{(45\,\text{kg})(0) + (80\,\text{kg})(2.5\,\text{m})}{(125\,\text{kg})} = 1.6\,\text{m}$$

If the boy moves a distance of 1 m, the center of mass is still at the same position, and we have

$$(1.6\,\text{m}) = \frac{(45\,\text{kg})(1\,\text{m}) + (80\,\text{kg})x_b}{(125\,\text{kg})}$$

That gives $x_b = 1.94$ m. Thus, the displacement of the center of mass of the boat is $(1.94\,\text{m}) - (2.5\,\text{m}) = -0.56$ m.

6.3.5   Velocity of the Center of Mass

The velocity of the center of mass of a system of particles that has a constant mass M is

$$\mathbf{v}_{cm} = \frac{d\mathbf{r}_{cm}}{dt} = \frac{1}{M}\frac{d}{dt}\left(\sum_{i=1}^{n} m_i \mathbf{r}_i\right) = \frac{1}{M}\sum_{i=1}^{n} m_i \dot{\mathbf{r}}_i$$

where $\dot{\mathbf{r}}_i = d\mathbf{r}_i/dt$, or

$$\mathbf{v}_{cm} = \sum_{i=1}^{n} \frac{m_i \mathbf{v}_i}{M} \tag{6.1}$$

where $\mathbf{v}_i$ is the ith particle velocity. The acceleration of the center of mass is given by

$$\mathbf{a}_{cm} = \frac{d\mathbf{v}_{cm}}{dt} = \frac{1}{M}\frac{d}{dt}\left(\sum_{i=1}^{n} m_i \mathbf{v}_i\right) = \frac{1}{M}\sum_{i=1}^{n} m_i \ddot{\mathbf{r}}_i$$

$$\mathbf{a}_{cm} = \frac{1}{M}\sum_{i=1}^{n} m_i \mathbf{a}_i \tag{6.2}$$

where $\mathbf{a}_i$ is the acceleration of the ith particle.

6.3.6   Momentum of a System of Particles

The total linear momentum of a system of particles is the vector sum of the linear momenta of the individual particles:

$$\sum_{i=1}^{n} m_i \mathbf{v}_i = \sum_{i=1}^{n} \mathbf{p}_i = \mathbf{p}_{tot} \tag{6.3}$$

By using Eq. 6.1

$$\mathbf{p}_{tot} = M\mathbf{v}_{cm} \tag{6.4}$$

Example 6.10 Two particles of masses $m_1 = 1$ kg and $m_2 = 2$ kg have position vectors given by $\mathbf{r}_1 = (2t\mathbf{i} - 4\mathbf{j})$ m and $\mathbf{r}_2 = (5t\mathbf{i} - 2t\mathbf{j})$ m respectively where t is time. Determine the velocity and linear momentum of the center of mass of the two- particle system at any time and at $t = 1$ s.

Solution 6.10

$$\mathbf{r}_{cm} = \frac{\sum_i m_i \mathbf{r}_i}{\sum_i m_i} = \frac{(1\,\text{kg})(2t\mathbf{i} - 4\mathbf{j}) + (2\,\text{kg})(5t\mathbf{i} - 2t\mathbf{j})}{(3\,\text{kg})}$$

That gives

$$\mathbf{r}_{cm} = \left(4t\mathbf{i} - \frac{4}{3}(t+1)\mathbf{j}\right)\,\text{m}$$

$$\mathbf{v}_{cm} = \frac{d\mathbf{r}_{cm}}{dt} = \left(4\mathbf{i} - \frac{4}{3}\mathbf{j}\right)\,\text{m/s}$$

The total linear momentum is

$$\mathbf{p}_{tot} = M\mathbf{v}_{cm} = (3\text{kg})\left(4\mathbf{i} - \frac{4}{3}\mathbf{j}\right) = (12\mathbf{i} - 4\mathbf{j})\,\text{kg.m/s}$$

at $t = 1$s

$$\mathbf{r}_{cm} = (4\mathbf{i} - \frac{8}{3}\mathbf{j})\,\text{m}$$

$$\mathbf{v}_{cm} = (4\mathbf{i} - \frac{4}{3}\mathbf{j})\,\text{m/s}$$

and

$$\mathbf{p}_{tot} = (12\mathbf{i} - 4\mathbf{j})\,\text{kg.m/s}$$

6.3.7   Motion of a System of Particles

From Newton's second law Eq. 6.2 can be written as

$$\mathbf{a}_{cm} = \frac{1}{M}\sum_{i=1}^{n}\mathbf{F}_i \qquad (6.5)$$

where $\mathbf{F}_i$ is the net force acting on the ith particle. If both the external forces on the system and the internal forces between the particles in the system are included, then $\mathbf{F}_i$ may be written as

$$\mathbf{F}_i = \mathbf{F}_{i(ext)} + \sum_{j}\mathbf{f}_{ij} \qquad (6.6)$$

Where $\mathbf{F}_{i(ext)}$ is the resultant external force acting on the ith particle. $\mathbf{f}_{ij}$ is the internal force exerted on the ith particle by the jth particle. Note that it is as-sumed that no force is exerted on the particle by itself, i.e., $\mathbf{f}_{ii} = 0$. Substituting Eq. 6.6 into Eq. 6.5 gives:

$$\mathbf{a}_{cm} = \frac{1}{M}\left(\sum_{i}\mathbf{F}_{i(ext)} + \sum_{i}\sum_{j}\mathbf{f}_{ij}\right) \qquad (6.7)$$

Now, from Newton's third law we have

$$\mathbf{f}_{ij} = -\mathbf{f}_{ji}$$

Therefore, the second term in Eq. 6.7 is equal to zero. Hence the net force acting on the system is due only to external forces. That gives

$$\mathbf{F}_{net} = \sum_{i}\mathbf{F}_{i(ext)} = M\mathbf{a}_{cm}$$

where $\mathbf{F}_{net}$ is the resultant external force on the center of mass, i.e.,

$$\mathbf{F}_{net} = \sum\mathbf{F}_{ext} = M\mathbf{a}_{cm}$$

By differentiating Eq. 6.4 with respect to time we have

$$M\mathbf{a}_{cm} = \frac{d\mathbf{p}_{tot}}{dt}$$

thus

$$\sum\mathbf{F}_{ext} = \frac{d\mathbf{p}_{tot}}{dt}$$

Thus, the net external force acting on a system of particles is equal to the time rate of change of the total linear momentum of the system.

6.3.8   Conservation of Momentum

For an isolated system of particles, we have

$$\sum\mathbf{F}_{ext} = 0$$

Thus

$$\frac{d\mathbf{p}_{tot}}{dt} = 0$$

and

$$\mathbf{p}_{tot} = M\mathbf{v}_{cm} = \text{constant}$$

Which is the law of conservation of linear momentum for a system of particles.

6.3.9   Angular Momentum of a System of Particles

The angular momentum $\mathbf{L}$ of a system of particles about a fixed point is the vector sum of angular momenta of the individual particles:

$$\mathbf{L} = \mathbf{L}_1 + \mathbf{L}_2 + \mathbf{L}_3 + + \mathbf{L}_n = \sum_{i=1}^{n}\mathbf{L}_i = \sum_{i=1}^{n}(\mathbf{r}_i \times \mathbf{p}_i) = \sum_{i=1}^{n}m_i(\mathbf{r}_i \times \mathbf{v}_i)$$

6.3.10   The Total Torque on a System

The total torque acting on a particle in a system is the sum of torques associated with the internal forces and of torques associated with external forces. Using Eq. 6.6 we have

$$\tau_i = \mathbf{r}_i \times \mathbf{F}_i = \mathbf{r}_i \times \left(\mathbf{F}_{iext} + \sum_j \mathbf{f}_{ij} \right) = \mathbf{r}_i \times \mathbf{F}_{iext} + \sum_j \mathbf{r}_i \times \mathbf{f}_{ij}$$

Summing over i we get

$$\sum_i \tau_i = \sum_i \mathbf{r}_i \times \mathbf{F}_i = \sum_i \mathbf{r}_i \times \mathbf{F}_{iext} + \sum_i \sum_j \mathbf{r}_i \times \mathbf{f}_{ij} \quad (6.8)$$

By using Newton's third law of action and reaction, the double sum in Eq. 6.8 has terms of the form

$$\mathbf{r}_i \times \mathbf{f}_{ij} + \mathbf{r}_j \times \mathbf{f}_{ji} = (\mathbf{r}_i - \mathbf{r}_j) \times \mathbf{f}_{ij}$$

Now, suppose that the internal forces between the two particles lie along the line joining the particles (i.e., the vectors $\mathbf{f}_{ij}$ and $(\mathbf{r}_i - \mathbf{r}_j)$ have the same direction). This condition is known as the strong law of action and reaction. It requires the internal forces to be central. If the internal forces are equal and opposite but not central, then they are said to satisfy the weak law of action and reaction. The force of gravity is an example of a force satisfying the strong law of action and reaction. Some forces such as the forces between two moving charges are not central. From this, it follows that the double summation in Eq. 6.8 is equal to zero.

$$\tau_{net} = \sum_i \tau_i = \sum_i \mathbf{r}_i \times \mathbf{F}_i = \sum_i \mathbf{r}_i \times \mathbf{F}_{iext}$$

Therefore, the total torque on the system about the origin is only the torque associated with external forces

$$\tau_{net} = \sum \tau_{ext} = \sum_{i=1}^n \mathbf{r}_i \times \mathbf{F}_{i(ext)} \quad (6.9)$$

6.3.11 The Angular Momentum and the Total External Torque

The angular momentum of the individual particles may change with time. This will change the total angular momentum of the system

$$\frac{d\mathbf{L}}{dt} = \sum_{i=1}^n \frac{d\mathbf{L}_i}{dt}$$

Eq. 6.9 may be written as

$$\tau_{net} = \sum \tau_{ext} = \sum_{i=1}^n \mathbf{r}_i \times \mathbf{F}_{i(ext)} = \frac{d}{dt} \left\{ \sum_{i=1}^n m_i (\mathbf{r}_i \times \mathbf{v}_i) \right\} = \frac{d}{dt} \left\{ \sum_{i=1}^n \mathbf{L}_i \right\} = \frac{d\mathbf{L}}{dt}$$

i.e., the net external torque about some origin exerted on a system of particles is equal to the time rate of change of the total angular momentum of the system.

6.3.12 Conservation of Angular Momentum

If

$$\sum \tau_{ext} = 0$$

$$\mathbf{L} = \sum_{i=1}^n m_i (\mathbf{r}_i \times \mathbf{v}_i) = \text{constant}$$

or

$$\mathbf{L}_i = \mathbf{L}_f$$

Hence, if the resultant external torque acting on a system is zero, the total angular momentum remains constant.

6.3.13 Kinetic Energy of a System of Particles

The total kinetic energy of a system of particles is the sum of the kinetic energies of the individual particles

$$K = \frac{1}{2} \sum_{i=1}^n m_i v_i^2$$

6.3.14 Work

Since the total force acting on the ith particle is given by

$$\mathbf{F}_i = \mathbf{F}_{i(ext)} + \sum_j \mathbf{f}_{ij}$$

then the total work done on such particle is given by

$$W_{12} = \sum_i \int_1^2 \mathbf{F}_i \cdot d\mathbf{s}_i$$

6.3.15 Work–Energy Theorem

The total work done in moving a system from one state to another is

$$W_{12} = \sum_i \int_1^2 \mathbf{F}_i \cdot d\mathbf{s}_i = \sum_i \int_1^2 \mathbf{F}_i \cdot \frac{d\mathbf{s}_i}{dt} dt = \sum_i \int_1^2 \mathbf{F}_i \cdot \mathbf{v}_i dt$$

$$= \sum_i \int_1^2 \mathbf{v}_i \cdot \mathbf{F}_i dt = \sum_i \int_1^2 \mathbf{v}_i \cdot \frac{d}{dt} (m_i \mathbf{v}_i) dt$$

Since

$$\mathbf{v}_i \frac{d}{dt} (m_i \mathbf{v}_i) = \frac{1}{2} \frac{d}{dt} (m_i (\mathbf{v}_i \cdot \mathbf{v}_i)) = \frac{1}{2} \frac{d}{dt} (m_i v_i^2)$$

it follows that

$$W_{12} = \frac{1}{2} \sum_i \int_1^2 \frac{d}{dt}(m_i v_i^2) dt = \frac{1}{2} \sum_i (m_i v_i^2)|_1^2 = K_2 - K_1$$

where $\frac{1}{2} \sum_i m_i v_i^2$ is the total kinetic energy of the system.

6.3.16 Potential Energy and Conservation of Energy of a System of Particles

Consider a system of particles in which the external and internal forces acting on the system are conservative. First, let us calculate the work done by the internal conservative forces. Suppose that $\mathbf{f}_{ij}$ is the conservative force acting on the ith particle due to the jth particle and $\mathbf{f}_{ji}$ is the force acting on the jth particle due to the ith particle. Note that $\mathbf{f}_{ij}$ and $\mathbf{f}_{ji}$ form an action and reaction pair, i.e., $\mathbf{f}_{ij} = -\mathbf{f}_{ji}$. Because these forces are conservative there is a potential energy associated with each force. That is,

$$\mathbf{f}_{ij} = -\nabla_i U_{ij}$$

and

$$\mathbf{f}_{ji} = -\nabla_j U_{ij}$$

From the law of action and reaction, U_{ij} is a function only of the distance between the particles. That is

$$U_{ij} = U_{ij}(|\mathbf{r}_i - \mathbf{r}_j|) = U_{ji}(|\mathbf{r}_i - \mathbf{r}_j|)$$

or

$$U_{ij}(r_{ij}) = U_{ji}(r_{ji})$$

where $|\mathbf{r}_i - \mathbf{r}_j| = r_{ij} = r_{ji}$ is the distance between the ith and jth particles. The work done by each pair of forces in displacing the ith and jth particles through $d\mathbf{r}_i$ and $d\mathbf{r}_j$, respectively, is

$$\mathbf{f}_{ij} \cdot d\mathbf{r}_i + \mathbf{f}_{ji} \cdot d\mathbf{r}_j = -\nabla_i U_{ij} \cdot d\mathbf{r}_i - \nabla_j U_{ij} \cdot d\mathbf{r}_j$$

$$= -\left[\frac{\partial U_{ij}}{\partial x_i} dx_i + \frac{\partial U_{ij}}{\partial y_i} dy_i + \frac{\partial U_{ij}}{\partial z_i} dz_i + \frac{\partial U_{ij}}{\partial x_j} dx_j + \cdots \cdots \right] = -dU_{ij}$$

Hence, the total work done by the internal conservative forces in moving the system from stage 1 to stage 2 is

$$W_{12(in,c)} = \sum_i \sum_j \int_1^2 \mathbf{f}_{ij} \cdot d\mathbf{r}_i = -\frac{1}{2} \sum_i \sum_j \int_1^2 dU_{ij}$$

$$= -\frac{1}{2} \sum_i \sum_j U_{ij}|_1^2 = U_{1(int)} - U_{2(int)} = -\Delta U_{(int)}$$

The factor 1/2 occurs since each term in the summation appears twice. Now, consider the total work done by the external conservative forces

$$W_{12(ext,c)} = \sum_i \int_1^2 \mathbf{F}_{i(ext)} . d\mathbf{s}_i = -\sum_i \int_1^2 \nabla_i U_i \cdot d\mathbf{s}_i = -\sum_i U_i|_1^2 = U_{1(ext)} - U_{2(ext)}$$

To show that energy is conserved when both the external and internal forces are conservative, we may define a total potential of the system as

$$U = \sum_i U_i + \frac{1}{2} \sum_i \sum_j U_{ij}$$

From the work–energy theorem, the work done by the total force F_i acting on the ith particle is equal to the change in the kinetic energy of that particle

$$W_{12} = \sum_i \int_1^2 \mathbf{F}_i \cdot d\mathbf{r}_i = K_2 - K_1$$

and since

$$W_{12} = W_{12(in,c)} + W_{12(ext,c)}$$

From this, we conclude that for a system of particles in which the internal and external forces are conservative, the total mechanical energy of the system is conserved

$$U_{1(int)} - U_{2(int)} + U_{1(ext)} - U_{2(ext)} = K_2 - K_1$$

or

$$U_1 - U_2 = K_2 - K_1$$

or

$$\Delta K = -\Delta U$$

Thus

$$\Delta K + \Delta U = 0$$

$$\Delta E = 0$$

6.3.17 Impulse

In Sect. 6.3.7, we have seen that the net external force on a system of particles is equal to the rate of change of the total linear momentum of the system

$$\mathbf{F}_{net} = \frac{d\mathbf{p}_{tot}}{dt}$$

The total linear impulse on the system as the system goes from one state to another is defined as

$$\mathbf{I} = \int_{t_1}^{t_2} \mathbf{F}_{net}dt = \int_{t_1}^{t_2} \frac{d\mathbf{p}_{tot}}{dt}dt = \mathbf{p}_{tot2} - \mathbf{p}_{tot1}$$

That is, the total linear impulse on the system is equal to the change in the total momentum of the system.

6.4 Motion Relative to the Center of Mass

The motion of a system of particles is sometimes described relative to the center of mass of the system. This method is used in some problems to simplify the analysis and add a particular symmetry to it.

6.4.1 The Total Linear Momentum of a System of Particles Relative to the Center of Mass

The position vector of the center of mass of the system with respect to an origin in an inertial frame of reference (for example, the lab frame) is given by

$$\mathbf{r}_{cm} = \frac{\Sigma_i^n m_i \mathbf{r}_i}{M} \tag{6.10}$$

From Fig. 6.12, the position vector ($\mathbf{r}_i'$) of the ith particle relative to the center of mass is

$$\mathbf{r}_i' = \mathbf{r}_i - \mathbf{r}_{cm}$$

or

$$\mathbf{r}_i = \mathbf{r}_i' + \mathbf{r}_{cm} \tag{6.11}$$

Where $\mathbf{r}_i$ is the position vector of the ith particle relative to the origin O. Substituting Eq. 6.11 into Eq. 6.10 gives

$$\mathbf{r}_{cm} = \frac{1}{M}\sum_{i=1}^n m_i(\mathbf{r}_i' + \mathbf{r}_{cm}) = \frac{1}{M}\sum_{i=1}^n m_i\mathbf{r}_i' + \frac{\sum_{i=1}^n m_i}{M}\mathbf{r}_{cm}$$

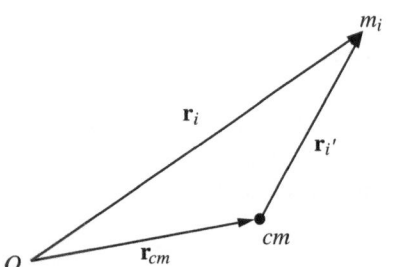

Fig. 6.12 The position vector ($\mathbf{r}_i'$) of the ith particle relative to the center of mass

$$= \frac{1}{M}\sum_{i=1}^n m_i\mathbf{r}_i' + \mathbf{r}_{cm}$$

therefore

$$\frac{1}{M}\sum_{i=1}^n m_i\mathbf{r}_i' = \mathbf{r}_{cm} - \mathbf{r}_{cm} = 0$$

That gives

$$\sum_{i=1}^n m_i\mathbf{r}_i' = \mathbf{0} \tag{6.12}$$

Differentiating Eq. 6.12 with respect to t gives

$$\sum_{i=1}^n m\mathbf{v}_i' = \mathbf{0} \tag{6.13}$$

or

$$\sum_{i=1}^n \mathbf{p}_i' = \mathbf{0}$$

or

$$\mathbf{p}' = \mathbf{0}$$

That is, the total linear momentum of the system is zero when observed from the center of mass frame.

6.4.2 The Total Angular Momentum About the Center of Mass

By differentiating Eq. 6.11 with respect to time gives

$$\mathbf{v}_i = \mathbf{v}_i' + \mathbf{v}_{cm} \tag{6.14}$$

where $\mathbf{v}_i$ and $\mathbf{v}_i'$ are the velocities of the particle relative to the origin O and the center of mass respectively $\mathbf{v}_{cm}$ is the velocity of the center of mass relative to O. The angular momentum of the system about the origin O is

$$\mathbf{L} = \sum_i m_i(\mathbf{r}_i \times \mathbf{v}_i) = \sum_i m_i\{(\mathbf{r}_i' + \mathbf{r}_{cm}) \times (\mathbf{v}_i' + \mathbf{v}_{cm})\}$$

$$= \sum_i m_i(\mathbf{r}_i' \times \mathbf{v}_i') + \sum_i m_i(\mathbf{r}_i' \times \mathbf{v}_{cm}) + \sum_i m_i(\mathbf{r}_{cm} \times \mathbf{v}_i') + \sum_i m_i(\mathbf{r}_{cm} \times \mathbf{v}_{cm})$$

The second and third terms are zero followed from Eqs. 6.12 and 6.13 where $\left(\sum_i m_i\mathbf{r}_i'\right) \times \mathbf{v}_{cm} = \mathbf{0}$ and $\mathbf{r}_{cm} \times \left(\sum_i m_i\mathbf{v}_i'\right) = \mathbf{0}$, hence

$$L = \sum_i m_i (\mathbf{r}'_i \times \mathbf{v}'_i) + \sum_i m_i (\mathbf{r}_{cm} \times \mathbf{v}_{cm})$$

Thus, the total angular momentum of the system of particles about an origin O equals the angular momentum of the system about the center of mass plus the angular momentum of the center of mass about O. Therefore, the total angular momentum $\mathbf{L}'$ about the center of mass is

$$\mathbf{L}' = \sum_i m_i (\mathbf{r}'_i \times \mathbf{v}'_i) = \sum_i m_i (\mathbf{r}_i \times \mathbf{v}_i) - M(\mathbf{r}_{cm} \times \mathbf{v}_{cm})$$

$$(6.15)$$

6.4.3 The Total Kinetic Energy of a System of Particles About the Center of Mass

The total kinetic energy of a system of particles relative to an origin in an inertial frame of reference is given by

$$K = \frac{1}{2} \sum_i m_i v_i^2 = \frac{1}{2} \sum_i m_i (\mathbf{v}_i \cdot \mathbf{v}_i)$$

From Eq. 6.14 we have

$$K = \frac{1}{2} \sum_i m_i ((\mathbf{v}'_i + \mathbf{v}_{cm}) \cdot (\mathbf{v}'_i + \mathbf{v}_{cm}))$$

$$= \frac{1}{2} \sum_i m_i (\mathbf{v}'_i \cdot \mathbf{v}'_i) + \sum_i m_i (\mathbf{v}'_i \cdot \mathbf{v}_{cm}) + \frac{1}{2} \sum_i m_i (\mathbf{v}_{cm} \cdot \mathbf{v}_{cm})$$

$$= \frac{1}{2} \sum_i m_i v_i'^2 + \mathbf{v}_{cm} \cdot \left(\sum_i m_i \mathbf{v}'_i \right) + \frac{1}{2} \left(\sum_i m_i \right) v_{cm}^2$$

From Eq. 6.13, the term in brackets in the second term is equal to zero. Hence

$$K = \frac{1}{2} \sum_i m_i v_i'^2 + \frac{1}{2} M v_{cm}^2$$

That is the total kinetic energy of a system of particles about an origin is equal to the kinetic energy of the system with respect to the center of mass plus the kinetic energy of the center of mass relative to the origin O. Therefore, the total kinetic energy of the system with respect to the center of mass is

$$K' = \frac{1}{2} \sum_i m_i v_i'^2 = \frac{1}{2} \sum_i m_i v_i^2 - \frac{1}{2} M v_{cm}^2$$

6.4.4 Total Torque on a System of Particles About the Center of Mass of the System

The total torque acting on a system of particles about the center of mass is (from theorem (5.6.1)) equal to the time rate of change of the angular momentum of the system about the center of mass. That is,

$$\boldsymbol{\tau}' = \frac{d\mathbf{L}'}{dt}$$

Example 6.11 Two particles of masses $m_1 = 1\,\text{kg}$ and $m_2 = 2\,\text{kg}$ are moving in the x-y plane. Their position vectors relative to the origin are $\mathbf{r}_1 = (t^2\mathbf{i} - 2t\mathbf{j})$ m and $\mathbf{r}_2 = (3t\mathbf{i} + \mathbf{j})$ m where t is time. Find: (a) the total angular momentum of the system; the total external torque acting on the system; and the total kinetic energy of the system all relative to the origin at any time; (b) repeat (a) relative to the center of mass.

Solution 6.11 (a)

$$\mathbf{v}_1 = \frac{d\mathbf{r}_1}{dt} = (2t\mathbf{i} - 2\mathbf{j})\,\text{m/s}$$

$$\mathbf{v}_2 = \frac{d\mathbf{r}_2}{dt} = (3\mathbf{i})\,\text{m/s}$$

The total angular momentum of the system relative to the origin is

$$\mathbf{L} = \sum_i m_i (\mathbf{r}_i \times \mathbf{v}_i) = (1)[(t^2\mathbf{i} - 2t\mathbf{j}) \times (2t\mathbf{i} - 2\mathbf{j})] + (2)[(3t\mathbf{i} + \mathbf{j}) \times (3)\mathbf{i}]$$

that gives

$$\mathbf{L} = ((2t^2 - 6)\mathbf{k})\,\text{kg.m}^2/\text{s}$$

The total kinetic energy of the system relative to O is

$$K = \frac{1}{2} \sum_{i=1}^n m_i v_i^2 = \frac{1}{2}(m_1 v_1^2 + m_2 v_2^2) = \frac{1}{2}[(1)(4t^2 + 4) + (2)(9)] = (2t^2 + 11)\,\text{J}$$

The net external torque about the origin is

$$\sum \boldsymbol{\tau}_{ext} = \frac{d\mathbf{L}}{dt} = ((4t)\mathbf{k})\,\text{N.m}$$

(b) To find the total angular momentum relative to the center of mass let's find first the total angular momentum of the center of mass relative to the origin

$$\mathbf{r}_{cm} = \frac{\sum_i m_i \mathbf{r}_i}{\sum_i m_i} = \frac{(1)(t^2\mathbf{i} - 2t\mathbf{j}) + (2)(3t\mathbf{i} + \mathbf{j})}{(3)}$$

$$= \left(\left(\frac{t^2}{3} + 2t\right)\mathbf{i} + \left(\frac{2}{3} - \frac{2}{3}t\right)\mathbf{j}\right) \text{m}$$

The velocity of the center of mass is

$$\mathbf{v}_{cm} = \left(\left(\frac{2}{3}t + 2\right)\mathbf{i} - \left(\frac{2}{3}\right)\mathbf{j}\right) \text{m/s}$$

and the total angular momentum of the center of mass relative to O is

$$\mathbf{L}_{cm} = M(\mathbf{r}_{cm} \times \mathbf{v}_{cm}) = (3)\left[\left(\left(\frac{t^2}{3} + 2t\right)\mathbf{i} + \left(\frac{2}{3} - \frac{2}{3}t\right)\mathbf{j}\right) \times \left(\left(\frac{2}{3}t + 2\right)\mathbf{i} - \left(\frac{2}{3}\right)\mathbf{j}\right)\right]$$

$$= \left(-\left(\frac{2}{3}t^2 + \frac{4}{3}t + 4\right)\mathbf{k}\right) \text{kg.m}^2/\text{s}$$

From Eq. 6.15, the total angular momentum relative to the center of mass is

$$\mathbf{L}' = \sum_i m_i(\mathbf{r}_i' \times \mathbf{v}_i') = \sum_i m_i(\mathbf{r}_i \times \mathbf{v}_i) - M(\mathbf{r}_{cm} \times \mathbf{v}_{cm})$$

$$= (2t^2 - 6)\mathbf{k} + \left(\frac{2t^2}{3} + \frac{4}{3}t + 4\right)\mathbf{k} = \left(\left(\frac{8}{3}t^2 + \frac{4}{3}t - 2\right)\mathbf{k}\right) \text{kg.m}^2/\text{s}$$

The net external torque about the center of mass is

$$\boldsymbol{\tau}' = \frac{d\mathbf{L}'}{dt} = \left(\left(\frac{16}{3}t + \frac{4}{3}\right)\mathbf{k}\right) \text{N.m}$$

The total kinetic energy of the system relative to the center of mass is

$$K' = \frac{1}{2}\sum_i m_i v_i'^2 = \sum_i m_i v_i^2 - \frac{1}{2}M v_{cm}^2$$

$$= (2t^2 + 11) - \frac{1}{2}(3)\left[\left(\frac{2}{3}t + 2\right)^2 + \frac{4}{9}\right] = \left(\frac{4t^2}{3} - 2t - \frac{13}{3}\right) \text{J}$$

Example 6.12 Two particles of equal mass m are rotating about their center of mass with a constant speed v as in Fig. 6.13. If they are separated by a distance $2d$, find the total angular momentum of the system.

Solution 6.12

$$L = mvd + mvd = 2mvd$$

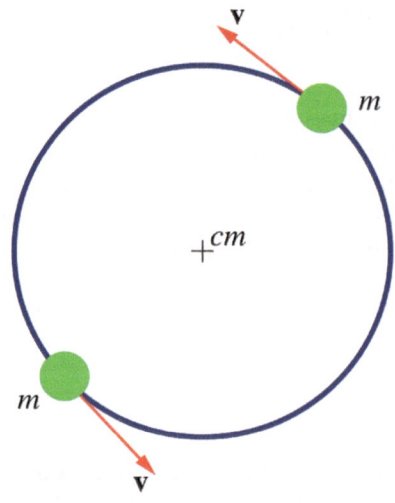

Fig. 6.13 Two particles rotating about their center of mass

6.4.5   Collisions and the Center of Mass Frame of Reference

In problems involving collisions, it is useful to use an inertial frame of reference that is attached to the center of mass to analyze the collision. This method is most commonly used in analyzing collisions between subatomic particles or atoms. In section (6.4.1), we proved that the total linear momentum of a system when observed from the center of mass frame is equal to zero.

$$\mathbf{p}_i' = \mathbf{p}_f' = \mathbf{0} \qquad (6.16)$$

Now consider a system consisting of two bodies undergoing a one-dimensional collision (see Fig. 6.14). Then from Eq. 6.16 we have

$$p_{1i}' = -p_{2i}'$$

and

$$p_{1f}' = -p_{2f}'$$

That is, when viewed from the center of mass frame the two objects approach each other with equal and opposite momenta and move away from each other with an equal and opposite momenta. Therefore, the center of mass frame simplifies the analysis since it exhibits a particular symmetry to the problem (see Fig. 6.15).

Example 6.13 A rocket is projected vertically upward and explodes into three fragments of equal mass when it reaches the top of its flight at an altitude of 40 m (see Fig. 6.16). If the two fragments land to the ground after 3 s from the explosion, find the time it takes the third fragment to hit the ground.

Solution 6.13 When the rocket reaches the top its velocity immediately before explosion is zero. Since $\mathbf{v}_1$, $\mathbf{v}_2$ and $\mathbf{v}_3$ are

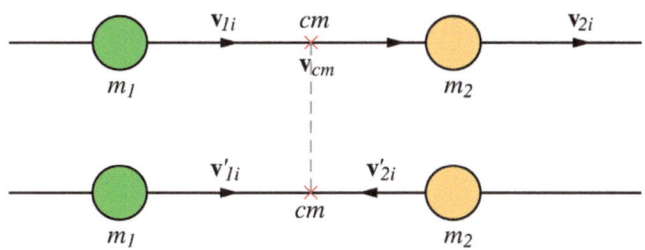

Fig. 6.14 Consider a system consisting of two bodies undergoing a one-dimensional collision

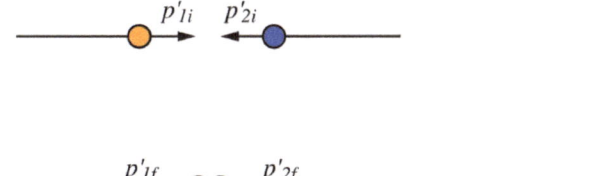

Fig. 6.15 The center of mass frame analysis of a collision

Fig. 6.16 A rocket is projected vertically upward and explodes into three fragments of equal mass when it reaches the top of its flight at an altitude of 40 m

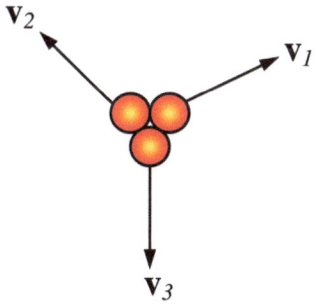

the velocities of the fragments immediately after explosion, we have from the conservation of momentum

$$m_1\mathbf{v}_1 + m_2\mathbf{v}_2 + m_3\mathbf{v}_3 = \mathbf{0}$$

Since $m_1 = m_2 = m_3$, then $v_1 + v_2 + v_3 = 0$. The first and second fragments land at the same time t' and hence they have the same vertical velocity initially which is equal to $-v_3/2$. Therefore

$$h = v_3 t + \frac{gt^2}{2}$$

and

$$h = \frac{-v_3 t'}{2} + \frac{gt'^2}{2}$$

That gives

$$v_3 = \frac{g(t'^2 - t^2)}{2t + t'}$$

and

$$h = \frac{gtt'(t + 2t')}{2(2t + t')}$$

Substituting the values of h and t' gives

$$29.4t^2 + 160t + 63.6 = 0$$

Thus, $t = 2.3\,\text{s}$.

Example 6.14 Find the center of mass of the Earth–Moon System and describe its motion around the sun.

Solution 6.14 As we shall see in Chap. 9, the center of mass of two bodies with different masses moving under gravity will trace an ellipse. Since the external forces on the sun can be neglected, we may consider it to be at rest in an inertial frame of reference and at the origin of a coordinate system (see Fig. 6.17). The center of mass of the Earth–Moon system is

$$\mathbf{r}_{cm} = \frac{M_E \mathbf{r}_E + M_M \mathbf{r}_M}{M_E + M_M}$$

where $\hat{\mathbf{r}}_E$ and $\hat{\mathbf{r}}_M$ are unit vectors in the direction of $\mathbf{r}_E$ and $\mathbf{r}_M$ respectively. The equation of motion of the center of mass is

$$\mathbf{F} = (M_E + M_M)\ddot{\mathbf{r}}_{cm}$$

The gravitational force on the Earth–Moon system exerted by the sun is

$$\mathbf{F} = -GM_S\left(\frac{M_E}{r_E^2}\hat{\mathbf{r}}_E + \frac{M_M}{r_M^2}\hat{\mathbf{r}}_M\right)$$

Since the distance between the earth and the moon is so small compared to their distance from the sun we may write $r_E \approx r_M \approx r_{cm}$

$$\mathbf{F} = -\frac{GM_S}{r_{cm}^2}(M_E + M_M)\hat{\mathbf{r}}_{cm} = (M_E + M_M)\ddot{\mathbf{r}}_{cm}$$

Hence, the center of mass of the Earth–Moon system moves as a single planet of mass $(M_E + M_M)$ about the sun as shown in Fig. 6.18.

Example 6.15 Describe the motion of a rocket in space using the law of conservation of momentum.

Fig. 6.17 The center of mass of the Earth-Moon system

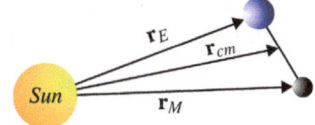

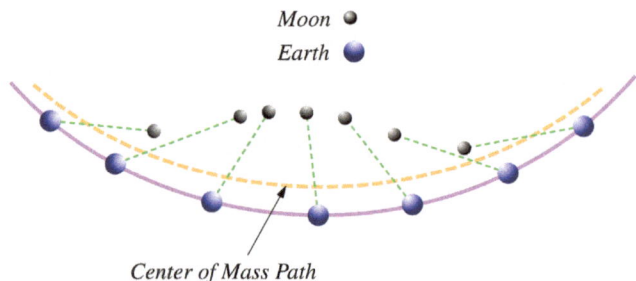

Fig. 6.18 The center of mass of the Earth-Moon system moves as a single planet of mass $(M_E + M_M)$ about the sun

Fig. 6.19 A rocket moving in space is a system with varying mass. Its motion is analyzed using the law of conservation of momentum

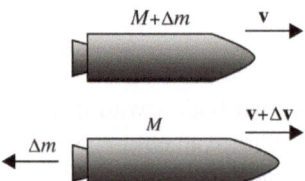

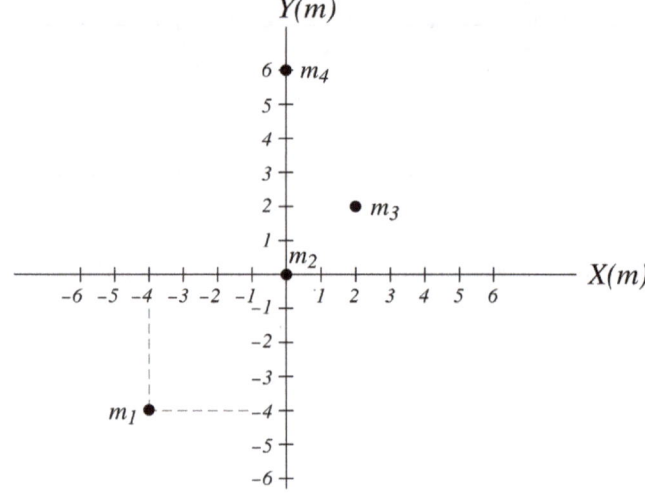

Fig. 6.20 A system of particles in x-y plane

Solution 6.15 A rocket moving in space is a system with varying mass. Its motion is analyzed using the law of conservation of momentum. In order for a rocket to move in space, its fuel is burned and gases are produced and ejected from its rear. This will cause the mass of the rocket to decrease continuously The ejected gases produce momentum in the backward direction and as a result the rocket receives a forward momentum and its velocity increases (see Fig. 6.19). Suppose at an instant t, the rocket has a mass M and velocity v relative to a stationary frame of reference. During a time interval t, a mass Δm of the fuel is expelled as gas with a velocity u relative to the rocket. The speed of the rocket increases to $v + \Delta v$ and the speed of the fuel relative to the stationary frame of reference is $v - u$. The initial momentum of the rocket is

$$\mathbf{p}(t) = (M + \Delta m)\mathbf{v}$$

and the final momentum is

$$\mathbf{p}(t + \Delta t) = M(\mathbf{v} + \Delta \mathbf{v}) + \Delta m \, (\mathbf{v} - \mathbf{u})$$

The change in the momentum is

$$\Delta \mathbf{p}(t + \Delta t) = \mathbf{p}(t + \Delta t) - \mathbf{p}(t) = M \Delta \mathbf{v} - (\Delta m)\mathbf{u}$$

Therefore, the force acting on the rocket is

$$\mathbf{F} = \frac{d\mathbf{p}}{dt} = \lim_{\Delta t \to 0} \frac{\Delta \mathbf{p}}{\Delta t} = M \frac{d\mathbf{v}}{dt} - \mathbf{u} \frac{dm}{dt}$$

Since the increase in the exhaust mass produce an equal decrease in the rocket mass, we have

$$dm = -dM$$

Thus

$$\mathbf{F} = M \frac{d\mathbf{v}}{dt} + \mathbf{u} \frac{dM}{dt}$$

If no external forces act on the rocket we have $\mathbf{F} = 0$ and

$$M \frac{d\mathbf{v}}{dt} = -\mathbf{u} \frac{dM}{dt}$$

hence

$$\int_{t_0}^{t} \frac{d\mathbf{v}}{dt} dt = -\mathbf{u} \int_{M_0}^{M} \frac{1}{M} \frac{dM}{dt} dt = -\mathbf{u} \int_{M_0}^{M} \frac{dM}{M}$$

That gives

$$\mathbf{v} - \mathbf{v}_0 = \mathbf{u} \ln \left(\frac{M_0}{M} \right)$$

Therefore, the final speed of the rocket depends on the exhaust speed and on the ratio of the initial and final masses.

Problems

1. Find the coordinate of the center of mass of the system shown in Fig. 6.20.
2. Find the center of mass of a uniform plate bounded by $y = -0.24x^2 + 6$ and the x-axis from $x = -5$ to $x = 5$ m.
3. Find the center of mass of the homogeneous sheet shown in Fig. 6.21.
4. Find the center of mass of the homogeneous sheet shown in Fig. 6.22.
5. Find the center of mass of a uniform solid circular cone of radius a and height h.
6. Find the center of mass of a uniform solid hemisphere of radius R.

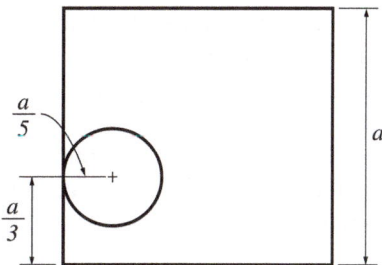

Fig. 6.21 A homogenous sheet with a hole

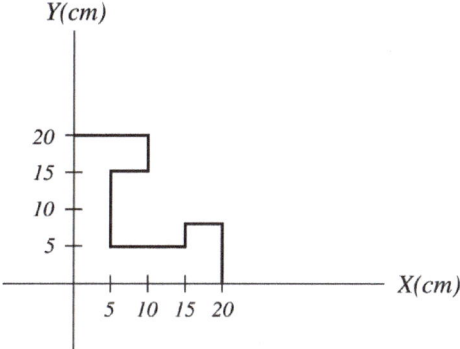

Fig. 6.22 A homogenous sheet in the x-y plane

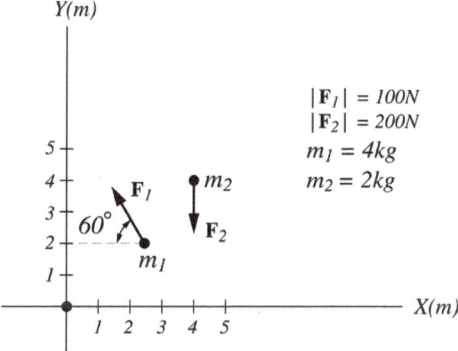

Fig. 6.23 The acceleration of the center of mass of two masses acted upon by different forces

7. Two masses initially at rest are located at the points shown in Fig. 6.23. If external forces act on the particles as in Fig. 6.23, find the acceleration of the center of mass.

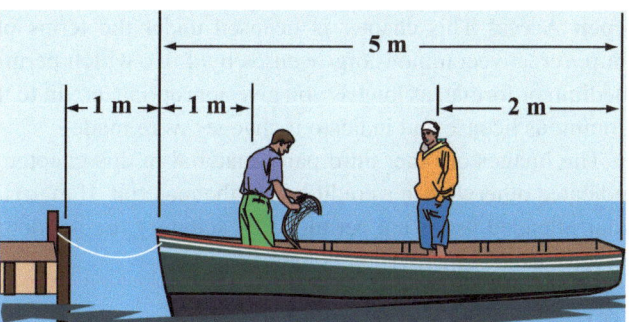

Fig. 6.24 By neglecting friction between the boat and water, the center of mass can be used to find the distance moved by the boat

8. A projectile of mass 15 kg is fired from the ground with an initial velocity of 12 m/s at an angle of 45° to the horizontal. 1 second later, the projectile explodes into two fragments A and B. If immediately after explosion, fragment A has a mass of 5 kg and a speed of 5 m/s at an angle of 30° to the horizontal, find the velocity of fragment B (assuming air resistance is neglected).

9. Two boys of masses 45 and 40 kg are standing on a boat of mass 150 kg and length 5 m as in Fig. 6.24. The boat is initially 1m from the pier. Assuming that there is no friction between the boat and the water, find the distance moved by the boat when the two meet at the middle of the boat.

10. Two particles of masses $m_1 = 3$ kg and $m_2 = 5$ kg are moving relative to the lab frame with velocities of 10 m/s along the y-axis and 15 m/s at an angle of 30° to the x-axis. Find (a) the velocity of their center of mass (b) the momentum of each particle in the center of mass frame (c) the total kinetic energy of the particles relative to the lab frame and relative to the center of mass frame.

11. Two particles of masses $m_1 = 1$ kg and $m_2 = 2$ kg are moving relative to the lab frame with velocities of $\mathbf{v}_1 = 2\mathbf{i} - 3\mathbf{j} + \mathbf{k}$ and $\mathbf{v}_2 = 7\mathbf{i} + \mathbf{j} - 2\mathbf{k}$. If at a certain instant they are located at $(-1, 1, 2)$ and $(3, 0, 1)$, find the angular momentum of the system relative to the origin and relative to the center of mass.

Open Access This chapter is licensed under the terms of the Creative Commons Attribution 4.0 International License (http://creativecommons.org/licenses/by/4.0/), which permits use, sharing, adaptation, distribution and reproduction in any medium or format, as long as you give appropriate credit to the original author(s) and the source, provide a link to the Creative Commons license and indicate if changes were made.

The images or other third party material in this chapter are included in the chapter's Creative Commons license, unless indicated otherwise in a credit line to the material. If material is not included in the chapter's Creative Commons license and your intended use is not permitted by statutory regulation or exceeds the permitted use, you will need to obtain permission directly from the copyright holder.

7.1 Rotational Motion

Rotational motion exists everywhere in the universe. The motion of electrons about an atom and the motion of the moon about the earth are examples of rotational motion. Objects cannot be treated as particles when exhibiting rotational motion since different parts of the object move with different velocities and accelerations. Therefore, it is necessary to treat the object as a system of particles.

7.2 The Plane Motion of a Rigid Body

When all parts of a rigid body move parallel to a fixed plane, then the motion of the object is referred to as plane motion. There are two types of plane motion, which are given as follows:

1. The pure rotational motion: The rigid body in such a motion rotates about a fixed axis that is perpendicular to a fixed plane. In other words, the axis is fixed and does not move or change its direction relative to an inertial frame of reference.
2. The general plane motion: The motion here can be considered as a combination of pure translational motion parallel to a fixed plane in addition to a pure rotational motion about an axis that is perpendicular to that plane. This chapter discusses the kinematics and dynamics of pure rotational motion.

7.2.1 The Rotational Variables

Suppose a rigid body of an arbitrary shape is in pure rotational motion about the z-axis (see Fig. 7.1). Let us analyze the motion of a particle that lies in a slice of the body in the x-y plane as in Fig. 7.2. This particle (at point P) will rotate in a circle of fixed radius r which represents the perpendicular distance from P to the axis of rotation. If you look at any other

particle in the object you will see that every particle will rotate in its own circle that has the axis of rotation at its center. In other words, different particles move in different circles but the center of all of these circles lies on the rotational axis. Suppose the particle moves through an arc length s starting at the positive x-axis. Its angular position is then given by

$$\theta = \frac{s}{r}$$

r and θ are the polar coordinates of a point in a plane (which was mentioned in Sect. 2.6) where θ is always measured from the positive x-axis. Because θ is the ratio of the arc length to the radius, it is a pure (dimensionless) number. The unit usually used to measure θ is the radians (rad). One radian is defined as the angle subtended by an arc of length that is equal to the radius of the circle. Since one rotation (360°) corresponds to $\theta = 2\pi r/r = 2\pi$ rad, it follows that:

$$1 \text{ rev} = 360° = 2\pi \text{ rad}$$

$$1 \text{ rad} = 57.3° = 0.159 \text{ rev}$$

Note that if the particle completes one revolution, θ will not become zero again, it is then equal to 2π rad. Thus for example for three revolutions the angular position is given by

$$\theta = (2\pi + 2\pi + 2\pi) \text{ rad} = 6\pi \text{ rad}$$

Suppose that the particle in Fig. 7.2 is at point P_1 at t_1 and at point P_2 at t_2 where it changes its angular position from θ_1 to θ_2 (see Fig. 7.3). Its angular displacement is then given by

$$\Delta\theta = \theta_2 - \theta_1$$

$\Delta\theta$ is positive for counterclockwise rotations (increasing θ) and negative for clockwise rotations (decreasing θ). If the particle undergoes this angular displacement during a time interval Δt, the average angular velocity $\overline{\omega}$ is then defined as

© The Author(s) 2019
S. Alrasheed, *Principles of Mechanics*, Advances in Science,
Technology & Innovation, https://doi.org/10.1007/978-3-030-15195-9_7

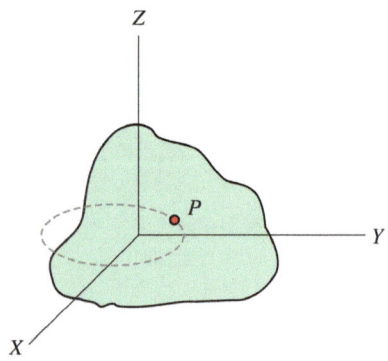

Fig. 7.1 A rigid body of an arbitrary shape is in pure rotational motion about the z-axis

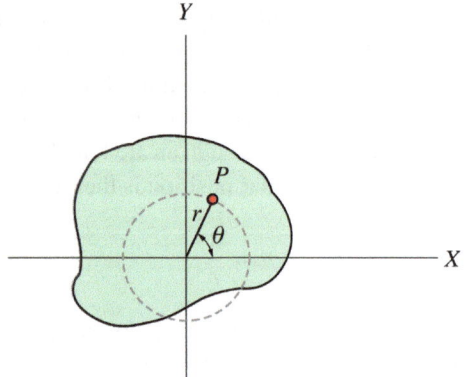

Fig. 7.2 The motion of a particle that lies in a slice of the body in the x-y plane

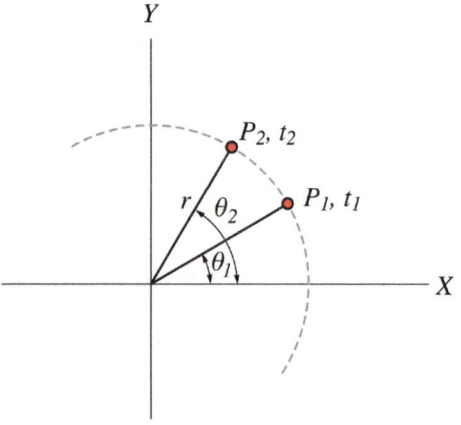

Fig. 7.3 The particle is at point P_1 at t_1 and at P_2 at t_2, where it changes its angular position from θ_1 to θ_2

$$\overline{\omega} = \frac{\theta_2 - \theta_1}{t_2 - t_1} = \frac{\Delta\theta}{\Delta t}$$

The instantaneous angular velocity is

$$\omega = \lim_{\Delta t \to 0} \frac{\Delta\theta}{\Delta t} = \frac{d\theta}{dt}$$

ω has units of rad/s or s^{-1}. The average angular acceleration is defined as

$$\overline{\alpha} = \frac{\omega_2 - \omega_1}{t_2 - t_1} = \frac{\Delta\omega}{\Delta t}$$

The instantaneous angular acceleration is

$$\alpha = \lim_{\Delta t \to 0} \frac{\Delta\omega}{\Delta t} = \frac{d\omega}{dt}$$

where α is in rad/s^2 or s^{-2}. Note that ω is positive for increasing θ and negative for decreasing θ, while α is positive for increasing ω and negative for decreasing ω. When a rigid body is in pure rotational motion, all particles in the body rotate through the same angle during the same time interval. Thus, all particles have the same angular velocity and the same angular acceleration. Therefore, ω and α describes the motion of the whole body In the case of pure rotational motion, the direction of ω is along the axis of rotation (also see Sect. 7.4), it can be determined by the right-hand rule or of advance of a right-handed screw as in Fig. 7.4. The direction of α is in the same direction of ω if ω is increasing or in the opposite direction if ω is decreasing.

The quantities θ, ω and α in pure rotational motion are the rotational analog of x, v and a in translational one-dimensional motion. The vectors ω and α are not used in the case of pure rotational motion, they are used in the general rotational motion when the axis of rotation changes its direction with time. Note that only the infinitesimal angular displacement $d\theta$ can be represented by a vector but not the finite angular displacement $\Delta\theta$. This is because the finite angular displacement $\Delta\theta$ does not obey the commutative law of vector addition (see Fig. 7.5) and therefore cannot be represented by a vector. Hence, the instantaneous angular velocity and acceleration (ω and α) can be represented by vectors but not their average values ($\overline{\omega}$ and $\overline{\alpha}$).

Example 7.1 Convert each of the following into the other angular units: 15°, 0.25 rev/s^2, 3 rad/s.

Solution 7.1

$$15° = (15 \text{ deg})\left(\frac{1 \text{ rev}}{360 \text{ deg}}\right) = 0.042 \text{ rev}$$

$$15° = (15 \text{ deg})\left(\frac{2\pi \text{ rad}}{360 \text{ deg}}\right) = 0.26 \text{ rad}$$

$$0.25 \text{ rev/s}^2 = \left(0.25 \frac{\text{rev}}{\text{s}^2}\right)\left(\frac{2\pi \text{ rad}}{1 \text{rev}}\right) = 1.57 \text{ rad/s}^2$$

$$0.25 \text{ rev/s}^2 = \left(0.25 \frac{\text{rev}}{\text{s}^2}\right)\left(\frac{360 \text{ deg}}{1 \text{ rev}}\right) = 90 \text{ deg/s}^2$$

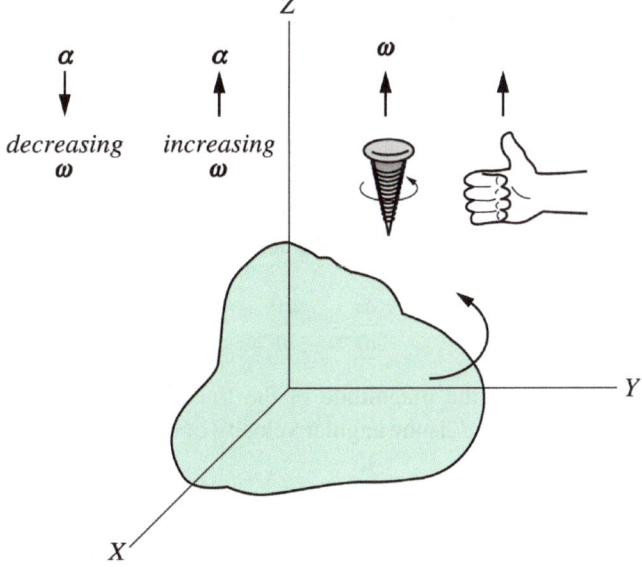

Fig. 7.4 The direction of ω is along the axis of rotation and can be determined by the right-hand rule or of advance of a right-handed screw

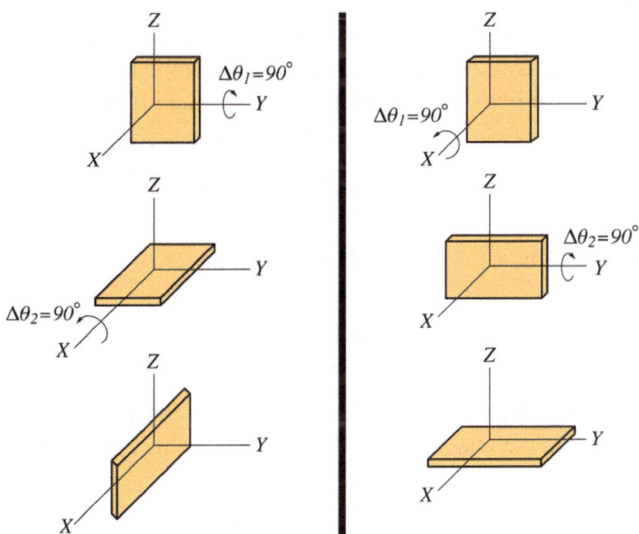

Fig. 7.5 Changing the order of addition will change the final result

$$3 \text{ rad/s} = \left(3 \frac{\text{rad}}{\text{s}}\right)\left(\frac{1 \text{ rev}}{2\pi \text{ rad}}\right) = 0.48 \text{ rev/s}$$

$$3 \text{ rad/s} = \left(3 \frac{\text{rad}}{\text{s}}\right)\left(\frac{360° \text{ deg}}{2\pi \text{ rad}}\right) = 172 \text{ deg/s}$$

Example 7.2 A rotating rigid object has an angular position given by $\theta(t) = ((0.3)t^2 + (0.4)t^3)$ rad. Determine: (a) the angular displacement of the object and the average angular velocity during the time interval from $t_1 = 1$s to $t_2 = 2$ s. (b) the instantaneous angular velocity and the instantaneous angular acceleration at $t = 5$ s.

Solution 7.2 (a)

$$\Delta\theta = \theta_2 - \theta_1$$

$$\theta_1 = ((0.3)(1 \text{ s})^2 + (0.4)(1 \text{ s})^3) = 0.7 \text{ rad}$$

and

$$\theta_2 = ((0.3)(2 \text{ s})^2 + (0.4)(2 \text{ s})^3) = 4.4 \text{ rad}$$

$$\Delta\theta = (4.4 \text{ rad}) - (0.7 \text{ rad}) = 3.7 \text{ rad}$$

$$\overline{\omega} = \frac{\Delta\theta}{\Delta t} = \frac{(3.7 \text{ rad})}{(1 \text{ s})} = 3.7 \text{ rad/s}$$

(b)

$$\omega = \frac{d\theta}{dt} = ((0.6)t + (1.2)t^2) \text{ rad/s}$$

at $t = 5$ s

$$\omega = (0.6)(5 \text{ s}) + (1.2)(5 \text{ s})^2 = 33 \text{ rad/s}$$

$$\alpha = \frac{d\omega}{dt} = ((0.6) + (2.4)t) \text{ rad/s}^2$$

at $t = 5$s

$$\alpha = (0.6) + (2.4)(5 \text{ s}) = 12.6 \text{ rad/s}^2$$

Example 7.3 A wheel is rotating with an angular acceleration that is given by $\alpha = (9 - 2t)$ rad/s^2. (a) Find the angular velocity and displacement at any time if at $t = 0$ the wheel has an angular velocity of 2 rad/s and an (initial) angular displacement of 3 rad; (b) at what angular displacement will the wheel reach its maximum angular velocity

Solution 7.3 (a)

$$\omega = \int \alpha dt = \int (9 - 2t)dt = 9t - t^2 + c_1$$

Since at $t = 0$ $\omega = 2$ rad/s, we have $c_1 = 2$ rad/s and hence

$$\omega = (9t - t^2 + 2) \text{ rad/s}$$

$$\theta = \int \omega dt = \int (9t - t^2 + 2)dt = \frac{9}{2}t^2 - \frac{1}{3}t^3 + 2t + c_2$$

Since at $t = 0$, $\theta = 3$ rad, then $c_2 = 3$ rad and

$$\theta = \left(\frac{9}{2}t^2 - \frac{1}{3}t^3 + 2t + 3\right) \text{ rad}$$

(b) The maximum velocity is when $\alpha = d\omega/dt = 0$, or $9 - 2t = 0$, i.e. at $t = 4.5$ s The angular displacement at that time is

$$\theta = \frac{9}{2}(4.5 \text{ s})^2 - \frac{1}{3}(4.5 \text{ s})^3 + 2(4.5 \text{ s}) + 3 = 72.8 \text{ rad}$$

7.3   Rotational Motion with Constant Acceleration

A pure rotational motion with constant angular acceleration is the rotational analogue of the pure translational motion with constant acceleration. The corresponding kinematic equations of pure rotational motion can be obtained by using the same method that is used for obtaining the kinematic equations of pure translational motion. To show this, consider a rigid object rotating with a constant angular acceleration during a time interval from t_1 to t_2 through an angle from θ_1 to θ_2. Let $t_1 = 0, t_2 = t, \omega_1 = \omega_0, \omega_2 = \omega, \theta_1 = \theta_0$, and $\theta_2 = \theta$. Because the angular acceleration is constant it follows that the angular velocity changes linearly with time and the average angular velocity is given by

$$\overline{\omega} = \frac{\omega_0 + \omega}{2}$$

Since

$$\alpha = \overline{\alpha} = \frac{\omega_2 - \omega_1}{t_2 - t_1} = \frac{\omega - \omega_0}{t}$$

we have

$$\omega = \omega_0 + \alpha t \qquad (7.1)$$

Furthermore

$$\overline{\omega} = \frac{\theta_2 - \theta_1}{t_2 - t_1} = \frac{\theta - \theta_0}{t} = \frac{\omega_0 + \omega}{2}$$

Hence

$$\theta = \theta_0 + \frac{1}{2}(\omega_0 + \omega)t \qquad (7.2)$$

Substituting Eq. 7.1 into Eq. 7.2 gives

$$\theta = \theta_0 + \frac{1}{2}(\omega_0 + \omega)t = \theta_0 + \frac{1}{2}(\omega_0 + \omega_0 + \alpha t)t$$

or

$$\theta = \theta_0 + \omega_0 t + \frac{1}{2}\alpha t^2 \qquad (7.3)$$

Finally solving for t from Eq. 7.1 and substituting into Eq. 7.2 gives

$$\theta = \theta_0 + \frac{1}{2}(\omega_0 + \omega)t = \theta_0 + \frac{1}{2}(\omega_0 + \omega)\left(\frac{\omega - \omega_0}{\alpha}\right)$$

or

$$\omega^2 = \omega_0^2 + 2\alpha(\theta - \theta_0) \qquad (7.4)$$

Note that as mentioned earlier, if a rigid object is in pure rotational motion, all particles in the object have the same angular velocity and angular acceleration. Different particles move in different circles but the center of these circles lies

at the axis of rotation. As the rigid body rotates, a particle in the body will move through a distance s along its circular path (see Fig. 7.6). The angular displacement of the particle is related to s by

$$s = r\theta$$

where r is the radius of the circle in which the particle is moving along. Differentiating the above equation with respect to t gives

$$\frac{ds}{dt} = r\frac{d\theta}{dt}$$

Since ds/dt is the magnitude of the linear velocity of the particle and $d\theta/dt$ is the angular velocity of the body we may write

$$v = r\omega \qquad (7.5)$$

Therefore, the farther the particle is from the rotational axis the greater its linear speed. The direction of the linear speed of the particles is always tangent to the path (as mentioned in Sect. 2.2.3). In Sect. 2.4.6 we have seen that a particle in nonuniform circular motion has both tangential and radial components of acceleration. The radial component is due to the change in the direction of the velocity and is given by

$$a_r = \frac{v^2}{r} \qquad (7.6)$$

Substituting Eq. 7.5 into Eq. 7.6 gives

$$a_r = \frac{v^2}{r} = r\omega^2$$

The tangential component of the acceleration is due to the change in the magnitude of the velocity and it is given by

$$a_t = \frac{dv}{dt} = r\frac{d\omega}{dt}$$

or

$$a_t = r\alpha$$

The total linear acceleration of the particle (see Fig. 7.7) is given by

$$\mathbf{a} = \mathbf{a}_t + \mathbf{a}_r$$

It's magnitude is given by

$$a = \sqrt{a_t^2 + a_r^2} = \sqrt{r^2\alpha^2 + r^2\omega^4} = r\sqrt{\alpha^2 + \omega^4}$$

Table. 7.1 shows the linear/rotational analogous equations.

Example 7.4 A disc of radius of 10 cm rotates from rest with a constant angular acceleration. If it requires 2 s for it to rotate through an angular displacement of 60°: (a) find the angular

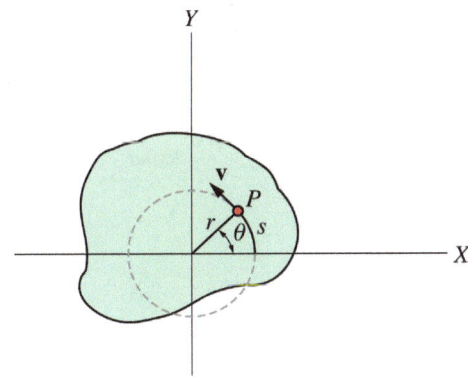

Fig. 7.6 As the rigid body rotates, a particle in the body will move through a distance s along its circular path

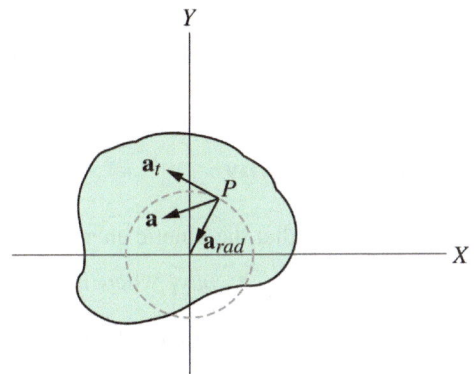

Fig. 7.7 The total acceleration of the particle

Table 7.1 Kinematic equations

Rotational motion about a fixed axis with constant α	Linear motion with constant a
$\omega = \omega_0 + \alpha t$	$v = v_0 + at$
$\theta = \theta_0 + \frac{1}{2}(\omega + \omega_0)t$	$x = x_0 + \frac{1}{2}(v + v)t$
$\theta = \theta_0 + \omega_0 t + \frac{1}{2}\alpha t^2$	$x = x_0 + v_0 t + \frac{1}{2}at^2$
$\omega^2 = \omega_0^2 + 2\alpha(\theta - \theta_0)$	$v^2 = v_0^2 + 2a(x - x_0)$

acceleration of the disc; (b) its angular velocity at $t = 2$s and at $t = 6$s, (c) the linear speed at $t = 2$s of a point that is at a distance of 7 cm from the center of the disc; (d) the distance that this point has moved during that time interval.

Solution 7.4 (a) We have $\omega_0 = 0$ and $\theta = (60 \text{ deg})(2\pi \text{ rad}/360 \text{ deg}) = 1.05$ rad. By choosing the reference position $\theta_0 = 0$ we have

$$\theta = \theta_0 + \omega_0 t + \frac{1}{2}\alpha t^2$$

$$\alpha = \frac{2\theta}{t^2} = \frac{2(1.05 \text{ rad})}{(2 \text{ s})^2} = 0.525 \text{ rad/s}^2$$

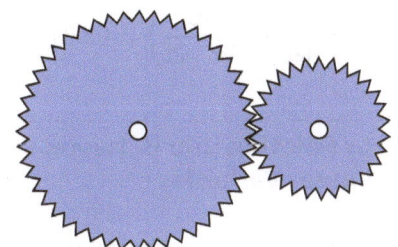

Fig. 7.8 Two sprockets connected at the rim

(b)

$$\omega = \omega_0 + \alpha t = (0.525 \text{ rad/s}^2)(2 \text{ s}) = 1.05 \text{ rad/s}$$

at $t = 6$ s

$$\omega = (0.525 \text{ rad/s}^2)(6\text{s}) = 3.15 \text{ rad/s}$$

(c)

$$v = r\omega = (0.07 \text{ m})(1.05 \text{ rad/s}) = 0.074 \text{ m/s}$$

(d)

$$s = r\theta = (0.07 \text{ m})(1.05 \text{ rad}) = 0.074 \text{ m}$$

Example 7.5 Two sprockets are attached to each other as in Fig. 7.8. There radii are $r_1 = 2$ cm and $r_2 = 5$ cm. If the angular velocity of the smaller sprocket is 2 rad/s, find the angular velocity of the other.

Solution 7.5 A point at the rim of one sprocket has the same linear speed as a point at the rim of the other sprocket since they are attached to each other, i.e.,

$$r_1\omega_1 = r_2\omega_2 = v$$

hence

$$\omega_2 = \frac{r_1}{r_2}\omega_1 = \frac{(2 \text{ cm})}{(5 \text{ cm})}(2 \text{ rad/s}) = 0.8 \text{ rad/s}$$

Example 7.6 Find the angular speed of the moon in its orbit about the earth in rev/day.

Solution 7.6 Assuming that the moon's orbit is circular, the linear speed of the moon is given by $v = 2\pi r/T$, where r is the mean distance from the earth to the moon and T is its period. Thus, the angular velocity of the moon is

$$\omega = rv = \frac{2\pi}{T} = \frac{2(3.14)}{(27.3 \text{ day})} = 0.23 \text{ rad/day}$$

or

$$\omega = \left(0.23 \frac{\text{rad}}{\text{day}}\right)\left(\frac{1 \text{ rev}}{2\pi \text{ rad}}\right) = 0.037 \text{ rev/day}$$

7.4 Vector Relationship Between Angular and Linear Variables

Consider a rigid body in pure rotational motion about a fixed axis (for example the z-axis). For any particle in the object, its linear velocity is given by

$$v = r\omega = R \sin \theta \omega$$

where R is the position vector of the particle from the origin (see Fig. 7.9) and θ is the angle between the position vector and the z-axis. As shown in Fig. 7.9, the direction of y is perpendicular to the plane formed by ω and R where it can be verified using the right-hand rule. Therefore, by using the definition of vector product we may write

$$\mathbf{v} = \boldsymbol{\omega} \times \mathbf{R} \qquad (7.7)$$

The total linear acceleration is

$$\mathbf{a} = \frac{d\mathbf{v}}{dt} = \frac{d}{dt}(\boldsymbol{\omega} \times \mathbf{R})$$

From Sect. 1.9.1 $(d/dt(\mathbf{A} \times \mathbf{B}) = \mathbf{A} \times d\mathbf{B}/dt + d\mathbf{A}/dt \times \mathbf{B})$ we have

$$\mathbf{a} = \frac{d\boldsymbol{\omega}}{dt} \times \mathbf{R} + \boldsymbol{\omega} \times \frac{d\mathbf{R}}{dt}$$

$$= \boldsymbol{\alpha} \times \mathbf{R} + \boldsymbol{\omega} \times \mathbf{v}$$

$$|\boldsymbol{\alpha} \times \mathbf{R}| = \alpha R \sin \theta = r\alpha = a_t$$

Furthermore, the direction of $\boldsymbol{\alpha} \times \mathbf{R}$ is tangent to the circular path of the particle at any instant (see Fig. 7.9). Thus the quantity $\boldsymbol{\alpha} \times \mathbf{R}$ is just the tangential component of the total acceleration

$$\mathbf{a_t} = \boldsymbol{\alpha} \times \mathbf{R} \qquad (7.8)$$

In addition

$$|\boldsymbol{\omega} \times \mathbf{v}| = \omega v \sin 90^\circ = \omega v = r\omega^2 = a_r$$

The direction of $\boldsymbol{\omega} \times \mathbf{v}$ is along the direction of r (radial direction). Hence, the quantity $\boldsymbol{\omega} \times \mathbf{v}$ is the radial component of the total acceleration

$$\mathbf{a_r} = \boldsymbol{\omega} \times \mathbf{v} \qquad (7.9)$$

Equations 7.7–7.9 are the vector relationship between angular and linear quantities.

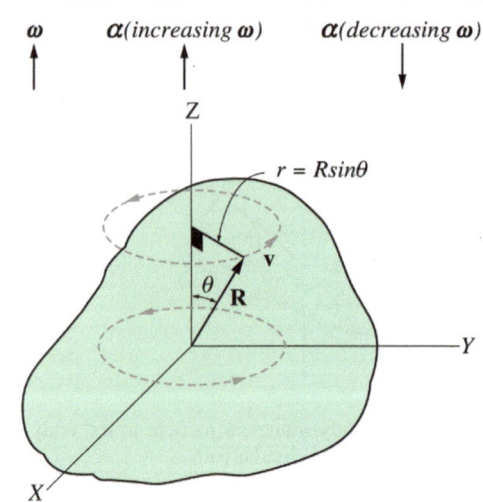

Fig. 7.9 A rigid body in pure rotational motion about a fixed axis (here the z-axis)

7.5 Rotational Energy

In Chap. 6 we have seen that the kinetic energy of a discrete system of particles is $K = \frac{1}{2} \sum_i m_i v_i^2$ where m_i and v_i are the mass and linear velocity of the ith particle respectively (see Fig. 7.10). From Eq. 7.5, we have

$$v_i = r_i \omega$$

where r_i is the perpendicular distance from the particle to the axis of rotation. Therefore the total kinetic energy of the system is

$$K_R = \frac{1}{2} \sum_i (m_i r_i^2) \omega^2$$

The quantity between brackets is known as the moment of inertia of the system

$$I = \sum_i m_i r_i^2$$

This quantity shows how the mass of the system is distributed about the axis of rotation. Thus, to find the rotational inertia, the axis of rotation must be specified. If the rotational axis changes its position or direction, I changes as well. The SI unit of the moment of inertia is kg m^2. The rotational kinetic energy can thus be written as

$$K_R = \frac{1}{2} I \omega^2$$

This quantity is the rotational analogue of the kinetic energy in translational motion. Note that this energy is not a new kind of energy; it is just the sum of the translational kinetic energies

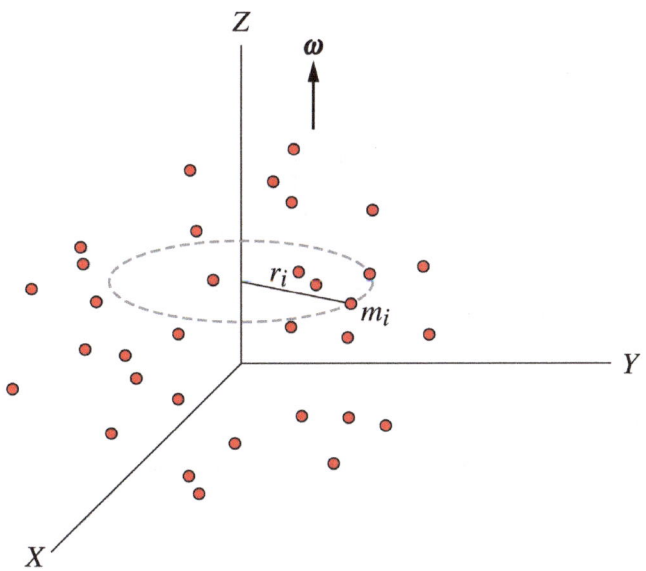

Fig. 7.10 A system of particles rotating about the z-axis

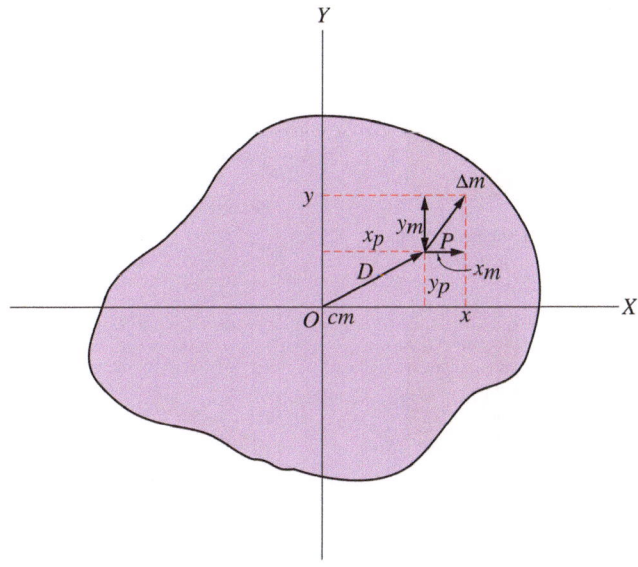

Fig. 7.11 The Parallel-axis Theorem

of the particles. For a rigid body which is a continuous system of particles, the sum is replaced by an integral

$$I = \lim_{\Delta m_i \to 0} \sum_i m_i r_i^2 = \int r^2 dm$$

In solving problems ρ, σ, and λ (see Sect. 6.3.4) are often used to express dm in terms of its position coordinates.

7.6 The Parallel-Axis Theorem

The parallel-axis theorem states that the moment of inertia I of a system about any axis that is parallel to an axis passing through the center of mass is

$$I = I_{cm} + MD^2$$

where I_{cm} is the moment of inertia about an axis passing through the center of mass, M is the total mass of the system, and D is the perpendicular distance between the two parallel axes.

Proof Consider an axis that is perpendicular to the page and passing through the center of mass of the object. Figure 7.11 shows a thin slice of the object that lies in the x-y plane. Because the origin is taken at the center of mass we have

$$z_{cm} = x_{cm} = y_{cm} = 0$$

The moment of inertia of the object about the center of mass axis is

$$I_{cm} = \int r^2 dm = \int (x^2 + y^2) dm$$

where x and y are the coordinates of the mass element dm from the center of mass (the origin). Now consider another axis that is parallel to the first axis and that passes through a point P as shown in Fig. 7.11. Suppose that the x and y coordinates of P from the center of mass are x_p and y_p. The moment of inertia about an axis passing through P is

$$I_P = \int [(x - x_P)^2 + (y - y_P)^2] dm$$

where $(x - x_P)$ and $(y - y_P)$ are coordinates of dm from point P Expanding this equation gives

$$I_P = \int (x^2 + y^2) dm - 2x_P \int x dm - 2y_P \int y dm + \int (x_P^2 + y_P^2) dm$$

Since $x_{cm} = y_{cm} = 0$ and since

$$x_{cm} = \frac{1}{M} \int x dm$$

and

$$y_{cm} = \frac{1}{M} \int y dm$$

it follows that the second and third terms are zero. Thus

$$I_P = I_{cm} + D^2 \int dm$$

where

$$D = \sqrt{(x_P^2 + y_P^2)}$$

is the perpendicular distance between the two parallel axes. Hence

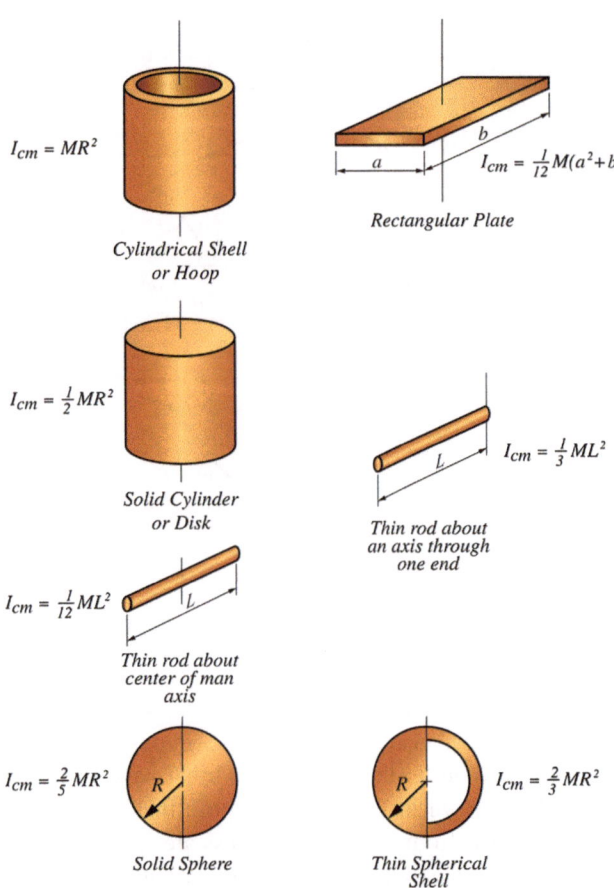

Fig. 7.12 The rotational inertia of various rigid bodies of uniform density

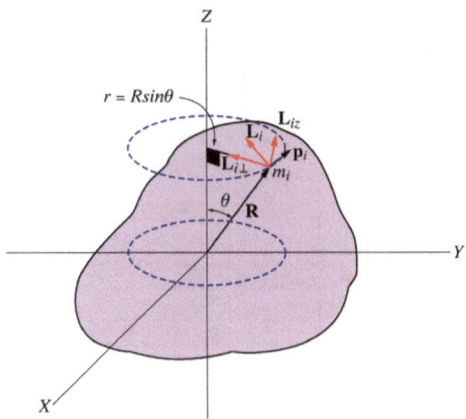

Fig. 7.13 A rigid body rotating about a fixed axis (the z-axis) with an angular speed ω

$$I_P = I_{cm} + MD^2 \quad \text{(Parallel–Axis Theorem)}$$

Special Moment of Inertia Fig. 7.12 gives the rotational inertia of various rigid bodies of uniform density.

7.7   Angular Momentum of a Rigid Body Rotating about a Fixed Axis

Consider a rigid body rotating about a fixed axis (the z-axis) with an angular speed ω as shown in Fig. 7.13. The angular momentum of the ith particle with respect to the origin is given by

$$\mathbf{L}_i = \mathbf{R}_i \times \mathbf{p}_i$$

Since the angle between $\mathbf{R}_i$ and $\mathbf{p}_i$ is 90, then $L_i = R_i p_i$. As seen from Fig. 7.13, $\mathbf{L}_i$ is not parallel to $\boldsymbol{\omega}$. $\mathbf{L}_i$ can be analyzed to two components, a component parallel to $\boldsymbol{\omega}$ written ($\mathbf{L}_{iz}$) and a component perpendicular to $\boldsymbol{\omega}$, ($\mathbf{L}_{i\perp}$). The magnitude of $\mathbf{L}_{iz}$ is given by

$$L_{iz} = L_i \sin\theta = R_i p_i \sin\theta = R_i(m_i v_i)\sin\theta$$

$$= R_i m_i(r_i\omega)\sin\theta = m_i r_i^2 \omega$$

where r_i is the radius of the circle in which the particle is moving along and $R_i = r_i \sin\theta$. Therefore, the total angular momentum of the rigid body along the z-direction is

$$L_z = \sum_i m_i r_i^2 \omega = \left(\sum_i m_i r_i^2\right)\omega$$

$$L_z = I\omega$$

where I is the moment of inertia of the rigid body about the rotational axis (z-axis). This equation can also be written in component form since $\mathbf{L}_z$ is parallel to $\boldsymbol{\omega}$, that is,

$$\mathbf{L}_z = I\boldsymbol{\omega} \tag{7.10}$$

Therefore, if a rigid body is rotating about a fixed axis (say the z-axis), the component of the angular momentum along that axis is given by Eq. 7.10. Now suppose that the rigid body is symmetric and homogeneous and that it is rotating about its symmetrical axis (see Fig. 7.14). For any two particles (1 and 2) opposing each other with an equal angular momenta $\mathbf{L}_1$ and $\mathbf{L}_2$, the perpendicular components, $\mathbf{L}_{1\perp}$ and $\mathbf{L}_{2\perp}$, of the angular momenta cancel each other out since they are in opposite directions. That leaves the parallel components $\mathbf{L}_{1z}$ and $\mathbf{L}_{2z}$ which add up since they have the same direction. For all particles in the object the total angular momentum is, therefore, given by

$$\mathbf{L} = \sum_i \mathbf{L}_{iz} = \mathbf{L}_z = I\boldsymbol{\omega}$$

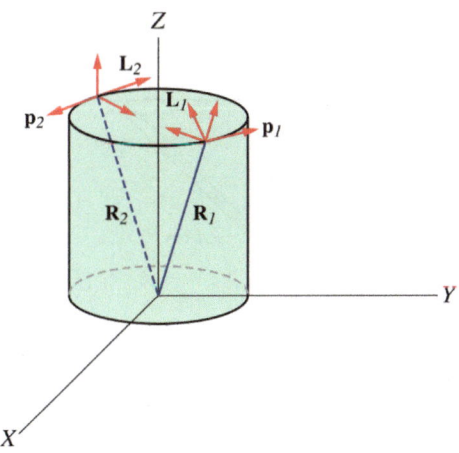

Fig.7.14 A homogenous symmetrical rigid body rotating about its symmetrical axis

Hence, the total angular momentum of a symmetrical homogeneous body in pure rotation about its symmetrical axis is given by

$$\mathbf{L} = I\boldsymbol{\omega} \qquad (7.11)$$

Note that Eq. 7.10 is valid for any rigid object in pure rotation where it only gives the component of the angular momentum that is parallel to the rotational axis. On the other hand, Eq. 7.11 is valid only for a symmetrical homogeneous rigid object rotating about its symmetrical axis, where the angular momentum in the equation is the total angular momentum and it is directed along the axis of rotation. The net external torque acing on the rigid object is equal to the rate of change of the total angular momentum of the object, i.e.,

$$\Sigma\boldsymbol{\tau}_{ext} = \frac{d\mathbf{L}}{dt}$$

In the case of any rigid object symmetrical or not, the net external torque acting on the object about the axis of rotation (say the z-axis) is equal to the rate of change of the component of angular momentum that is along that axis

$$\Sigma\boldsymbol{\tau}_{extz} = \frac{d\mathbf{L}_z}{dt} = \frac{d(I\omega)}{dt} = I\alpha$$

However, if the object is symmetric and homogeneous in pure rotation about its symmetrical axis we may write

$$\Sigma\boldsymbol{\tau}_{ext} = \frac{d\mathbf{L}}{dt} = \frac{d(I\omega)}{dt} = I\alpha$$

Example 7.7 A 5 kg wheel of radius of 0.1 m decelerates from an angular speed of 5 rad/s to rest after going through an angular displacement of 10 rev If a frictional force causes the wheel to decelerate, find the torque due to this force.

Solution 7.7 The angular displacement is

$$\Delta\theta = (10 \text{ rev})\left(\frac{2\pi \text{ rad}}{1 \text{ rev}}\right) = 62.8 \text{ rad}$$

The angular acceleration of the wheel is

$$\alpha = \frac{\omega^2 - \omega_0^2}{2\Delta\theta} = \frac{0 - (5 \text{ rad/s})^2}{2(62.8 \text{ rad})} = -0.2 \text{ rad/s}$$

The external torque is

$$\tau = I\alpha = MR^2\alpha = (5 \text{ kg})(0.1 \text{ m})^2(-0.2 \text{ rad/s}^2) = -0.01 \text{ N m}$$

Example 7.8 Three masses are connected by massless rods as in Fig. 7.15. If $m = 0.1$ kg, find the moment of inertia of the system and the corresponding kinetic energy if it rotates with an angular speed of 5 rad/s about: (a) the z-axis; (b) the y-axis and; (c) the x-axis ($a = 0.2$ m).

Solution 7.8 (a)

$$I_z = \sum_i m_i r_i^2 = 2ma^2 + \frac{m}{2}a^2 + ma^2 = \frac{7}{2}ma^2$$
$$= \frac{7}{2}(0.1 \text{ kg})(0.2 \text{ m})^2 = 0.014 \text{ kg m}^2$$

$$K_R = \frac{1}{2}I_z\omega^2 = \frac{1}{2}(0.014 \text{ kg m}^2)(5 \text{ rad/s})^2 = 0.175 \text{ J}$$

(b)

$$I_y = \frac{m}{2}a^2 + 2ma^2 = \frac{5}{2}ma^2 = \frac{5}{2}(0.1 \text{ kg})(0.2 \text{ m})^2 = 0.01 \text{ kg m}^2$$

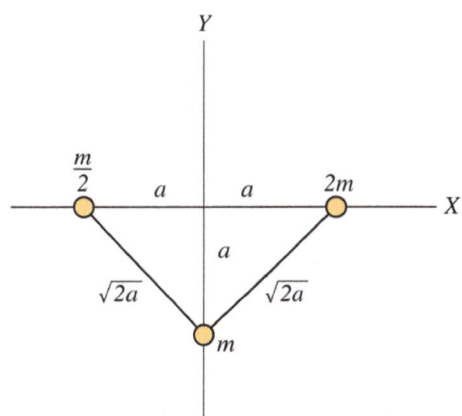

Fig. 7.15 Three masses connected by massless rods

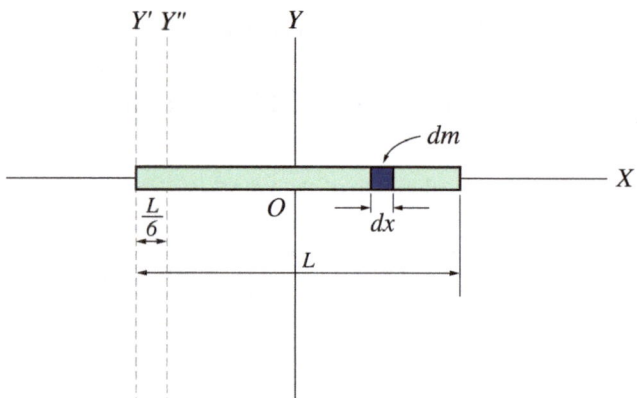

Fig. 7.16 A uniform thin rod of mass M and length L

$$K_R = \frac{1}{2}I_y\omega^2 = \frac{1}{2}(0.01 \text{ kg m}^2)(5 \text{ rad/s})^2 = 0.125 \text{ J}$$

(c)

$$I_x = ma^2 = (0.1 \text{ kg})(0.2 \text{ m})^2 = 4 \times 10^{-3} \text{ kg m}^2$$

$$K_R = \frac{1}{2}I_x\omega^2 = \frac{1}{2}(4 \times 10^{-3} \text{ kg m}^2)(5 \text{ rad/s})^2 = 0.05 \text{ J}$$

Example 7.9 Fig. 7.16 shows a uniform thin rod of mass M and length L. Find the moment of inertia of the rod about an axis that is perpendicular to it and passing through: (a) the center of mass; (b) at one end; (c) at a distance of $L/6$ from one end.

Solution 7.9 (a) The mass dm of an element in the rod is

$$dm = \lambda dx = \left(\frac{M}{L}\right)dx$$

$$I_{cm} = I_y = \int r^2 dm = \int_{x=-\frac{L}{2}}^{\frac{L}{2}} x^2\left(\frac{M}{L}\right)dx = \frac{M}{L}\left(\frac{x^3}{3}\right)\Big|_{-L/2}^{L/2} = \frac{1}{12}ML^2$$

(b)

$$I_{y'} = I_{cm} + MD^2 = \frac{1}{12}ML^2 + M\left(\frac{L}{2}\right)^2 = \frac{1}{3}ML^2$$

(c)

$$I_{y''} = I_{cm} + MD^2 = \frac{1}{12}ML^2 + M\left(\frac{L}{2} - \frac{L}{6}\right)^2 = \frac{7}{36}ML^2$$

Example 7.10 Fig. 7.17 shows a uniform thin plate of mass M and surface density σ. Find the moment of inertia of the plate about an axis passing through its center of mass if its length is b and its width is a (the z-axis).

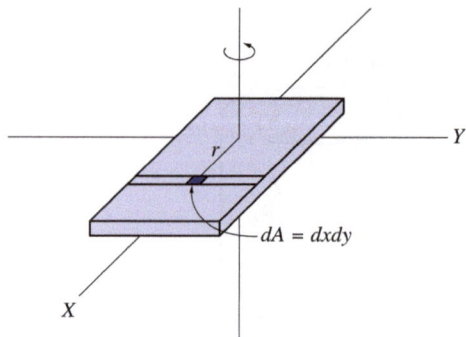

Fig. 7.17 A uniform thin plate of mass M and surface density σ

Solution 7.10 A mass element dm has an area $dxdy$ and is at a distance $r = \sqrt{x^2 + y^2}$ from the axis of rotation. Therefore, we have

$$I_{cm} = \int r^2 dm = \int r^2 \sigma dA = \int_{y=-a/2}^{a/2}\int_{y=-b/2}^{b/2}(x^2+y^2)\left(\frac{M}{ab}\right)dxdy$$

$$= \frac{M}{ab}\int_{y=-a/2}^{a/2}\left(\frac{x^3}{3}+xy^2\right)\Big|_{x=-b/2}^{b/2}dy = \frac{M}{ab}\int_{y=-a/2}^{a/2}\left(\frac{b^3}{12}+by^2\right)dy$$

$$= \frac{M}{ab}\left(\frac{b^3y}{12}+\frac{y^3b}{3}\right)\Big|_{x=-a/2}^{a/2} = \frac{M}{ab}\left[\frac{ab^3}{12}+\frac{ab^3}{12}\right] = \frac{1}{12}M\left(a^2+b^2\right)$$

Example 7.11 Find the moment of inertia of a uniform solid cylinder of radius R, length L and mass M about its axis of symmetry.

Solution 7.11 Method 1: Using a single integration by dividing the cylinder into thin cylindrical shells each of radius r, length L and thickness dr as in Fig. 7.18, then each volume element is given by

$$dV = 2\pi r dr L$$

and

$$dm = \rho dV = \rho(2\pi r dr L)$$

$$I = \int r^2 dm = \int_0^R r^2(\rho 2\pi r L dr) = 2\pi\rho L\int_0^R r^3 dr = \frac{\pi\rho L}{2}R^4$$

Since

$$\rho = \frac{M}{\pi R^2 L}$$

then

$$I = \frac{1}{2}MR^2$$

Method 2: Using double integration: dividing the cylinder into thin rods each of mass

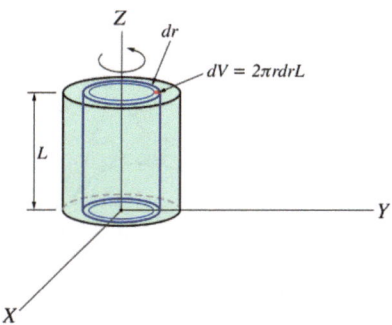

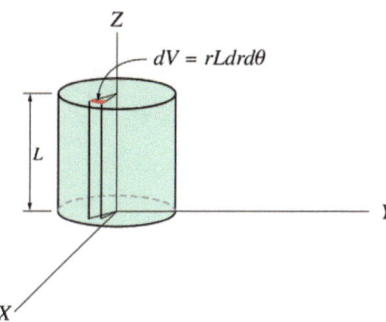

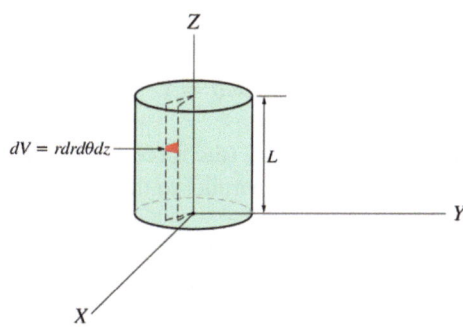

Fig. 7.18 Calculating the moment of inertia of a uniform solid cylinder with the volume element defined in different ways

$$dm = \rho dV = \rho L r dr d\theta$$

$$I = \int r^2 dm = \int_0^{2\pi} \int_{r=0}^R r^3 \rho L dr d\theta = \rho \frac{L}{4} R^4 \int_{\theta=0}^{2\pi} d\theta = \frac{\pi \rho L R^4}{2}$$

Since

$$\rho = \frac{M}{\pi R^2 L}$$

We have

$$I = \frac{1}{2} MR^2$$

Method 3: Using triple integration Dividing the cylinder into small cubes each of mass given by

$$dm = \rho r dr d\theta dz$$

$$I = \int r^2 dm = \int_{\theta=0}^{2\pi} \int_{r=0}^R \int_{z=0}^L \rho r^3 dr d\theta dz = \rho L \frac{R^4}{4} \int_{\theta=0}^{2\pi} d\theta = \frac{\pi \rho L R^4}{2}$$

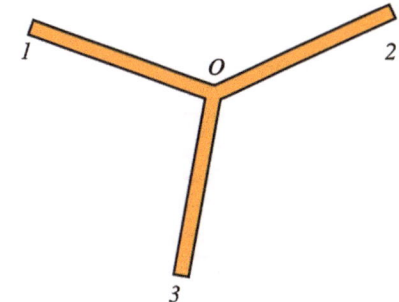

Fig. 7.19 Three rods of length L and mass M are connected together

Since

$$\rho = \frac{M}{\pi R^2 L}$$

Therefore,

$$I = \frac{1}{2} MR^2$$

Example 7.12 Three rods of length L and mass M are connected together as in Fig. 7.19. Determine the moment of inertia of the system about an axis passing through O and perpendicular to the page (the rods lie in the same plane).

Solution 7.12 The moment of inertia of a thin rod about an axis that is perpendicular to it and passing through one end is $1/3 ML^2$. The total moment of inertia at O is the sum of the moment of inertias of the rods, i.e.,

$$I = I_1 + I_2 + I_3 = 3 \left(\frac{1}{3} ML^2 \right) = ML^2$$

Example 7.13 Find the moment of inertia of a spherical shell of radius R and mass M about an axis passing through its center of mass.

Solution 7.13 Let us divide the spherical shell into thin rings each of area (see Fig. 7.20) given by

$$dA = 2\pi R \sin \theta R d\theta = 2\pi R^2 \sin \theta d\theta$$

$$I = \int r^2 dm = \int R^2 \sin^2 \theta \sigma 2\pi R^2 \sin \theta d\theta$$

since $\sigma = M / 4\pi R^2$, we have

$$I = \frac{M}{2} R^2 \int_{\theta=0}^{\pi} \sin^3 \theta d\theta = \frac{M}{2} R^2 \int_{\theta=0}^{\pi} (1 - \cos^2 \theta) \sin \theta d\theta$$

$$= \frac{M}{2} R^2 \left[- \cos \theta + \frac{\cos^3 \theta}{3} \right]_{\theta=0}^{\pi} = \frac{2}{3} MR^2$$

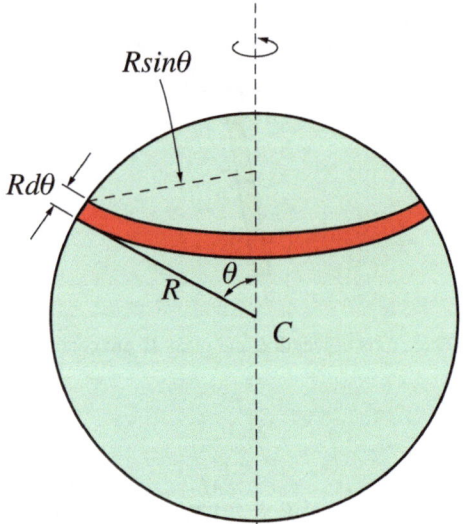

Fig. 7.20 A spherical shell divided into thin rings

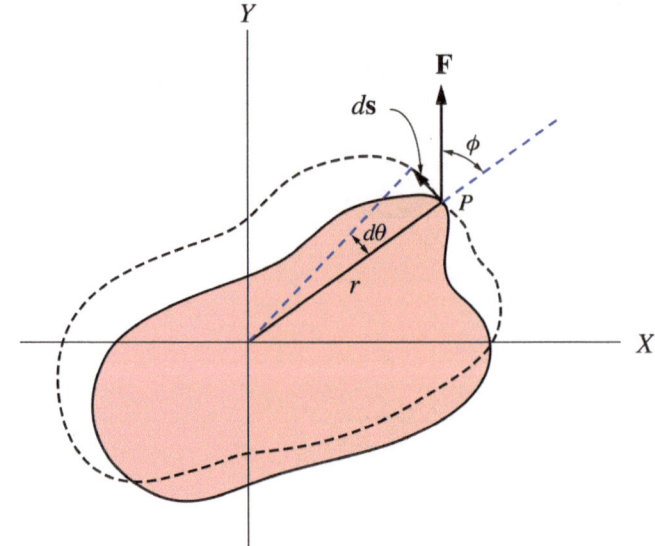

Fig. 7.21 A rigid body rotating about a fixed axis

7.8   Conservation of Angular Momentum of a Rigid Body Rotating About a Fixed Axis

In Chap. 5 we have seen that if the net external torque acting on a system of particles relative to an origin is zero then the total angular momentum of the system about that origin is conserved

$$\mathbf{L}_i = \mathbf{L}_f = \text{constant (isolated system)}$$

In the case of a rigid object in pure rotational motion, if the component of the net external torque about the rotational axis (say the z-axis) is zero then the component of angular momentum along that axis is conserved, i.e., if

$$\tau_z = \frac{dL_z}{dt} = 0$$

then

$$I_i \omega_i = I_f \omega_f$$

That is, the angular momentum is not necessarily conserved in all directions. It is conserved in the direction where the net external torque is equal to zero.

7.9   Work and Rotational Energy

Consider a rigid body rotating about a fixed axis as in Fig. 7.21. If a force that lies in the x-y plane is applied to the body at P, then the work done on the body if it rotates through an angle $d\theta$ is

$$dW = \mathbf{F} \cdot d\mathbf{s} = \mathbf{F} \cdot \frac{d\mathbf{s}}{dt} dt = \mathbf{F} \cdot \mathbf{v} dt = \mathbf{F} \cdot (\boldsymbol{\omega} \times \mathbf{r}) dt$$

$$= (\mathbf{r} \times \mathbf{F}) \cdot \boldsymbol{\omega} dt = \boldsymbol{\tau} \cdot \boldsymbol{\omega} dt$$

Since $\boldsymbol{\tau}$ and $\boldsymbol{\omega}$ are parallel, (the force lies in the x-y plane therefore the total torque is parallel to the z-axis) we have

$$dW = \tau \omega dt = \tau \frac{d\theta}{dt} dt = \tau d\theta$$

Therefore, the total work done in displacing the body from θ_1 to θ_2 is

$$W = \int_{\theta_1}^{\theta_2} \tau d\theta \qquad (7.12)$$

If this torque is constant we have

$$W = \tau(\theta_2 - \theta_1) = \tau \triangle \theta$$

The Work–Energy Theorem The work–energy theorem states that the work done by an external force while a rigid object rotate from θ_1 to θ_2 is equal to the change in the rotational energy of the object. This follows from Eq. 7.12 and by using the fact that along the axis of rotation the torque is given by $\tau_z = I\alpha$ (see Sect. 7.7), thus

$$W = \int_{\theta_1}^{\theta_2} \tau d\theta = \int_{\theta_1}^{\theta_2} I\alpha d\theta = \int_{\omega_1}^{\omega_2} I\omega \frac{d\omega}{dt} dt = \int_{\omega_1}^{\omega_2} I\omega d\omega = \frac{1}{2} I\omega_2^2 - \frac{1}{2} I\omega_1^2$$

$$W = \triangle K = \frac{1}{2} I\omega_2^2 - \frac{1}{2} I\omega_1^2$$

Table 7.2 Analogous Equations in linear Motion and Rotational Motion about a Fixed Axis

Rotational motion	Linear motion
$\tau = I\alpha$	$F = ma$
$W = \int_{\theta_0}^{\theta} \tau \, d\theta$	$W = \int_{x_0}^{x} F \, dx$
$K_R = \frac{1}{2}I\omega^2$	$K = \frac{1}{2}mv^2$
$P = \tau\omega$	$P = Fv$

7.10 Power

The instantaneous power delivered to rotate an object about a fixed axis is found from

$$P = \frac{dW}{dt} = \frac{\tau_z d\theta}{dt} = \tau_z\omega_z$$

Table. 7.2 shows analogous equations in linear motion and rotational motion about a fixed axis

Example 7.14 A disc of radius $R = 0.08$ m and mass of 5 kg is rotating about its central axis with an angular speed of 170 rev/min. Find: (a) the rotational kinetic energy of the disc; (b) Suppose that the same disc rotate using a motor that delivers an instantaneous of power 0. 2hp, find in that case the torque applied to the disc.

Solution 7.14 (a) Since the rotational axis is the axis of symmetry of the disc, then the moment of inertia is

$$I = \frac{1}{2}MR^2 = \frac{1}{2}(5 \text{ kg})(0.08 \text{ m})^2 = 0.016 \text{ kg m}^2$$

The angular velocity of the disc is

$$\omega = \left(\frac{170 \text{ rev}}{\text{min}}\right)\left(\frac{2\pi \text{ rad}}{1 \text{ rev}}\right)\left(\frac{1 \text{ min}}{60 \text{ s}}\right) = 17.8 \text{ rad/s}$$

$$K = \frac{1}{2}I\omega^2 = \frac{1}{2}(0.016 \text{ kg m}^2)(17.8 \text{ rad/s})^2 = 2.5 \text{ J}$$

(b)

$$P = (0.2 \text{ hp})\left(\frac{746 \text{ W}}{1\text{hp}}\right) = 149.2 \text{ W}$$

and

$$\tau = \frac{P}{\omega} = \frac{(149.2 \text{ W})}{(17.8 \text{ rad/s})} = 8.4 \text{ N m}$$

Example 7.15 Consider a light rope wrapped around a uniform cylindrical shell of mass 30 kg and radius of 0.2 m as in Fig. 7.22. Suppose that the cylinder is free to rotate about its central axis and that the rope is pulled from rest with a constant force of magnitude of 35 N. Assuming that the rope does not slip, find: (a) the torque applied to the cylinder about

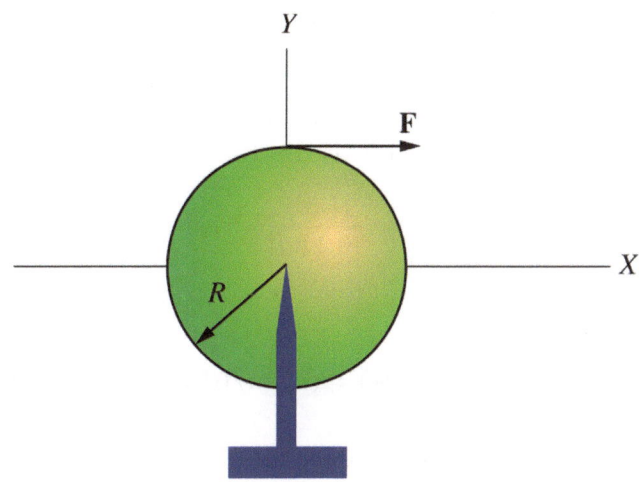

Fig. 7.22 A light rope wrapped around a uniform cylindrical shell

its central axis; (b) the angular acceleration of the cylinder; (c) the acceleration of a point in the unwinding rope; (d) the number of revolutions made by the cylinder when it reaches an angular velocity of 12 rad/s, (e) the work done by the applied force when the rope is pulled a distance of 1m, (f) the work done using the work–energy theorem.

Solution 7.15 (a) Because the line of action of both the weight and the normal forces passes through the central axis of the cylinder, they produce no torque. Hence, the total torque acting on the cylinder is

$$\tau = FR = (35 \text{ N})(0.2 \text{ m}) = 7 \text{ N/m}$$

(b) The moment of inertia of the cylinder is

$$I = MR^2 = (30 \text{ kg})(0.2 \text{ m})^2 = 1.2 \text{ kg m}^2$$

and

$$\alpha = \frac{\tau}{I} = \frac{(7 \text{ N m})}{(1.2 \text{ kg m}^2)} = 5.8 \text{ rad/s}^2$$

(c) The acceleration of a point in the unwinding rope is the same as the acceleration of a point at the rim of the cylinder, i.e.,

$$a = R\alpha = (0.2 \text{ m})(5.8 \text{ rad/s}^2) = 1.2 \text{ m/s}^2$$

(d)

$$\omega^2 = \omega_0^2 + 2\alpha\theta$$

Since $\omega_0 = 0$,

$$\theta = \frac{(12 \text{ rad/s})^2}{2(5.8 \text{ rad/s}^2)} = 12.4 \text{ rad}$$

or

$$\theta = (12.4 \text{ rad})\left(\frac{1 \text{ rev}}{2\pi \text{ rad}}\right) = 2 \text{ rev}$$

(e) If the rope has moved a distance of 1m, the angular displacement of the cylinder is

$$\theta = \frac{s}{R} = \frac{(1 \text{ m})}{(0.2 \text{ m})} = 5 \text{ rad}$$

the work done is

$$W = \int_{\theta_0}^{\theta} \tau d\theta = \tau(\theta - \theta_0) = (7 \text{ N m})((5 \text{ rad}) - 0) = 35 \text{ J}$$

(f) The final angular speed when $\theta = 5$ rad is

$$\omega^2 = \omega_0^2 + 2\alpha\theta = 0 + 2(5.8 \text{ rad/s}^2)(5 \text{ rad})$$

That gives $\omega = 7.6$ rad/s. From the work–energy theorem we have

$$W = \Delta K = \frac{1}{2}I\omega^2 - \frac{1}{2}I\omega_0^2 = \frac{1}{2}(1.2 \text{ kg m}^2)(7.6 \text{ rad/s})^2 - 0 = 35 \text{ J}$$

Example 7.16 A uniform rod of mass $M = 0.75$ kg and length $L = 1$m is hinged at one end and is free to rotate in a vertical plane as in Fig. 7.23. If the rod is released from rest at an angle $\theta = 30°$ to the horizontal, find; (a) the initial angular acceleration of the rod when it is released; (b) the initial acceleration of a point at the end of the rod; (c) from conservation of energy find the angular speed of the rod at its lowest position (Neglect friction at the pivot).

Solution 7.16 (a) Since the normal force exerted by the pin on the rod passes through O, then the only force that contributes to the torque is the force of gravity This force acts at the center of gravity which is at the center of mass (see Sect. 8.4). Therefore the net external torque is

$$\tau = \frac{MgL}{2}\cos\theta = \frac{(0.75 \text{ kg})(9.8 \text{ m/s}^2)(1 \text{ m})}{2}\cos 30° = 3.2 \text{ N m}$$

The moment of inertia about the rotational axis is

$$I = \frac{1}{3}ML^2 = \frac{(0.75 \text{ kg})(1 \text{ m})^2}{3} = 0.25 \text{ kg m}^2$$

and hence

$$\alpha = \frac{\tau}{I} = \frac{(3.2 \text{ N m})}{(0.25 \text{ kg m}^2)} = 12.8 \text{ rad/s}^2$$

(b) The acceleration of a point at the end of the rod is

$$a_t = r\alpha = L\alpha = (1 \text{ m})(12.8 \text{ rad/s}^2) = 12.8 \text{ m/s}^2$$

(c) When the rod reaches its lowest position, the potential energy of its center of mass is transformed into rotational kinetic energy of the rod. From conservation of energy we have $K_i + U_i = K_f + U_f$. Taking the potential energy to be zero at the lowest position, gives

$$0 + Mg\frac{L}{2}(\sin\theta + 1) = \frac{1}{2}I\omega^2 + 0$$

That gives

$$\omega = \sqrt{Mg\frac{L}{I}(\sin\theta + 1)} = \sqrt{\frac{(0.75 \text{ kg})(9.8 \text{ m/s}^2)(1\text{m})}{(0.25 \text{ kg m}^2)}(\sin 30° + 1)} = 6.64 \text{ rad/s}$$

Example 7.17 Find the net torque on the system shown in Fig. 7.24 where $r_1 = 5$ cm, $r_2 = 15$ cm, $F_1 = 10$ N, $F_2 =$

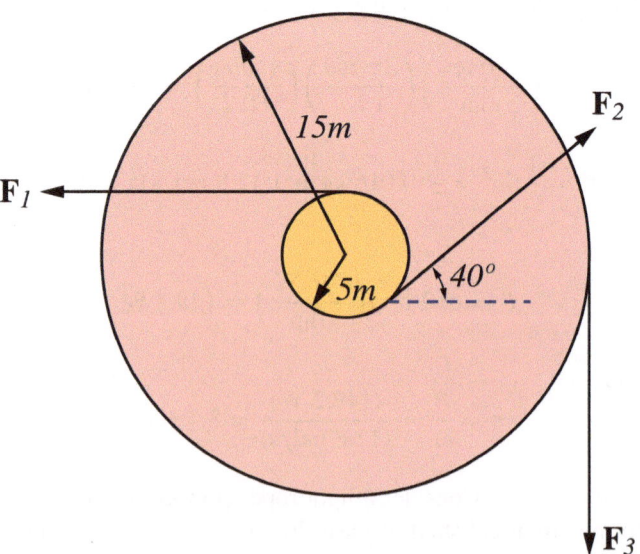

Fig. 7.24 A cylinder with a core section is free to rotate about its center. Ropes wrapped around the inner and outer sections exert different forces

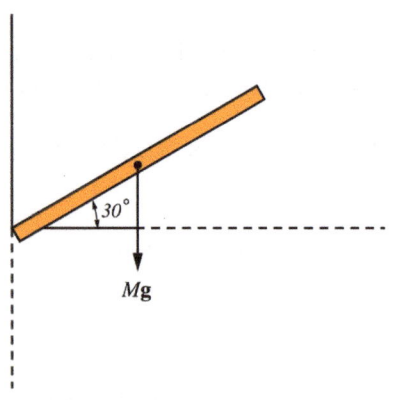

Fig. 7.23 A uniform rod free to rotate at one end

Fig. 7.25 A block of mass m is attached to a light string that is wrapped around the rim of a uniform solid disk of radius R and mass M

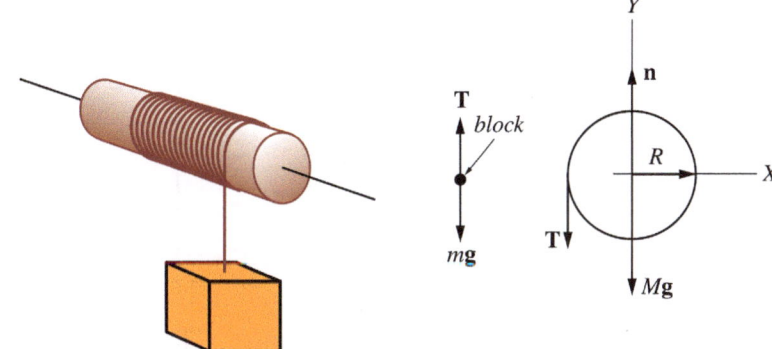

20 N and $F_3 = 15$ N. Neglect the mass and friction of the ropes and pulleys.

Solution 7.17 Since all forces lie in the same plane the net torque is

$$\tau_{net} = \tau_1 + \tau_2 + \tau_3 = (10 \text{ N})(0.05 \text{ m}) + (20 \text{ N})(0.05 \text{ m})$$
$$- (15 \text{ N})(0.15 \text{ m}) = -0.75 \text{ N m}$$

Example 7.18 A block of mass m is attached to a light string that is wrapped around the rim of a uniform solid disc of radius R and mass M as in Fig. 7.25. Assuming that the string does not slip and that the disc rotates without friction, find: (a) the acceleration of the block; (b) the angular acceleration of the disc, and; (c) the tension in the string when the system is released from rest.

Solution 7.18 The free-body diagrams of the disc and the block are shown in Fig. 7.25. Applying Newton's second law to the block gives

$$T - mg = -ma$$

or

$$a = \frac{mg - T}{m} \qquad (7.13)$$

where positive y is chosen to be directed upwards. Applying Newton's second law in angular form to the disc gives

$$\tau = RT = I\alpha$$

or

$$\alpha = \frac{RT}{I}$$

Since the acceleration of the block is equal to the (tangential) acceleration of a point at the rim of the disc we have

$$a = R\alpha = \frac{TR^2}{I} \qquad (7.14)$$

Equating Eqs. 7.13 and 7.14 gives

$$\frac{TR^2}{I} = \frac{mg - T}{m}$$

$$T = \frac{g}{1/m + R^2/I} = \frac{g}{1/m + 2R^2/MR^2}$$

that gives

$$T = \frac{mg}{1 + 2m/M}$$

Substituting this into Eq. 7.14

$$a = \frac{TR^2}{I} = \frac{2TR^2}{MR^2}$$

gives

$$a = \frac{g}{1 + M/2m}$$

Finally

$$\alpha = \frac{a}{R} = \frac{g}{R(1 + M/2m)}$$

Example 7.19 A homogeneous solid sphere of mass 4.7 kg and radius of 0.05 m rotate from rest about its central axis with a constant angular acceleration of 3 rad/s². Find: (a) the torque that produces this angular acceleration; (b) the work done on the sphere after 7 revolutions; (c) the work done after 7 revolutions using the work–energy theorem.

Solution 7.19 (a)

$$\tau = I\alpha = \frac{2}{5}MR^2\alpha = \frac{2}{5}(4.7 \text{ kg})(0.05 \text{ m})^2(3 \text{ rad/s}^2) = 0.014 \text{ N}$$

(b)

$$\theta = (7 \text{ rev})\left(\frac{2\pi \text{ rad}}{1 \text{ rev}}\right) = 44 \text{ rad}$$

and

$$W = \tau\Delta\theta = (0.014 \text{ N/m})(44 \text{ rad}) = 0.6 \text{ J}$$

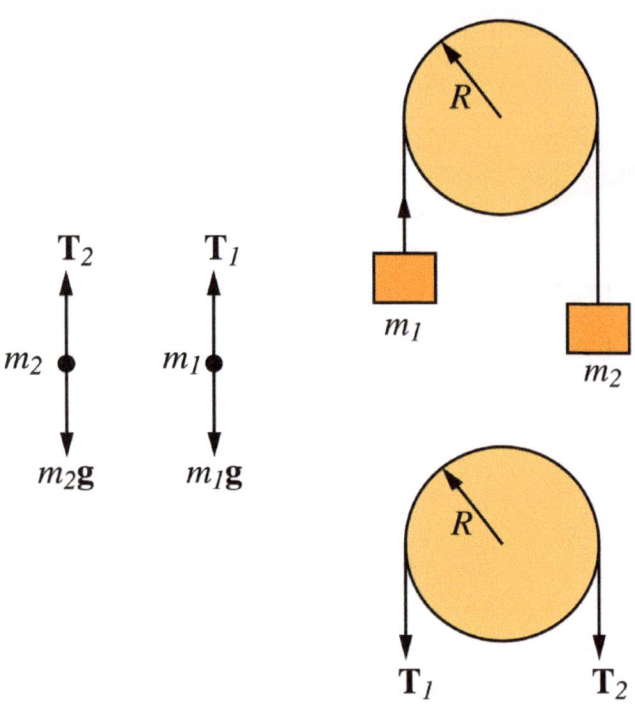

Fig. 7.26 AtwoodÕs machine

assuming $\theta_0 = 0$.

(c) After seven revolutions the angular velocity is

$$\omega^2 = \omega_0^2 + 2\alpha(\theta - \theta_0)$$

Since $\omega_0 = 0$, we have

$$\omega^2 = 2\alpha\theta = 2(3 \text{ rad/s}^2)(44 \text{ rad})$$

that gives $\omega = 16.24$ rad/s. Hence

$$W = \frac{1}{2}I\omega^2 - \frac{1}{2}I\omega_0^2 = \frac{1}{2}(4.7 \times 10^{-3} \text{ kg m}^2)(16.24 \text{ rad/s}^2)^2 - 0 = 0.6 \text{ J}$$

Example 7.20 Fig. 7.26 shows Atwood's machine when the mass of the pulley is considered. If the system is released from rest (and assuming that the string does not stretch or slip) and that the friction of the pulley is negligible, find linear acceleration of the blocks and the angular acceleration of the pulley.

Solution 7.20 Fig. 7.26 shows the free-body diagram for each block and for the pulley Applying Newton's second law gives

$$T_1 - m_1 g = m_1 a$$

$$T_2 - m_2 g = -m_2 a$$

$$\tau = (T_1 - T_2)R = -I\alpha$$

and

$$n - T_1 - T_2 - Mg = 0$$

The torque is negative because the pulley rotates in the clockwise direction. Therefore we have

$$T_1 - T_2 + g(m_2 - m_1) = a(m_1 + m_2)$$

and

$$T_2 - T_1 = \frac{I\alpha}{R} = \frac{Ia}{R^2}$$

That gives

$$a = \frac{g(m_2 - m_1)}{(m_1 + m_2 + I/R^2)}$$

If the pulley is a uniform solid disc then

$$I = \frac{1}{2}MR^2$$

and

$$a = \frac{g(m_2 - m_1)}{(m_1 + m_2 + M/2)}$$

$$\alpha = \frac{g(m_2 - m_1)}{R(m_1 + m_2 + M/2)}$$

Example 7.21 A uniform solid cylinder of radius of 0.2 m and mass of 10 kg is rotating about its central axis. If the angular speed of the cylinder is 5 rad/s:(a) calculate the angular momentum of the cylinder about its central axis; (b) Suppose the cylinder accelerates at a constant rate of 0.5 rad/s^2, find the angular momentum of the cylinder at $t = 3$s(c) find the applied torque; (d) find the work done after 3s.

Solution 7.21 (a) The moment of inertia of the cylinder is

$$I = \frac{1}{2}MR^2 = \frac{1}{2}(10 \text{ kg})(0.2 \text{ m})^2 = 0.2 \text{ kg m}^2$$

for homogeneous symmetrical objects the total angular momentum is

$$L = I\omega = (0.2 \text{ kg m}^2)(5 \text{ rad/s}) = 1 \text{ kg m}^2/\text{s}$$

(b) At $t = 3$ s

$$\omega = \omega_0 + \alpha t = (5 \text{ rad/s}) + (0.5 \text{ rad/s}^2)(3 \text{ s}) = 6.5 \text{ rad/s}$$

at that instant

$$L = I\omega = (0.2 \text{ kg m}^2)(6.5 \text{ rad/s}) = 1.3 \text{ kg m}^2/\text{s}$$

(c)

$$\tau = I\alpha = (0.2 \text{ kg m}^2)(0.5 \text{ rad/s}^2) = 0.1 \text{ N m}$$

Fig. 7.27 A uniform solid sphere rotating about an axis tangent to the sphere

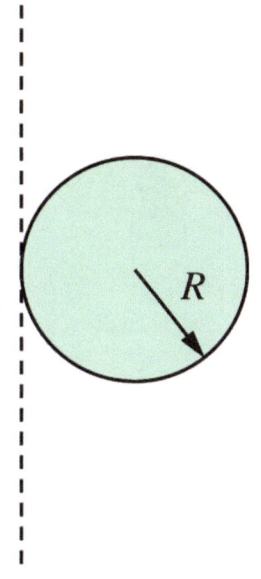

(d)

$$W = \frac{1}{2}I\omega^2 - \frac{1}{2}I\omega_0^2 = \frac{1}{2}(0.2\,\text{kg m}^2)((6.5\,\text{rad/s})^2 - (5\,\text{rad/s})^2) = 1.72\,\text{J}$$

Example 7.22 A uniform solid sphere of radius of 5 cm and mass of 4.7 kg is rotating about an axis that is tangent to the sphere (see Fig. 7.27). If its angular acceleration is given by $\alpha = (4t)\,\text{rad/s}^2$ and if at $t = 0$, $\omega_0 = 0$, find the angular momentum of the sphere and the applied torque as a function of time.

Solution 7.22

$$\omega = \int \alpha dt = \int 4t dt = 2t^2 + c$$

since at $t = 0$, $\omega_0 = 0$ then $c = 0$ and

$$\omega = (2t^2)\,\text{rad/s}$$

The moment of inertia of the sphere is

$$I = \frac{2}{5}MR^2 + MR^2 = \frac{7}{5}MR^2 = \frac{7}{5}(4.7\,\text{kg})(0.05\,\text{m})^2 = 0.016\,\text{kg m}^2$$

and

$$L = I\omega = (0.016\,\text{kg m}^2)((2t^2)\,\text{rad/s}) = (0.03t^2)\,\text{kg m}^2/\text{s}$$

$$\tau = \frac{dL}{dt} = (0.06t)\,\text{N m}$$

Example 7.23 In Example 7.8 find the angular momentum in each case.

Solution 7.23 (a)

$$L = I_z\omega = (0.014\,\text{kg m}^2)(5\,\text{rad/s}) = 0.07\,\text{kg m}^2/\text{s}$$

(b)

$$L = I_y\omega = (0.01\,\text{kg m}^2)(5\,\text{rad/s}) = 0.05\,\text{kg m}^2/\text{s}$$

(c)

$$L = I_x\omega = (4 \times 10^{-3}\,\text{kg m}^2)(5\,\text{rad/s}) = 0.02\,\text{kg m}^2/\text{s}$$

Example 7.24 A uniform solid sphere of radius of 0.2 m is rotating about its central axis with an angular speed of 5 rad/s. If an impulsive force that has an average value of 100 N acts at the rim of the sphere at the center level for a short time of 2 ms:(a) find the angular impulse of the force; (b) the final angular speed of the sphere.

Solution 7.24 (a)

$$\Delta L = \int_{t_1}^{t_2} \tau dt = \tau_{ave}\Delta t = \overline{F}Rt = (100\,\text{N})(0.2\,\text{m})(2 \times 10^{-3}\,\text{s}) = 0.04\,\text{kg m}^2/\text{s}$$

(b)

$$\Delta L = I(\omega_f - \omega_i)$$

$$(0.04\,\text{kg m}^2/\text{s}) = (0.2\,\text{kg m}^2)(\omega_f - (5\,\text{rad/s}))$$

That gives $\omega_f = 5.2\,\text{rad/s}$.

Example 7.25 A man stands on a platform that is free to rotate without friction about a vertical axis as in Fig. 7.28. If the system is initially rotating with an angular speed of 0.3 rev/s: (a) find the final angular speed of the system if the man draws the weights in; (b) find the increase in the kinetic energy of the system and its source. ($I_i = 15\,\text{kg m}^2$ And $I_f = 3\,\text{kg m}^2$).

Solution 7.25 Because the resultant external torque on the system is zero, it follows that the total angular momentum of the system is conserved. That is

$$L_i = L_f$$

$$I_i\omega_i = I_f\omega_f$$

hence

$$\omega_f = \frac{I_i}{I_f}\omega_i = \frac{(15\,\text{kg m}^2/\text{s})}{(3\,\text{kg m}^2/\text{s})}(0.3\,\text{rev/s}) = 1.5\,\text{rev/s}$$

(b)

$$\omega_i = \left(0.3\,\frac{\text{rev}}{\text{s}}\right)\left(\frac{2\pi\,\text{rad}}{1\,\text{rev}}\right) = 1.9\,\text{rad/s}$$

Fig. 7.28 A man stands on a platform that is free to rotate without friction about a vertical axis

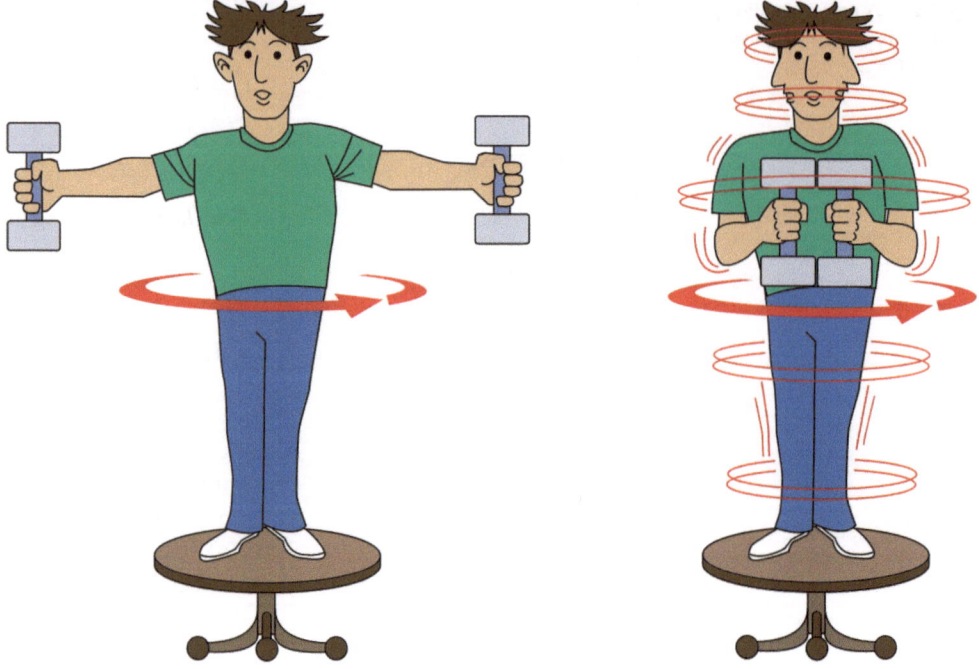

$$\omega_f = \left(1.5 \; \frac{\text{rev}}{\text{s}}\right)\left(\frac{2\pi \; \text{rad}}{1 \; \text{rev}}\right) = 9.4 \; \text{rad/s}$$

$$K_i = \frac{1}{2}I_i\omega_i^2 = \frac{1}{2}(15 \; \text{kg m}^2)(1.9 \; \text{rad/s})^2 = 27 \; \text{J}$$

$$K_f = \frac{1}{2}I_f\omega_f^2 = \frac{1}{2}(3 \; \text{kg m}^2)(9.4 \; \text{rad/s})^2 = 132.5 \; \text{J}$$

This increase in the kinetic energy is because the man does work when he moves the dumbbells inwards.

Example 7.26 A uniform disc of moment of inertia of 0.1 kg m^2 is rotating without friction with an angular speed of 3 rad/s about an axle passing through its center of mass as in Fig. 7.29. When another disc of moment of inertia of 0.05 kg m^2 that is initially at rest is dropped on the first, the two will eventually rotate with the same angular speed due to friction between them. Determine (a) the final angular speed; (b) the change in the kinetic energy of the system.

Solution 7.26 (a) Since the net external torque acting on the system is zero, it follows that the total angular momentum of the system is conserved, i.e.,

$$L_i = L_f$$

or

$$I_1\omega_1 = (I_1 + I_2)\omega$$

hence

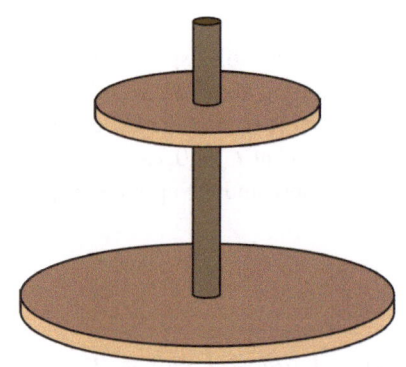

Fig. 7.29 A uniform disc rotating without friction. Another disc that is initially at rest is dropped on the first, the two will eventually rotate with the same angular speed due to friction between them

$$\omega = \frac{I_1\omega_1}{(I_1 + I_2)} = \frac{(0.1 \; \text{kg m}^2)(3 \; \text{rad/s})}{(0.15 \; \text{kg m}^2)} = 2 \; \text{rad/s}$$

(b)

$$K_i = \frac{1}{2}I_1\omega_1^2 = \frac{1}{2}(0.1 \; \text{kg m}^2)(3 \; \text{rad/s})^2 = 0.45 \; \text{J}$$

$$K_f = \frac{1}{2}(I_1 + I_2)\omega^2 = \frac{1}{2}(0.15 \; \text{kg m}^2)(2 \; \text{rad/s})^2 = 0.3 \; \text{J}$$

This decrease in kinetic energy is due to the internal nonconservative (frictional) force that acts within the system.

Problems

1. A wheel is initially rotating at 60 rad/s in the clockwise direction. If a counterclockwise torque acts on the wheel

producing a counterclockwise angular acceleration $\alpha = 2t$ rad/s^2, find the time required for the wheel to reverse its direction of motion.

2. If the angular position of a point on a rotating wheel is given by $\theta = 2t + 5t^2$ rad, find the angular speed and angular acceleration of the point at $t = 2$ s.

3. A wheel of radius of 0.5 m rotates from rest at a constant angular acceleration of 2.5 rad/s^2. At $t = 2$ s Find (a) the angular speed of the wheel (b) the angle in radians through which the wheel rotates (c) the tangential and radial acceleration of a point at the rim of the wheel.

4. Find the angular speed in radians per second of the earth about (a) its axis (b) the sun.

5. An L-shaped bar rotates counterclockwise with an angular acceleration of ω (see Fig. 7.30). Find (in vector form) the linear velocity and acceleration of the point P on the bar.

6. Four masses are connected by light rigid rods as in Fig. 7.31. Calculate the moment of inertia of the system about (a) the x-axis (b) the y-axis (c) the z-axis.

7. Find the moment of inertia of a uniform solid sphere of radius R and mass M about an axis passing through its center of mass.

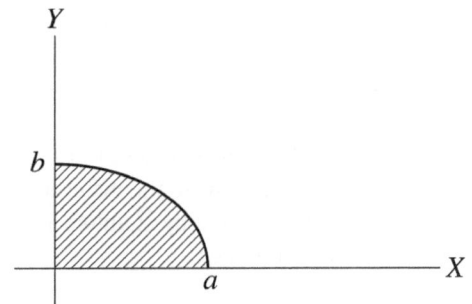

Fig. 7.32 An elliptical quadrant

Fig. 7.33 A uniform rod of length L and mass M is pivoted at O. A projectile of mass m moving at velocity v collides with the rod and sticks to it

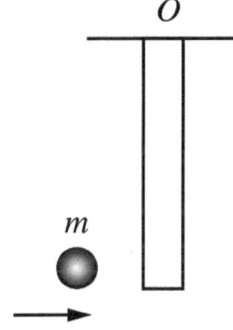

8. Find the moment of inertia of an elliptical quadrant about the y-axis (see Fig. 7.32).

9. A 5 kg uniform solid cylinder of radius 0.2 m rotate about its center of mass axis with an angular speed of 10 rev/min. Find (a) its rotational kinetic energy (b) its angular momentum.

10. A wheel of mass of 20 kg and radius of 0.75 m is initially rotating at 120 rev/min. If its angular speed is increased to 300 rev/min in 20 s, find (a) the work done on the wheel (b) the average power delivered to the wheel.

11. A wheel of mass 10 kg and radius 0.4 m accelerates uniformly from rest to an angular speed of 800 rev/min in 20 s. Find (a) the torque applied to the wheel (b) the work done on the wheel (c) the work done using the work–energy theorem.

12. A uniform rod of length L and mass M is pivoted at O (see Fig. 7.33). If a projectile of mass m moving at velocity v collide with the rod and stick to it, find the angular momentum of the system immediately before and immediately after the collision.

13. A disc of radius 2.2 m and mass of 120 kg rotate about a frictionless vertical axle that passes through its center. A man of mass 65 kg walks slowly from the rim of the disc towards the center. Find the angular speed of the disc when the man is at a distance of 0.7 m from the center if its angular speed when the man starts walking is 1.6 rad/s.

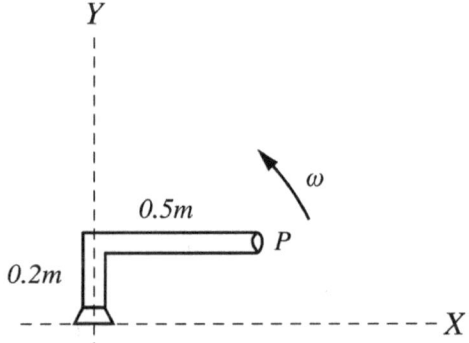

Fig. 7.30 An L-shaped bar rotating counterclockwise

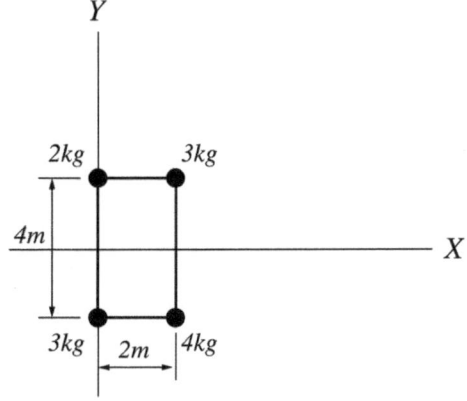

Fig. 7.31 Four masses connected by light rigid rods

Open Access This chapter is licensed under the terms of the Creative Commons Attribution 4.0 International License (http://creativecommons.org/licenses/by/4.0/), which permits use, sharing, adaptation, distribution and reproduction in any medium or format, as long as you give appropriate credit to the original author(s) and the source, provide a link to the Creative Commons license and indicate if changes were made.

The images or other third party material in this chapter are included in the chapter's Creative Commons license, unless indicated otherwise in a credit line to the material. If material is not included in the chapter's Creative Commons license and your intended use is not permitted by statutory regulation or exceeds the permitted use, you will need to obtain permission directly from the copyright holder.

8.1 Rolling Motion

Rolling motion represents the general plane motion of a rigid body It can be considered as a combination of pure translational motion parallel to a fixed plane plus a pure rotational motion about an axis that is perpendicular to that plane. The axis of rotation usually passes through the center of mass. In Sect. 6.4, we've seen that the motion of an object (or a system of particles) can always be considered as a combination of the motion of the object relative to its center of mass plus the motion of its center of mass relative to some origin O. From Sect. 6.4.3, the kinetic energy of an object relative to the origin is

$$K = \frac{1}{2}\sum_i m_i v_i'^2 + \frac{1}{2}M v_{cm}^2 \qquad (8.1)$$

where v_{cm} is the velocity of the center of mass of the object relative to the origin O, m_i is the mass of the ith particle and v_i' is the linear velocity of the ith particle relative to the center of mass. In the case of the general plane motion of a rigid body, the motion can be considered as a combination of pure translational motion of the center of mass plus pure rotational motion about an axis passing through the center of mass and perpendicular to the plane of motion. Therefore, the first term in Eq. 8.1 can be written as

$$v_i' = \omega r_i'$$

where r_i' is the perpendicular distance from the ith particle to the center of mass axis. Hence

$$K = \frac{1}{2}\left(\sum_i m_i r_i'^2\right)\omega^2 + \frac{1}{2}M v_{cm}^2$$

$$K = \frac{1}{2}I_{cm}\omega^2 + \frac{1}{2}M v_{cm}^2$$

Thus, the total kinetic energy of a rolling object is the sum of the translational kinetic energy of its center of mass and the rotational kinetic energy about its center of mass.

8.2 Rolling Without Slipping

An important special case of the general plane motion is rolling without slipping. Such motion occurs if a perfectly rigid body rolls on a perfectly rigid surface. As the object rolls without slipping, the instantaneous s' point of contact between the object and the surface is at rest relative to the surface since there is no slipping. Now, consider a wheel of radius R rolling without slipping along the straight track shown in Fig. 8.1. The center of mass of the wheel moves along a straight line, while a point on the rim such as P moves in a cycloid path. As the wheel rotates through an angle θ, its center of mass moves through a distance equal to the arc length s (see Fig. 8.2) given by

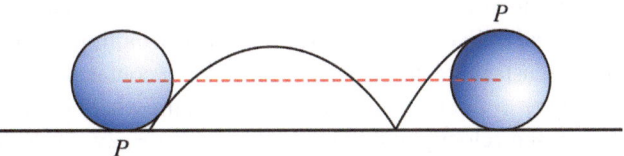

Fig. 8.1 A wheel of radius R rolling without slipping along the straight track

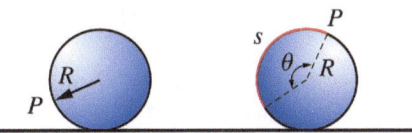

Fig. 8.2 As the wheel rotates through an angle θ, its center of mass moves through a distance equal to the arc length s

© The Author(s) 2019
S. Alrasheed, *Principles of Mechanics*, Advances in Science,
Technology & Innovation, https://doi.org/10.1007/978-3-030-15195-9_8

Fig. 8.3 The combination of pure rotational and translational motions

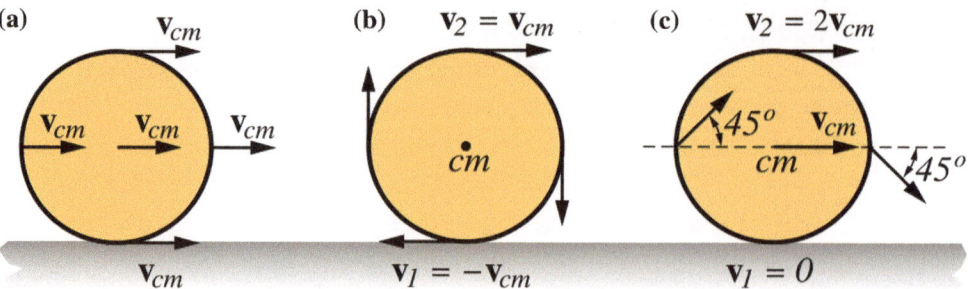

$$s = R\theta$$

Hence, the speed of the center of mass is

$$v_{cm} = \frac{ds}{dt} = R\frac{d\theta}{dt} = R\omega$$

The acceleration of the center of mass is given by

$$a_{cm} = \frac{dv_{cm}}{dt} = R\frac{d\omega}{dt} = R\alpha$$

The combination of pure rotational and translational motions is viewed in Fig. 8.3. In the pure translational motion (see Fig. 8.3 part a) every particle in the wheel moves with the velocity $\mathbf{v}_{cm}$. In pure rotational motion (see Fig. 8.3 part b), each particle moves with an angular speed ω about the center of mass axis and the linear speed of any particle at the rim is

$$v_{cm} = R\omega \qquad (8.2)$$

The resulting motion of these two combined motions is shown in Fig. 8.3 part c, where the linear velocity of each particle is the vector sum of its linear velocity in pure translational motion and its linear velocity in pure rotational motion. Therefore, the instantaneous velocity of the point of contact is equal to zero ($\mathbf{v}_1 = 0$) and of a point at the top of the wheel is equal to twice the velocity of the center of mass ($\mathbf{v}_2 = 2\mathbf{v}_{cm}$). Note that Eq. 8.2 is valid only in the special case of rolling without slipping; in the general rolling motion this equation does not hold. The total kinetic energy of a rigid object rolling without slipping is therefore given by

$$K = \frac{1}{2}I_{cm}\omega^2 + \frac{1}{2}Mv_{cm}^2$$

$$= \frac{1}{2}I_{cm}\omega^2 + \frac{1}{2}MR^2\omega^2$$

Another way to view rolling without slipping is to consider the wheel to be in pure rotational motion about an instantaneous axis that passes through the point of contact P (see Fig. 8.4). In that case, the velocity of the point of contact P is zero and

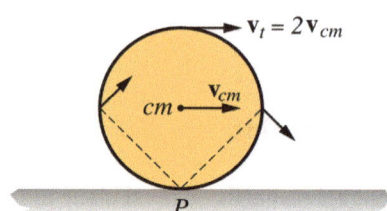

Fig. 8.4 Another way to view rolling without slipping is to consider the wheel to be in pure rotational motion about an instantaneous axis that passes through the point of contact P

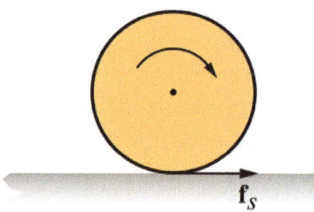

Fig. 8.5 A statistical frictional force acts on it at the instantaneous point of contact producing a torque about the center

the velocity of the center of mass is $v_{cm} = R\omega$ (since it is at a distance R from the axis of rotation) and the velocity of a point at the top is $v_t = 2R\omega = 2v_{cm}$. Note that the angular velocity ω of the wheel is the same as its angular velocity if the axis of rotation is at the center of mass.

For simplicity, only homogeneous symmetrical objects will be considered here such as hoops, cylinders, and spheres. When a rigid body rolls without slipping with a constant speed, there will be no frictional force acting on the body at the instantaneous point of contact. However, if the object is accelerating, then a statistical frictional force acts on it at the instantaneous point of contact producing a torque about the center (see Fig. 8.5). This will cause the object to rotate about its center of mass. The direction of the statistical force opposes the tendency of the object to slide. For example, if a wheel is rolling down an incline, the direction of the frictional force will be opposing the downward motion.

In most situations, the body and the surface are not perfectly rigid. As a result, the normal force would not be a single force; rather it would be a number of forces that are distributed over the area of contact (see Fig. 8.6). Therefore, each normal force will exert an opposing torque since its line of action will

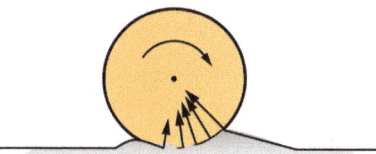

Fig. 8.6 If the body and the surface are not perfectly, the normal force would not be a single force; rather it would be a number of forces that are distributed over the area of contact

not pass through the center of mass. Furthermore, as the object rolls over the surface, both the object and the surface undergo deformation resulting in a loss in the mechanical energy.

Example 8.1 A uniform solid hoop of mass of 32 kg and radius of 1.2 m rolls without slipping on a horizontal track where the center of mass speed is 2 m/s. Find: (a) the total energy of the hoop and compare it with its total energy if it would slide without rolling; (b) the speed of the hoop at its top and bottom.

Solution 8.1 (a) the total energy is given by

$$K = \frac{1}{2}I_{cm}\omega^2 + \frac{1}{2}Mv_{cm}^2$$

$$= \frac{1}{2}(MR^2)\left(\frac{v_{cm}}{R}\right)^2 + \frac{1}{2}Mv_{cm}^2 = Mv_{cm}^2 = (32\ \text{kg})(2\ \text{m/s})^2 = 128\ \text{J}$$

If the hoop slides without rolling its total kinetic energy is $\frac{1}{2}Mv_{cm}^2$, that is, its value is half of that if the hoop were to roll without slipping.

(b)

$$v_{\text{top}} = 2v_{cm} = 2(2\ \text{m/s}) = 4\ \text{m/s}$$

$$v_{\text{bottom}} = 0$$

Example 8.2 A uniform solid cylinder, sphere, and hoop roll without slipping from rest at the top of an incline (see Fig. 8.7). Find out which object would reach the bottom first.

Solution 8.2 For each object, we have

$$K_i + U_i = K_f + U_f$$

$$0 + Mgh = \frac{1}{2}Mv_{cm}^2 + \frac{1}{2}I_{cm}\left(\frac{v_{cm}}{R}\right)^2$$

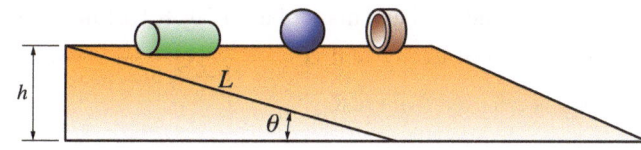

Fig. 8.7 A uniform solid cylinder, sphere and hoop roll without slipping from rest at the top of an incline

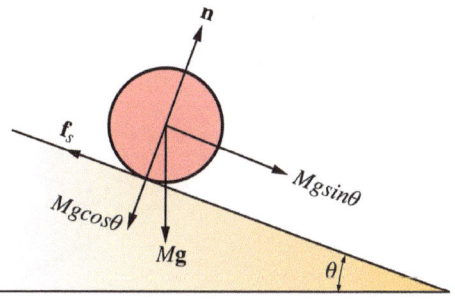

Fig. 8.8 A marble ball of radius R and mass M rolls without slipping down the incline

$$v_{cm} = \sqrt{\frac{2gh}{1 + I_{cm}/MR^2}}$$

Hence, the speed of the center of mass of any object at the bottom of the incline does not depend on its mass or size; it depends only on its shape. Therefore, all objects of the same shape such as spheres (of any mass or size) have the same speed at the bottom. That is, the smaller the ratio I_{cm}/MR^2 the faster the object moves since less of its energy goes to rotational kinetic energy and more goes to translational kinetic energy The ratio I_{cm}/MR^2 is equal to 0.4, 0.5, and 1 for a sphere, cylinder, and hoop, respectively Therefore, these objects will finish in the order of any sphere, any cylinder, and any hoop.

Example 8.3 A marble ball of radius R and mass M rolls without slipping down the incline shown in Fig. 8.8. Find: (a) its acceleration; (b) the minimum coefficient of static friction that is required to prevent slipping.

Solution 8.3 (a) Applying Newton's second law in both linear and angular form (see Fig. 8.7) we have

$$\sum F_x = Mg\ \sin\theta - f_s = Ma_{cm} \qquad (8.3)$$

$$\sum F_y = n - Mg\cos\theta = 0$$

and

$$\sum \tau = f_s R = I_{cm}\alpha = \left(\frac{2}{5}MR^2\right)\left(\frac{a_{cm}}{R}\right)$$

that gives

$$f_s = \frac{2}{5}Ma_{cm} \qquad (8.4)$$

Substituting Eq. 8.4 into Eq. 8.3 gives

$$Mg\sin\theta - \frac{2}{5}Ma_{cm} = Ma_{cm}$$

hence

Fig. 8.9 A string wrapped around a uniform solid cylinder of radius of R and mass of M

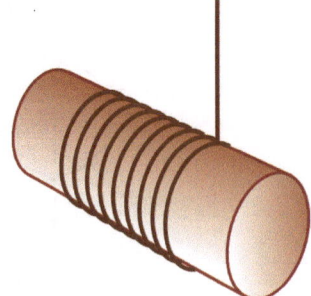

$$a_{cm} = \frac{5}{7} g \sin \theta$$

and

$$f_s == \frac{2}{7} Mg \sin \theta$$

(b) At the verge of slipping, the statistical frictional force is a maximum given by

$$f_{s\,max} = \mu_s n = \frac{2}{7} Mg \sin \theta$$

Hence, the coefficient of static friction must be at least as great as $\mu_s = \frac{2}{7} \tan \theta$ in order for the ball not to slip.

Example 8.4 A string is wrapped around a uniform solid cylinder of radius of R and mass of M as in Fig. 8.9. If the cylinder is released from rest while the string is fixed in place and assuming that the string does not slip at the cylinder's surface, find: (a) the acceleration of the center of mass using Newton's laws (b) the acceleration of the center of mass using energy methods if the cylinder descends a distance h (c) the tension in the string.

Solution 8.4 (a) Applying Newton's second law in both the linear and angular form gives

$$\sum F_y = T - Mg = -Ma_{cm} \qquad (8.5)$$

$$\sum \tau = TR = I_{cm}\alpha = \frac{1}{2} MR^2 (\frac{a_{cm}}{R})$$

hence

$$T = \frac{1}{2} Ma_{cm} \qquad (8.6)$$

Substituting Eq. 8.6 into Eq. 8.5 gives

$$-M_9 + \frac{1}{2} Ma_{cm} = -Ma_{cm}$$

that gives

$$a_{cm} = \frac{2}{3} g$$

(b) Energy Method

$$K_i + U_i = K_f + U_f$$

$$0 + Mgh = \frac{1}{2} Mv_{cm}^2 + \frac{1}{2} I_{cm} \omega^2$$

$$0 + Mgh = \frac{1}{2} Mv_{cm}^2 + \frac{1}{2} \left(\frac{1}{2} MR^2\right)\left(\frac{v_{cm}}{R}\right)^2$$

that gives

$$v_{cm} = \sqrt{\frac{4}{3} gh}$$

From the expression $v^2 = v_0^2 + 2a_{cm}h$, and since $v_0 = 0$ we have

$$a_{cm} = \frac{v_{cm}^2}{2h} = \frac{4gh}{3(2h)} = \frac{2}{3} g$$

(b) From Eq. 8.6,

$$T = \frac{1}{2} Ma_{cm} = \frac{1}{2} M\left(\frac{2}{3} g\right) = \frac{1}{3} Mg$$

Example 8.5 A uniform solid sphere of radius R and mass M is released from rest at the top of an incline at a distance h above the ground. If it rolls without slipping, find the speed of the center of mass at the bottom of the incline.

Solution 8.5

$$K_i + U_i = K_f + U_f$$

$$0 + Mgh = \frac{1}{2} Mv_{cm}^2 + \frac{1}{2} I_{cm} \omega^2$$

$$0 + Mgh = \frac{1}{2} Mv_{cm}^2 + \frac{1}{2} \left(\frac{2}{5} MR^2\right)\left(\frac{v_{cm}}{R}\right)^2$$

That gives

$$v_{cm} = \sqrt{\frac{10}{7} gh}$$

Example 8.6 A block of mass m is attached to a light string that passes over a light pulley and is connected to a uniform solid sphere of radius R and mass M as in Fig. 8.10. Show that the acceleration of the system is $a = \dfrac{g}{1 + 7/5(M/m)}$ when the block is released from rest.

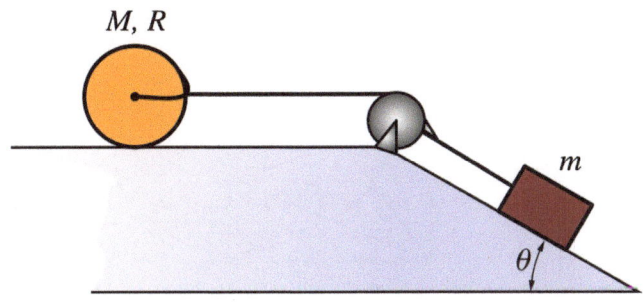

Fig. 8.10 A block of mass m is attached to a light string that passes over a light pulley connected to a uniform solid sphere of radius R and mass M

Solution 8.6 From conservation of energy, we have

$$mgh = \frac{1}{2}Mv_{cm}^2 + \frac{1}{2}I_{cm}\omega^2 + \frac{1}{2}mv^2$$

Since the block and the sphere are connected, they have the same speed, therefore

$$mgh = \frac{1}{2}Mv^2 + \frac{1}{2}\left(\frac{2}{5}MR^2\right)\left(\frac{v^2}{R}\right)^2 + \frac{1}{2}mv^2$$

Therefore, the speed of the system when the block is at the bottom of the incline is

$$v = \sqrt{\frac{2gh}{1 + 7M/5m}}$$

The acceleration of the system is

$$v^2 - v_0^2 = 2ah$$

or

$$a = \frac{v^2}{2h} = \frac{2gh}{2h(1 + 7/5(M/m))}$$

that gives

$$a = \frac{g}{(1 + 7/5(M/m))}$$

8.3 Static Equilibrium

An extended object is said to be in equilibrium if two conditions are satisfied. First, the net external force acting on the object must be equal to zero. Second, the net external torque on the object about any origin must also be equal to zero. In other words, an object is in equilibrium if its total linear momentum and its total angular momentum (about any origin) are constants. Only the first condition is necessary if the object can be treated as a particle. Thus, the conditions of equilibrium may be written as

$$\sum \mathbf{F} = 0 \ (Translational \ Equilibrium) \qquad (8.7)$$

$$\sum \boldsymbol{\tau} = 0 \ (Rotational \ Equilibrium) \qquad (8.8)$$

In terms of components, we may write

$$\sum F_x = 0, \ \sum F_y = 0, \ \sum F_z = 0 \qquad (8.9)$$

$$\sum \tau_x = 0, \ \sum \tau_y = 0, \ \sum \tau_z = 0 \qquad (8.10)$$

An object is said to be in static equilibrium if it is at rest (there isn't any kind of motion with respect to our inertial frame of reference). Now consider the case in which all external forces acting on the object lie in the same plane (for example the x–y plane). Such forces are called coplanar forces. The net external torque due to these forces is then perpendicular to the x–y plane and parallel to the z-axis. Equations 8.9 and 8.10 are, therefore, reduced to

$$\sum F_x = 0, \ \sum F_y = 0, \ \sum \tau_z = 0$$

Next, we will prove that if the object is in translational equilibrium where ($\Sigma \mathbf{F} = \mathbf{0}$) and the net external torque on the object is equal to zero about some origin, it is also equal to zero about any other origin. Note that the origin may be chosen anywhere inside or outside the object. Suppose that a number of forces $\mathbf{F}_1, \mathbf{F}_2, \mathbf{F}_3, \ldots \mathbf{F}_n$ are acting on a rigid object at different points (see Fig. 8.11) and that the object is in translational equilibrium. The point of application of $\mathbf{F}_1$ relative to O is r_1 and of $\mathbf{F}_2$ is r_2 and so on. The net external torque about O is given by

$$\sum \boldsymbol{\tau}_0 = \boldsymbol{\tau}_1 + \boldsymbol{\tau}_2 + \cdots + \boldsymbol{\tau}_n = \mathbf{r}_1 \times \mathbf{F}_1 + \mathbf{r}_2 \times \mathbf{F}_2 + + \cdots \mathbf{r}_n \times \mathbf{F}_n$$

The net external torque about O' (see Fig. 8.12) is

$$\sum \boldsymbol{\tau}_{0'} = \boldsymbol{\tau}_1' + \boldsymbol{\tau}_2' + + \cdots \boldsymbol{\tau}_n' = \mathbf{r}_1' \times \mathbf{F}_1 + \mathbf{r}_2' \times \mathbf{F}_2 + + \cdots \mathbf{r}_n' \times \mathbf{F}_n$$

$$= (\mathbf{r}_1 - \mathbf{r}_{0'}) \times \mathbf{F}_1 + (\mathbf{r}_2 - \mathbf{r}_{0'}) \times \mathbf{F}_2 + \cdot + \cdots (\mathbf{r}_n - \mathbf{r}_{0'}) \times \mathbf{F}_n$$

$$= \mathbf{r}_1 \times \mathbf{F}_1 + \mathbf{r}_2 \times \mathbf{F}_2 + . + \cdots \mathbf{r}_n \times \mathbf{F}_n - (\mathbf{r}_{0'} \times \mathbf{F}_1 + \mathbf{r}_{0'} \times \mathbf{F}_2 + + \cdots \mathbf{r}_{0'} \times \mathbf{F}_n)$$

$$= \sum \boldsymbol{\tau}_0 - (\mathbf{r}_{0'} \times (\mathbf{F}_1 + \mathbf{F}_2 + + \mathbf{F}_n)) = \sum \boldsymbol{\tau}_0 - (\mathbf{r}_{0'} \times \sum \mathbf{F})$$

Since $\Sigma \mathbf{F} = \mathbf{0}$ we have

$$\sum \boldsymbol{\tau}_{0'} = \sum \boldsymbol{\tau}_0$$

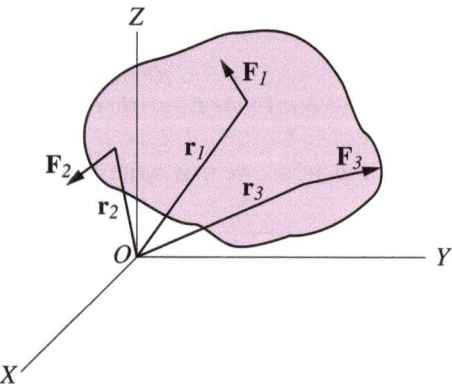

Fig. 8.11 A number of forces $\mathbf{F}_1, \mathbf{F}_2, \mathbf{F}_3, ..\mathbf{F}_n$ act on a rigid object at different points

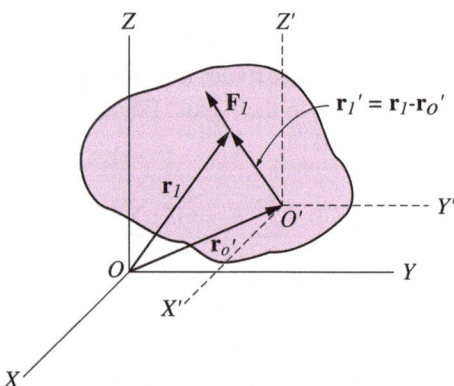

Fig. 8.12 The net external torque on the object about O$\breve{o}$

8.4   The Center of Gravity

The resultant gravitational force acting on an object is the resultant of the individual gravitational forces acting on different mass elements of the object (see Fig. 8.13), i.e.,

$$\sum \mathbf{F} = \sum m_i \mathbf{g} \tag{8.11}$$

This force can be replaced by a single force that is equal to the weight of the object ($M\mathbf{g}$) and that acts at a single point called the center of gravity Now consider an object that is near the earth's surface where the force of gravity is assumed to be constant over that range. Equation 8.11 becomes

$$\sum \mathbf{F} = \sum m_i \mathbf{g} = \mathbf{g} \sum m_i = M\mathbf{g} = \mathbf{w}$$

To locate the center of gravity, let us calculate the net torque acting on an object about an origin due to gravity This torque is the vector sum of the individual torques acting on different mass elements. That is,

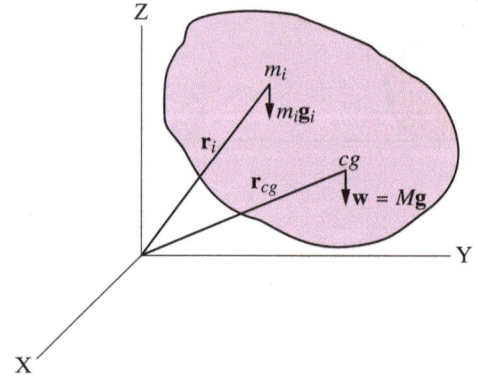

Fig. 8.13 The resultant gravitational force acting on an object is the resultant of the individual gravitational forces acting on different mass elements of the object

$$\boldsymbol{\tau} = \sum_i \boldsymbol{\tau}_i = \sum_i (\mathbf{r}_i \times m_i \mathbf{g}) = \left(\sum_i m_i \mathbf{r}_i \right) \times \mathbf{g}$$

$$\boldsymbol{\tau} = \frac{\left(\sum_i m_i \mathbf{r}_i \right)}{M} \times M\mathbf{g} = \mathbf{r}_{cm} \times \mathbf{w}$$

$$\boldsymbol{\tau} = \mathbf{r}_{cm} \times \mathbf{w}$$

Therefore, we conclude that if the gravitational field (g) is constant over the body, the center of gravity of the object coincides with its center of mass.

Example 8.7   Two blocks of masses $m_2 = 20$ kg and $m_1 = 10$ kg are supported by a uniform horizontal beam of length $L = 1.5$m and mass $M = 6$ kg (see Fig. 8.14). Find: (a) the normal force exerted by the fulcrum (supporting point) on the beam if it is placed under the center of gravity of the beam; (b) the distance x in which m_2 must be placed in order for the system to be balanced.

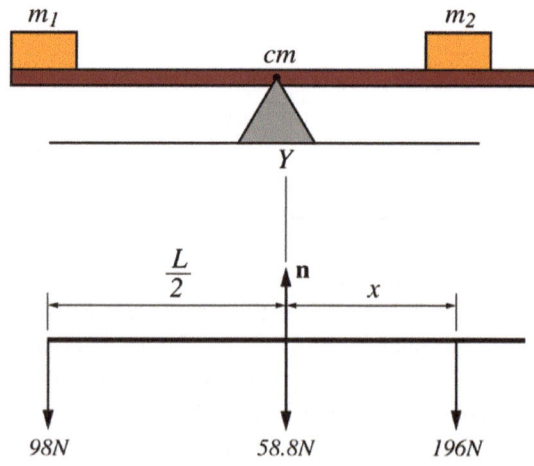

Fig. 8.14 Two blocks supported by a uniform horizontal beam

Fig. 8.15 The free-body diagram of a ladder of length L and mass $M = 20$ kg resting against a smooth vertical wall

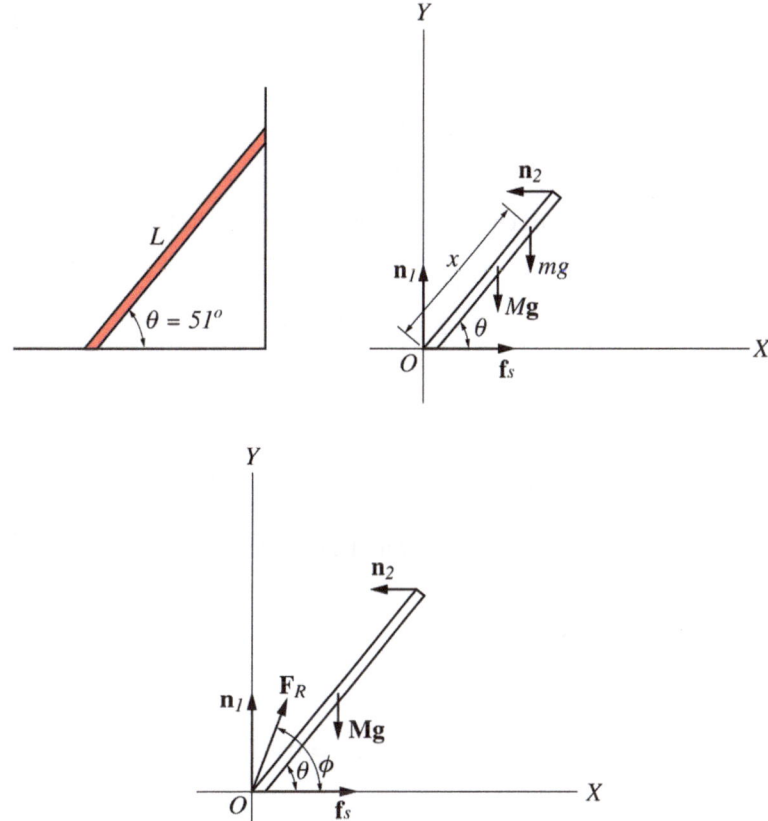

Solution 8.7 (a) The free-body diagram of the system in shown in Fig. 8.14 where $w_1 = 196$ N, $w_2 = 98$ N, and $w = 58.8$ N. Applying Newton's second law to the beam gives

$$\sum F_y = n - (59 \text{ N}) - (98 \text{ N}) - (196 \text{ N}) = 0$$

and

$$n = 353 \text{ N}$$

(b) The net external torque about an axis passing through the center of the beam and perpendicular to the page is

$$\sum \tau_z = (98 \text{ N})(0.75 \text{ m}) - (196 \text{ N})x = 0$$

$$x = 0.37 \text{ m}$$

Example 8.8 A ladder of length L and mass $M = 20$ kg rests against a smooth vertical wall as shown in Fig. 8.15. If the center of gravity of the ladder is at a distance of $L/3$ from the base, determine: (a) the minimum coefficient of static friction such that the ladder does not slip; (b) the magnitude and direction of the resultant of the contact forces acting on the ladder at the base; (c) if a man of mass of 70 kg climbs up

the ladder, what is the maximum distance the man can climb before the ladder slips if $\mu_s = 0.4$.

Solution 8.8 (a) Figure 8.15 shows the free-body diagram of the ladder. Applying Newton's second law to the ladder gives

$$\sum F_x = f_s - n_2 = 0$$

$$f_s = n_2$$

and

$$\sum F_y = n_1 - Mg = 0$$

$$n_1 = Mg$$

Applying Newton's second law in angular form about O (the point must be chosen to give minimum unknowns) we have

$$\sum \tau_z = n_2 L \sin \theta - \frac{1}{3} M g L \cos \theta = 0 \qquad (8.12)$$

If the ladder is at the verge of slipping the statistical frictional force is maximum $f_s = \mu_s n_1$. From Eq. 8.12, we have

$$n_2 = \frac{Mg}{3\tan\theta} = \frac{(196\text{ N})}{3\tan(51°)} = 53\text{ N} = f_s$$

hence

$$\mu_s = \frac{f_s}{n_1} = \frac{(53\text{ N})}{(196\text{ N})} = 0.27$$

(b) The resultant of the contact forces on the ladder at the base is

$$F_R = \sqrt{f_s^2 + n_1^2} = \sqrt{(53\text{ N})^2 + (196)^2}\text{ N} = 203\text{ N}$$

the direction of F_R is

$$\phi = \tan^{-1}\frac{n_1}{f_s} = \tan^{-1}\frac{(196\text{ N})}{(52.9\text{ N})} = 75°$$

(c) The free-body diagram is shown in Fig. 8.15. From the equilibrium condition, we have

$$\sum F_x = f_s - n_2 = 0$$

and

$$\sum F_y = n_1 - mg - Mg = 0$$

or

$$f_s = n_2$$

and

$$n_1 = (m + M)g$$

Furthermore, the resultant external torque about O is

$$\sum \tau_z = n_2 L \sin\theta - \frac{1}{3}MgL\cos\theta - mgx\cos\theta = 0$$

thus

$$n_2 = \frac{g}{\tan\theta}\left(\frac{M}{3} + m\left(\frac{x}{L}\right)\right)$$

at the verge of slipping

$$f_s = \mu_s n_1 = \mu_s g(M + m) = (0.4)(9.8\text{ m/s}^2)(90\text{ kg}) = 353\text{ N} = n_2$$

Hence

$$x = 0.54 L$$

Example 8.9 A uniform beam of weight w and length L is held by two supports as in Fig. 8.16. A block of weight w_1 is resting on the beam at a distance of $L/6$ from the center of gravity of the beam. Find the magnitude of the forces exerted by the supports on the beam.

Solution 8.9 The free-body diagram of the system is shown in Fig. 8.16. Because the beam has a uniform density its cen-

ter of mass and gravity are located at its geometrical center. Applying Newton's second law gives

$$\sum F_y = 0$$

$$F_2 + F_1 - w - w_1 = 0 \tag{8.13}$$

Taking the torque about an axis passing through one end (at F_1) gives

$$\sum \tau_z = 0$$

$$F_2 L - \frac{2}{3}Lw_1 - \frac{L}{2}w = 0 \tag{8.14}$$

From Eqs. 8.13 and 8.14 we have

$$F_2 = \frac{2}{3}w_1 + \frac{w}{2}$$

and

$$F_1 = \frac{w_1}{3} + \frac{w}{2}$$

Example 8.10 A man of mass of 80 kg is standing at the end of a uniform beam of mass of 30 kg and length of 12 m as shown in Fig. 8.17. Find the tension in the rope and the reaction force exerted by the hinge on the beam.

Solution 8.10 (a) The free-body diagram is shown in Fig. 8.17. Applying Newton's second law to the beam gives

$$\sum F_y = T\sin 50° + F_R \sin\theta - (294\text{ N}) - (784\text{ N}) = 0$$

$$\sum F_x = F_R \cos\theta - T\cos 50° = 0$$

The resultant torque about an axis passing through O is

$$\sum \tau_z = T\sin 50° L - L(784\text{ N}) - \frac{L}{2}(294\text{ N}) = 0$$

That gives $T = 1215.3\text{ N}$. Hence

$$F_R \cos\theta = T\cos 50° = (1215.3\text{ N})(0.64) = 781.2\text{ N} \tag{8.15}$$

and

$$F_R \sin\theta = -T\sin 50° + (294\text{ N}) + (784\text{ N})$$
$$= -(1215.3\text{ N})(0.76) + (294\text{ N}) + (784\text{ N}) = 147\text{ N} \tag{8.16}$$

Dividing Eq. 8.16 by Eq. 8.15 gives

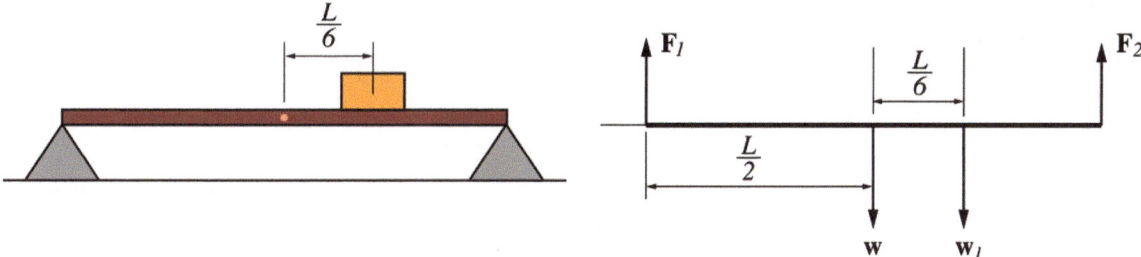

Fig. 8.16 A uniform beam of weight w and length L balanced by two supports

Fig. 8.17 A man standing at the end of a uniform beam

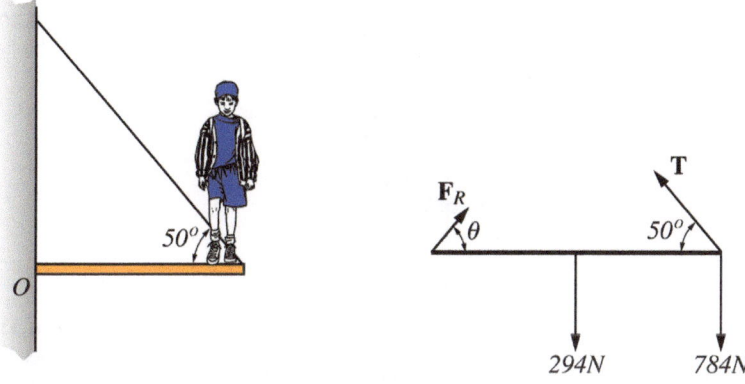

$$\tan \theta = \frac{(147 \text{ N})}{(781.2 \text{ N})} = 0.2$$

$$\theta = 10.6°$$

and

$$F_R = \sqrt{(147)^2 + (7812)^2} = 795 \text{ N}$$

Example 8.11 A uniform beam of weight of 120 N and length of L is in horizontal static equilibrium as in Fig. 8.18. Neglecting the masses of the ropes, find the tension in each string. (The center of mass is at $L/3$ from one end).

Solution 8.11 The free-body diagram is shown in Fig. 8.18. Applying Newton's second law to the beam gives

$$\sum F_y = T_1 \cos \theta + T_2 \cos 30° - (120 \text{ N}) = 0$$

or

$$T_1 \cos \theta + T_2(0.87) = (120 \text{ N}) \quad (8.17)$$

Also

$$\sum F_x = T_1 \sin \theta - T_2 \sin 30° = 0$$

or

$$T_1 \sin \theta = T_2 \sin 30° \quad (8.18)$$

Taking the resultant torque on the beam about one end (at T_1) gives

$$\sum \tau = (120 \text{ N})\frac{L}{3} - LT_2 \cos 30° = 0$$

or

$$T_2 = 46.2 \text{ N}$$

Substituting T_2 into Eqs. 8.18 and 8.17 gives

$$T_1 \sin \theta = (46.2 \text{ N}) \sin 30° = 23.1 \text{ N}$$

and

$$T_1 \cos \theta + (46.2 \text{ N})(0.87) = (120 \text{ N})$$

$$T_1 \cos \theta = 80 \text{ N}$$

Hence

$$\tan \theta = \frac{(23.1 \text{ N})}{(80 \text{ N})} = 0.3$$

That gives $\theta = 16.7°$ and $T_1 = (23.1 \text{ N})/\sin 16.7° = 80.3 \text{ N}$.

Example 8.12 A solid sphere of mass of 12 kg is in static equilibrium inside the wedge shown in Fig. 8.19. If the surface of the wedge is frictionless, find the forces that the wedge exerts on the sphere.

Fig. 8.18 A uniform beam held by ropes in static equilibrium

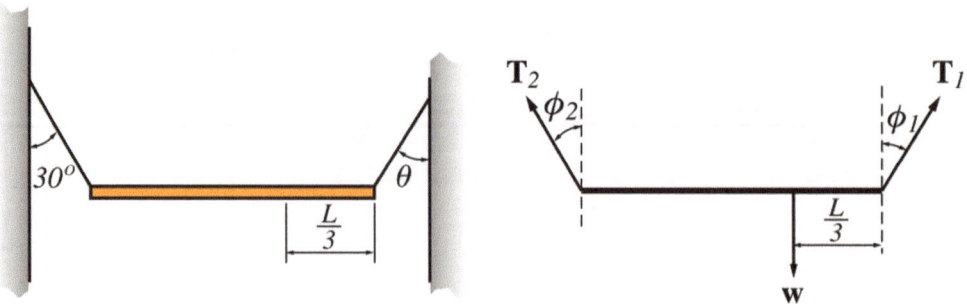

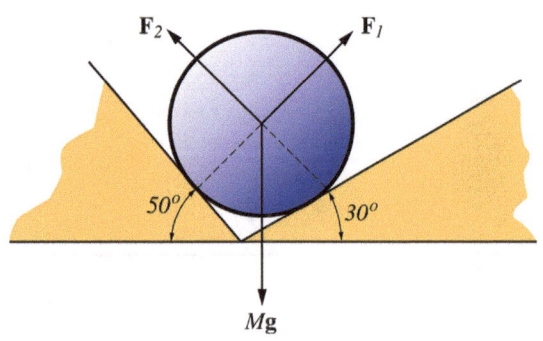

Fig. 8.19 A solid sphere in static equilibrium inside a wedge

Fig. 8.20 A block suspended by a cable attached to a uniform rod

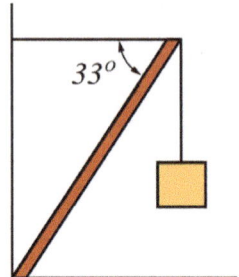

Fig. 8.21 A uniform sphere suspended by a light string and leaning on a frictionless wall

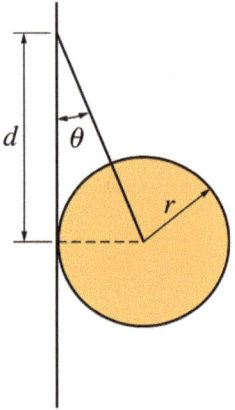

Solution 8.12 Applying Newton's second law gives

$$\sum F_x = F_1 \sin 50° - F_2 \sin 30° = 0$$

or

$$F_1 = 0.65 F_2$$

Also we have

$$\sum F_y = F_1 \cos 50° + F_2 \cos 30° - Mg = 0$$

or

$$0.65 F_2 \cos 50° + F_2 \cos 30° - Mg = 0$$

That gives $F_2 = 91.6$ N. Therefore

$$F_1 = 0.65 F_2 = 0.65(91.6 \text{ N}) = 59.5 \text{ N}$$

Problems

1. A uniform cylinder of mass 3 kg and radius of 0.05 m rolls without slipping along a horizontal surface. Find the total energy of the cylinder at the instant its speed is 2 m/s.
2. A uniform solid cylinder of mass 10 kg and radius of 0.2 m rolls up the incline of angle 45° with an initial velocity of 15 m/s. Find the height in which the cylinder will stop.

3. A wheel of mass 2 kg and radius of 0.05 m rolls without slipping with an angular speed of 3 rad/s on a horizontal surface. How much work is required to accelerate the wheel to an angular speed of 15 rad/s.
4. A block weighing 1000 N is held by a cable that is attached to a uniform rod of weight 500 N (see Fig. 8.20). Find (a) the tension in the cable, (b) the horizontal and vertical components of the force exerted on the base of the rod.
5. A uniform sphere of radius r and mass m is held by a light string and leans on a frictionless wall as in Fig. 8.21. If the string is attached a distance d above the center of the sphere, find (a) the tension in the string, (b) the reaction force exerted by the wall on the sphere.
6. Find the minimum force applied at the top of a wheel of mass M and radius R to raise it over a step of height h as in Fig. 8.22. Assume that the wheel does not slip on the step.

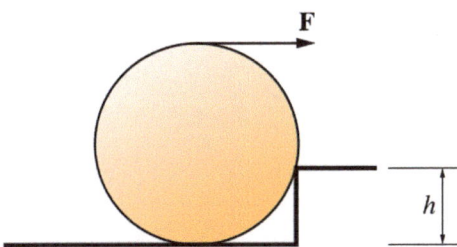

Fig. 8.22 A wheel raised over a step

7. Three identical uniform blocks each of length L are on top of each other as in Fig. 8.23. Find the maximum value of h in order for the stack to be in equilibrium.

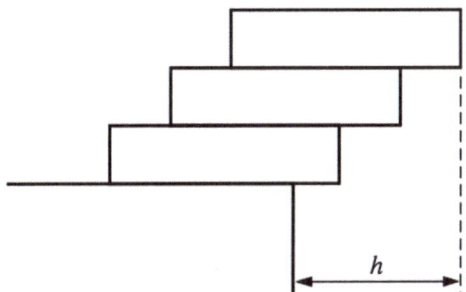

Fig. 8.23 Three identical uniform blocks on top of each other

Open Access This chapter is licensed under the terms of the Creative Commons Attribution 4.0 International License (http://creativecommons.org/licenses/by/4.0/), which permits use, sharing, adaptation, distribution and reproduction in any medium or format, as long as you give appropriate credit to the original author(s) and the source, provide a link to the Creative Commons license and indicate if changes were made.

The images or other third party material in this chapter are included in the chapter's Creative Commons license, unless indicated otherwise in a credit line to the material. If material is not included in the chapter's Creative Commons license and your intended use is not permitted by statutory regulation or exceeds the permitted use, you will need to obtain permission directly from the copyright holder.

Central Force Motion

9.1 Motion in a Central Force Field

A force is said to be central under two conditions. First, the direction of the force must always be toward or away from a fixed point (see Fig. 9.1). This point is known as the center of the force. Second, the magnitude of the force should only be proportional to the distance r between the particle and the center of the force. The central force may be written as

$$\mathbf{F} = f(r)\mathbf{r}_1 \qquad (9.1)$$

where $\mathbf{r}_1$ is a unit vector in the direction of $\mathbf{r}$. Therefore, if $f(r) < 0$, then the central force is an attractive force since it is directed toward the center of the force O (as shown in Fig. 9.1) and if $f(r) > 0$, the force is repulsively directed away from O.

Example 9.1 Which of the following forces are repulsive and which are attractive? (a)$\mathbf{F} = \dfrac{-3}{\sqrt{r}}\mathbf{r}_1$ (b)$\mathbf{F} = 4r^2\mathbf{r}_1$ (c)$\mathbf{F} = r(r-2)\mathbf{r}_1$.

Solution 9.1 (a) Attractive, (b) repulsive, and (c) attractive if $0 < r < 2$ and repulsive if $r > 2$.

9.1.1 Properties of a Central Force

1. The resulting motion of the particle takes place in a plane. To show that we have from Eq. 9.1

$$\mathbf{F} = f(r)\mathbf{r}_1 = m\mathbf{a}$$

thus, a is parallel to $\mathbf{r}(\mathbf{r} = r\mathbf{r}_1)$ and we may write

$$\mathbf{r} \times \mathbf{a} = 0$$

Hence,

$$\mathbf{r} \times \frac{d\mathbf{v}}{dt} = 0$$

or

$$\frac{d}{dt}(\mathbf{r} \times \mathbf{v}) = 0$$

Thus,

$$\mathbf{r} \times \mathbf{v} = \mathbf{h} = \text{constant} \qquad (9.2)$$

where $\mathbf{h}$ is a constant vector. Therefore, $\mathbf{r}$ and $\mathbf{v}$ always lie in the same plane where $\mathbf{h}$ is perpendicular to that plane for every value of t. As a result, the path of the particle takes place in a plane.

2. The angular momentum of the particle is conserved. From Eq. 9.2, we have

$$m(\mathbf{r} \times \mathbf{v}) = m\mathbf{h}$$

or

$$\mathbf{L} = m\mathbf{h} = \text{constant}$$

Thus, the angular momentum is equal to a constant at all times (conserved).

3. The position vector $\mathbf{r}$ of the particle with respect to the center of force sweeps out equal areas in equal times or in other words, the areal velocity is constant. To show that, consider the plane of motion to be the x–y plane. During an infinitesimally small time interval dt, the radius vector $\mathbf{r}$ sweeps out an area equal to dA. From Fig. 9.2, this area is equal to half of the area of a parallelogram with sides r and dr. That is,

$$d\mathbf{A} = \frac{1}{2}|\mathbf{r} \times d\mathbf{r}|$$

or

$$d\mathbf{A} = \frac{1}{2}|\mathbf{r} \times \mathbf{v}dt|$$

© The Author(s) 2019
S. Alrasheed, *Principles of Mechanics*, Advances in Science,
Technology & Innovation, https://doi.org/10.1007/978-3-030-15195-9_9

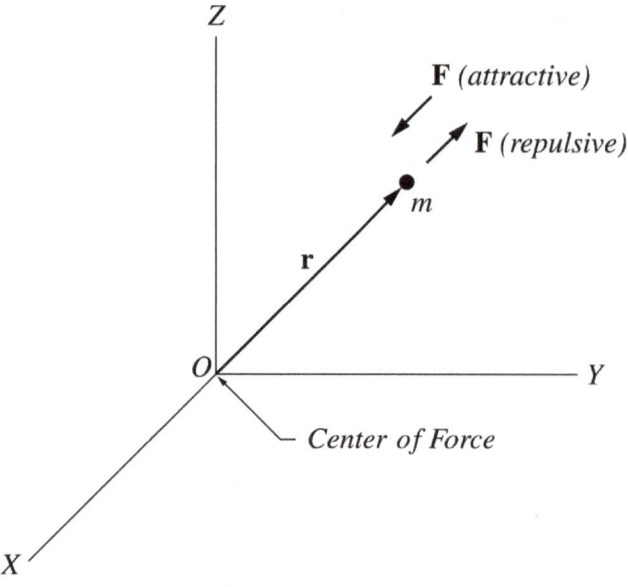

Fig. 9.1 The central force

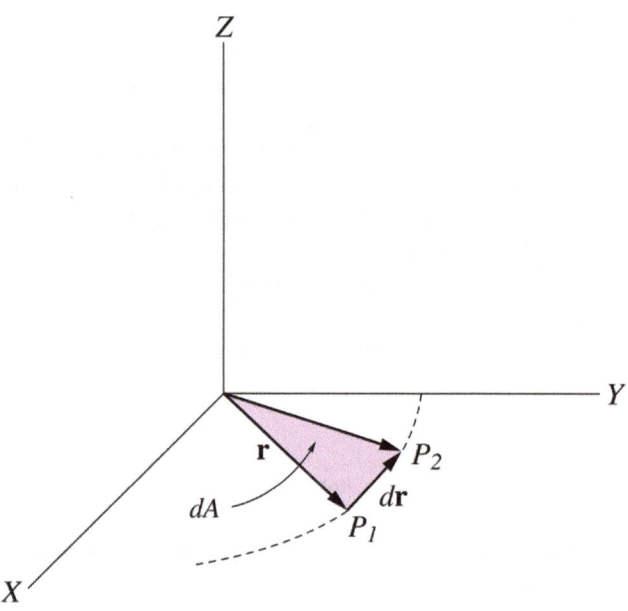

Fig. 9.2 During an infinitesimally small time interval dt, the radius vector $\mathbf{r}$ sweeps out an area equal to dA

or

$$\frac{d\mathbf{A}}{dt} = \frac{1}{2}|\mathbf{r} \times \mathbf{v}|$$

Thus,

$$\frac{dA}{dt} = \frac{h}{2} = \text{constant}$$

9.1.2 Equations of Motion in a Central Force Field

The most convenient coordinate system to describe the motion of a particle, under the influence of a central force, is the polar coordinate system. This convenience lies in the fact that the central force is in the r-direction. In Sect. 2.6, it has been shown that the acceleration of a particle in a plane, in terms of its polar coordinates, is given by

$$\mathbf{a} = (\ddot{r} - r\dot{\theta}^2)\mathbf{r}_1 + (r\ddot{\theta} + 2\dot{r}\dot{\theta})\boldsymbol{\theta}_1$$

Applying Newton's second law to the particle gives

$$\mathbf{F} = m\mathbf{a}$$

$$f(r)\mathbf{r}_1 = m[(\ddot{r} - r\dot{\theta}^2)\mathbf{r}_1 + (r\ddot{\theta} + 2\dot{r}\dot{\theta})\boldsymbol{\theta}_1]$$

That gives

$$f(r) = m(\ddot{r} - r\dot{\theta}^2) \tag{9.3}$$

$$m(r\ddot{\theta} + 2\dot{r}\dot{\theta}) = 0 \tag{9.4}$$

In Sect. 2.6, we've also seen that the velocity of a particle in polar coordinates is given by

$$\mathbf{v} = \dot{r}\mathbf{r}_1 + r\dot{\theta}\boldsymbol{\theta}_1$$

Therefore, we have

$$\mathbf{r} \times \mathbf{v} = r\mathbf{r}_1 \times (\dot{r}\mathbf{r}_1 + r\dot{\theta}\boldsymbol{\theta}_1) = r\dot{r}\,(\mathbf{r}_1 \times \mathbf{r}_1) + r^2\dot{\theta}(\mathbf{r}_1 \times \boldsymbol{\theta}_1)$$

$$= \mathbf{0} + r^2\dot{\theta}(\mathbf{r}_1 \times \boldsymbol{\theta}_1) = \mathbf{h}$$

Taking the plane of motion to be the x–y plane, then $\mathbf{r}_1 \times \boldsymbol{\theta}_1$ is parallel to the z-direction and we have

$$\mathbf{h} = r^2\dot{\theta}\mathbf{k} = h\mathbf{k}$$

Hence,

$$r^2\dot{\theta} = h \tag{9.5}$$

and Eq. 9.2 can be written as

$$\frac{d}{dt}(r^2\dot{\theta}) = 0$$

or

$$r^2\dot{\theta} = \text{constant}$$

Substituting Eq. 9.5 into Eq. 9.3 gives

$$f(r) = m\left(\ddot{r} - \frac{h^2}{r^3}\right) \tag{9.6}$$

Let $u = 1/r$, then $\dot{r} = -\dot{u}(1/u^2)$. Since $r^2\dot{\theta} = h$, we have $u^2 = \dot{\theta}/h$. Thus

$$\dot{r} = -h\left(\frac{\dot{u}}{\dot{\theta}}\right) = -h\left(\frac{du/dt}{d\theta/dt}\right) = -h\left(\frac{du}{d\theta}\right) \tag{9.7}$$

And

$$\ddot{r} = \frac{d}{dt}\left(-h\frac{du}{d\theta}\right) = \frac{d}{d\theta}\left(-h\frac{du}{d\theta}\right)\frac{d\theta}{dt}$$

$$\ddot{r} = -h\left(\frac{d^2u}{d\theta^2}\right)\dot{\theta} = -h^2u^2\left(\frac{d^2u}{d\theta^2}\right) \tag{9.8}$$

Substituting Eq. 9.8 into Eq. 9.6 gives

$$f(1/u) = m\left(-h^2u^2\left(\frac{d^2u}{d\theta^2}\right) - h^2u^3\right)$$

or

$$\frac{d^2u}{d\theta^2} + u = \frac{-1}{mh^2u^2}f(1/u) \tag{9.9}$$

This is the equation of path in a central force field.

9.1.3 Potential Energy of a Central Force

Consider a particle moving from point P_1 to P_2 (see Fig. 9.3) while a central force that has its center at the origin acts on it. The path of the particle may be considered as a combination of radial and curved segments. The central force is always acting in the direction of the radial segments and is perpendicular to the displacement along any of the curved segments. Thus, the work done by the central force along any curved segment is zero and the total work done in moving the particle along any path is equal to the work done along a radial line from r_i to r_f (see Fig. 9.4). That is, the work done by a central force is independent of path. It depends only on the initial and final positions of the particle.

From this, we conclude that the central force is a conservative force. You may also prove that $\nabla \times \mathbf{F} = \mathbf{0}$. Hence, there exists a potential energy and the work done by the gravitational force may be written as

$$W = -\triangle U$$

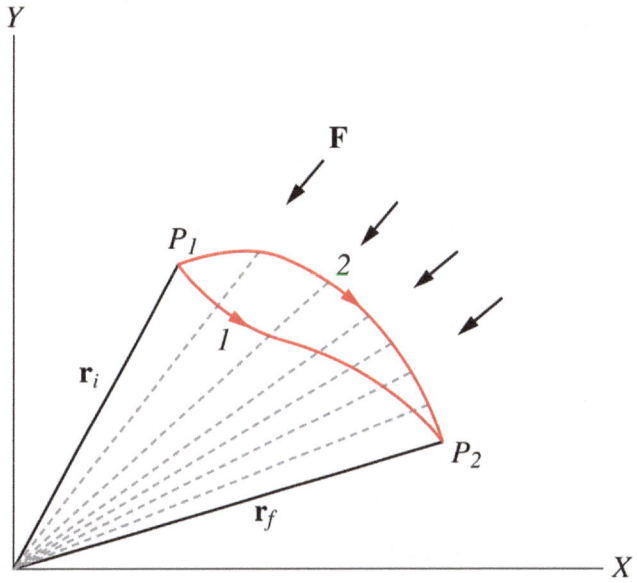

Fig. 9.3 A particle moving from point P_1 to P_2, while a central force that has its center at the origin acts on it

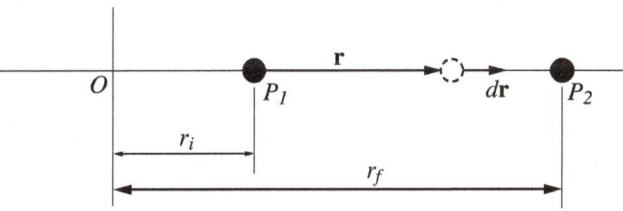

Fig. 9.4 The central force is always acting in the direction of the radial segments and is perpendicular to the displacement along any of the curved segments. Therefore, the total work done in moving the particle along any path is equal to the work done along a radial line from r_i to r_f

The work done in moving the particle from P_1 to P_2 is

$$W = \int_{P_1}^{P_2}\mathbf{F}\cdot d\mathbf{r} = \int_{r_i}^{r_f}f(r)\mathbf{r}_1\cdot d\mathbf{r} = \int_{r_i}^{r_f}f(r)\frac{\mathbf{r}}{r}\cdot d\mathbf{r}$$

Since $\mathbf{r}\cdot d\mathbf{r} = rdr$, we have

$$W = \int_{r_i}^{r_f}f(r)dr$$

or

$$\triangle U = U_f - U_i = -\int_{r_i}^{r_f}f(r)dr \tag{9.10}$$

9.1.4 The Total Energy

Since F is a conservative force, it follows that the total energy is conserved (constant), that is,

$$E = \frac{1}{2}mv^2 + U(r)$$

Since

$$v^2 = \mathbf{v} \cdot \mathbf{v} = \dot{r}^2 + r^2 \dot{\theta}^2$$

we have

$$E = \frac{1}{2} m (\dot{r}^2 + r^2 \dot{\theta}^2) + U(r) \qquad (9.11)$$

Substituting Eqs. 9.5 and 9.7 into Eq. 9.11 gives

$$E = \frac{1}{2} m \left(h^2 \left(\frac{du}{d\theta} \right)^2 + \left(\frac{1}{u^2} \right) (h u^2)^2 \right) + U$$

or

$$\left(\frac{du}{d\theta} \right)^2 + u^2 = \frac{2(E - U)}{m h^2} \qquad (9.12)$$

9.2   The Law of Gravity

In 1687, Isaac Newton made a remarkable discovery. Newton stated that the force that holds planets in their orbit is the same force that makes an apple fall from a tree. Newton's law of gravity states that *every particle in the universe attracts every other particle with a force that is directly proportional to the product of the masses of the particles and inversely proportional to the square of the distance between them*. The magnitude of this gravitational force is given by

$$F = \frac{G m_1 m_2}{r^2}$$

where m_1 and m_2 are the masses of the particles, r is the distance between them, and G is the universal gravitational constant. G has the same value if the particles (or objects) are located anywhere in the universe and it is given by

$$G = 6.672 \times 10^{-11} \, \text{N} \cdot \text{m}^2/\text{kg}^2$$

The gravitational force is effective when one or both the masses are very large. This is because G is a very small number. Note that, the gravitational force is not a contact force; it is a field force that can act through any medium. The direction of the gravitational force is along the line joining the two particles.

Therefore, the gravitational force is a central force since its magnitude is proportional only to the distance between the two particles (where one of the particles can be considered as the center of force), and its direction is along the line joining them (toward the center of force).

Figure 9.5 shows two particles of masses m_1 and m_2. Each particle exerts a gravitational force on the other. Let the gravitational force exerted on m_2 by m_1 to be $\mathbf{F}_{21}$, and that exerted on m_1 by m_2 to be $\mathbf{F}_{12}$. From Newton's third law of action and reaction, we have

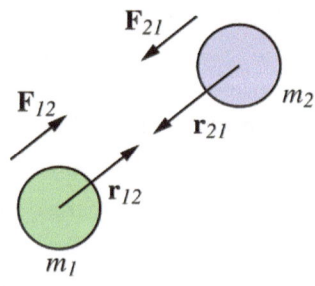

Fig. 9.5 Two particles of masses m_1 and m_2. Each particle exerts a gravitational force on the other

$$\mathbf{F}_{12} = -\mathbf{F}_{21}$$

That is, the two forces form an action and reaction pair. In terms of unit vectors, we may write

$$\mathbf{F}_{21} = -\frac{G m_1 m_2}{r_{12}^2} \mathbf{r}_{12}$$

and

$$\mathbf{F}_{12} = -\frac{G m_1 m_2}{r_{21}^2} \mathbf{r}_{21}$$

where $\mathbf{r}_{12}$ is a unit vector that is directed along the line joining the two particles (directed from m_1 to m_2) and $\mathbf{r}_{21}$ is a unit vector directed from m_2 to m_1. The negative sign indicates that the force is attractive. That is, the force exerted on m_1 by m_2 will move m_1 in the direction opposite of $\mathbf{r}_{21}$, i.e., toward m_2. Where the force exerted on m_2 by m_1 will move m_2 opposite to $\mathbf{r}_{12}$ (toward m_1). If particle P of mass of m_P interacts with a system of particles, the resultant gravitational force $\mathbf{F}_P$ exerted on particle P due to all particles in the system is the vector sum of the individual forces that each particle in the system exerts on particle P:

$$\mathbf{F}_P = \sum_{i=1}^{n} \mathbf{F}_{Pi} = \sum_{i=1}^{n} \frac{-G m_P m_i}{r_{iP}^2} \mathbf{r}_{iP}$$

where $\mathbf{r}_{iP}$ is a unit vector directed from the ith particle in the system toward the particle P and $\mathbf{F}_{Pi}$ is the force exerted on particle P by the ith particle. If particle P of mass m interacts with an extended body of mass M, the resultant gravitational force $\mathbf{F}_P$ exerted on particle P is the vector sum of the individual forces $d\mathbf{F}$ exerted on particle P due to each mass element dM in the object, but in this case, the sum is replaced by an integral

$$\mathbf{F}_P = \int d\mathbf{F} = -G m \int \frac{dM}{r^2} \mathbf{r}_1$$

where $\mathbf{r}_1$ is a unit vector directed from the mass element dM to the particle as shown in Fig. 9.6. The force of gravity gives planets and other heavy celestial bodies their spherical shape. That is because as the mass of the body becomes larger the force of gravity becomes stronger and all particles from all

Fig. 9.6 A particle P of mass m interacting with an extended body of mass M

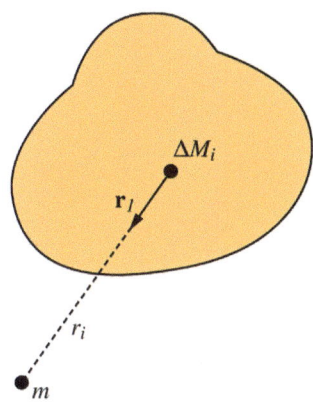

sides are attracted evenly toward the center. As a result, the body tends to have a spherical shape.

Example 9.2 Two particles of masses $m_1 = 0.2$ kg and $m_2 = 0.3$ kg are separated by a distance of 0.05 m. Find (a) the gravitational force that each particle exerts on the other; (b) at what distance a third particle $m_3 = 0.5$ kg must be placed at the other side of m_1 such that the net gravitational force on m_1 is zero. (All particles lie on a straight line).

Solution 9.2 (a)

$$F_{12} = F_{21} = \frac{Gm_1m_2}{r_{12}^2} = \frac{(6.67 \times 10^{-11}\,\text{N m}^2/\text{kg}^2)(0.2\,\text{kg})(0.3\,\text{kg})}{(0.05\,\text{m})^2} = 1.6 \times 10^{-9}\text{N}$$

(b)

$$F_{13} = \frac{Gm_1m_3}{r_{31}^2}$$

$$F_{12} = \frac{Gm_1m_2}{r_{21}^2}$$

If the net force on m_1 is zero, we have

$$\sum F_1 = F_{13} - F_{12} = 0$$

or

$$F_{13} = F_{12}$$

$$\frac{Gm_1m_3}{r_{31}^2} = \frac{Gm_1m_2}{r_{21}^2}$$

that gives

$$r_{31}^2 = \frac{m_3}{m_2}r_{21}^2 = \frac{(0.5\,\text{kg})}{(0.3\,\text{kg})}(0.05\,\text{m})^2$$

$$r_{31} = 0.064\,\text{m}$$

9.2.1 The Gravitational Force Between a Particle and a Uniform Spherical Shell

Case I: A Particle outside the Shell Consider a particle of mass m located outside a uniform spherical shell at point P as in Fig. 9.7. Imagine this shell to be made of a large number of thin rings each of outer thickness $Rd\theta$ and inner thickness l. The ring is so thin (since $d\theta$ is used) that every particle in the ring is at a distance s from P Furthermore, each particle in the ring exerts a gravitational force on the particle at P.

From the symmetry of the ring, if a particle (1) on the upper side exerts a gravitational force $\mathbf{F}_1$ on m, there is always another particle (2) at the opposite side of the ring exerting another force ($\mathbf{F}_2$) on the particle. Because $\mathbf{F}_1$ and $\mathbf{F}_2$ are equal in magnitude, then their y components cancel each other out and their x components add up (see Fig. 9.7). Thus, the resultant force exerted on m due to all particles of the sphere is the sum of the x components of their forces. Therefore the resultant force on m is along the x direction (toward the center of the shell). The gravitational force exerted on m by a thin ring of mass dM is

$$dF_g = \frac{GmdM}{s^2}\cos\phi$$

To express dM in terms of the density of the ring, we find the volume of the thin ring

$$dV = (2\pi R \sin\theta)(Rd\theta)l = 2\pi lR^2 \sin\theta d\theta$$

Since the shell has a uniform volume density ρ, dM is given by

$$dM = \rho dV = \rho 2\pi lR^2 \sin\theta d\theta$$

Thus,

$$dF_g = \frac{2\pi\rho lmGR^2 \cos\phi \sin\theta d\theta}{s^2} \tag{9.13}$$

From Fig. 9.7,

$$\cos\phi = \frac{r - R\cos\theta}{s} \tag{9.14}$$

From the cosines law, we have

$$s^2 = R^2 + r^2 - 2Rr\cos\theta \tag{9.15}$$

Substituting Eqs. 9.14 and 9.15 into Eq. 9.13 gives

$$dF_g = \frac{2\pi\rho lmGR^2(r - R\cos\theta)\sin\theta d\theta}{(r^2 + R^2 - 2rR\cos\theta)^{3/2}} \tag{9.16}$$

From Eq. 9.15, we have

Fig. 9.7 Because $\mathbf{F}_1$ and $\mathbf{F}_2$ are equal in magnitude, then their y components cancel each other out and their x components add up

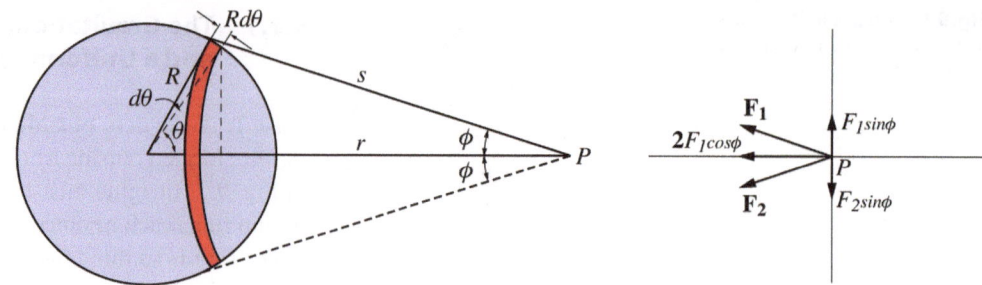

$$2sds = 2rR \sin\theta d\theta$$

To integrate over all rings, θ will change from $\theta = 0$ to π. From Eq. 9.15, we have at $\theta = 0$, $s = r - R$ since $(r \geq R)$, and at $\theta = \pi$, $s = r + R$. Also, we have from Eq. 9.15

$$\cos\theta = \frac{R^2 + r^2 - s^2}{2rR}$$

Thus

$$r - R\cos\theta = \frac{r^2 + s^2 - R^2}{2r}$$

Substituting this into Eq. 9.16 gives

$$F_g = \frac{\pi G\rho lRm}{r^2}\int_{r-R}^{r+R}\left(1 + \frac{r^2 - R^2}{s^2}\right)ds = \frac{4\pi G\rho lR^2m}{r^2} \tag{9.17}$$

Since $4\pi R^2\rho l = M$, it follows that

$$F_g = \frac{GMm}{r^2}$$

That is, the spherical shell behaves as a particle of mass M located at its center.

Case II: A Particle inside the Shell If a particle is inside a uniform spherical shell, the derivation of the gravitational force exerted on the particle by the spherical shell is the same as if the particle were outside the shell, except that the lower integration limit is different. At $\theta = 0$, $s = R - r$ since $r < R$. Thus, we have

$$F_g = \frac{\pi G\rho lRm}{r^2}\int_{R-r}^{r+R}\left(1 + \frac{r^2 - R^2}{s^2}\right)ds = 0$$

where $r < R$. That is, if the particle is inside the shell, the gravitational force exerted on it by the shell is zero. However, objects outside the shell may still exerts forces on the particle. In summary, we have

$$F_g = \frac{GMm}{r^2} \ (r \geq R)$$

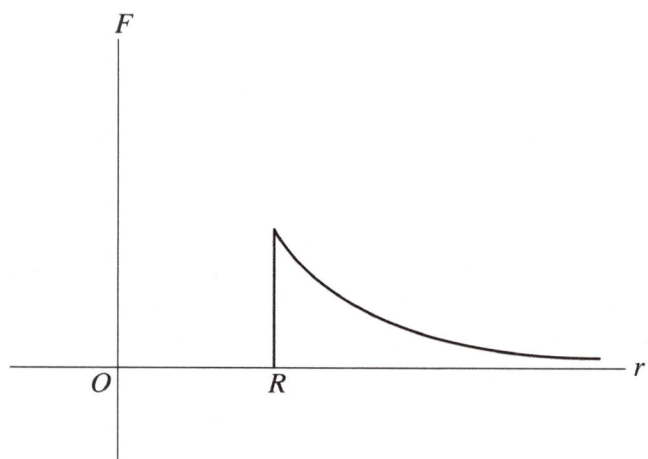

Fig. 9.8 The force exerted on a particle as a function of its r

$$F_g = 0 \ (r < R)$$

Figure 9.8 shows the force exerted on a particle as a function of its location.

9.2.2 The Gravitational Force between a Particle and a Uniform Solid Sphere

Case I: A Particle outside the Sphere Consider a particle of mass m located outside a uniform solid sphere. The sphere may be considered to be made of a series of concentric spherical shells. The force exerted on the particle by each shell is given by

$$dF_g = \frac{GdMm}{r^2}$$

The mass of each shell is $dM = \rho dV = \rho 4\pi a^2 da$. Where ρ is the volume density of the sphere and a is the distance from the shell to the center of the sphere and da is the thickness of the shell, Hence,

$$dF_g = \frac{Gm\rho 4\pi a^2 da}{r^2}$$

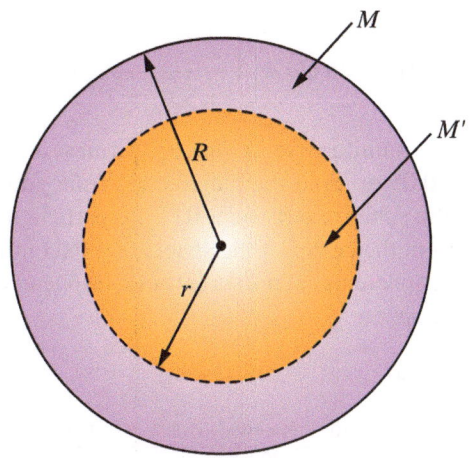

Fig. 9.9 If a particle of mass m is located inside a uniform solid sphere of mass M, then the gravitational force exerted on the particle is due only to the part of the sphere of radius $r < R$ and of mass of M

The total force exerted on m by the sphere is

$$F_g = \frac{Gm\rho 4\pi}{r^2} \int_0^R a^2 \, da$$

$$F_g = \frac{G(\rho^4/3\pi R^3)m}{r^2}$$

$$F_g = \frac{GMm}{r^2} \tag{9.18}$$

Thus, the solid sphere behaves as a particle of mass M located at the center of the sphere.

Case II: A Particle inside the Sphere If a particle of mass m is located inside a uniform solid sphere of mass M, then the gravitational force exerted on the particle is due only to the part of the sphere of radius $r < R$ and of mass of M (see Fig. 9.9). The remaining part of the sphere is a spherical shell which exerts no force on the particle since the particle is located inside it. From Eq. 9.18, the gravitational force exerted on the particle due to a sphere of radius r and mass M_1 is given by

$$F_g = \frac{GM_1 m}{r^2} \tag{9.19}$$

Since the sphere has a uniform density, we have

$$\rho = \frac{M_1}{V_1} = \frac{M}{V}$$

or

$$\frac{M_1}{M} = \frac{V_1}{V} = \frac{4/3\pi r^3}{4/3\pi R^3} = \frac{r^3}{R^3}$$

or

$$M_1 = M\frac{r^3}{R^3} \tag{9.20}$$

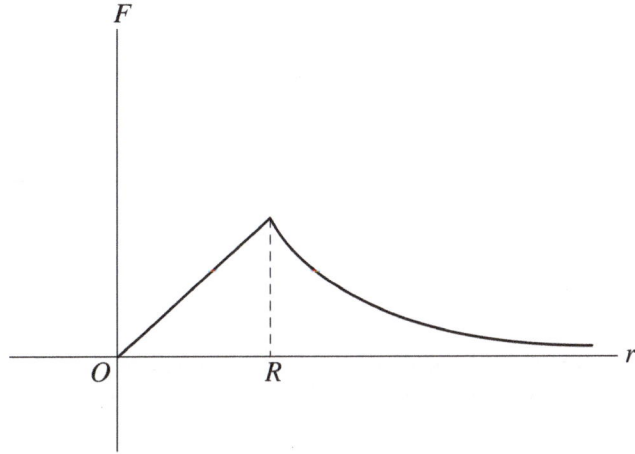

Fig. 9.10 The force exerted on a particle as a function of its r

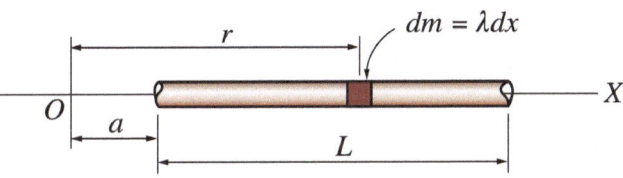

Fig. 9.11 The force exerted on a particle of mass m that is at a distance of a from a thin rod of mass M and length L

Substituting Eq. 9.20 into Eq. 9.19 gives

$$F_g = \frac{GmMr}{R^3}$$

where $r < R$. Therefore at the center of the sphere,

$$F_g = 0$$

Figure 9.10 shows the force exerted on a particle as a function of its location.

Example 9.3 (a) Find the gravitational force exerted on a particle of mass m that is at a distance of a from a thin rod of mass M and length L as in Fig. 9.11; (b) find the force in (a) if $a \gg L$.

Solution 9.3 (a)

$$dF = \frac{Gm \, dM}{x^2}$$

since the rod is uniform we have

$$dM = \lambda \, dx = \frac{M}{L} dx$$

Thus

$$dF = \frac{GmM}{Lx^2} dx$$

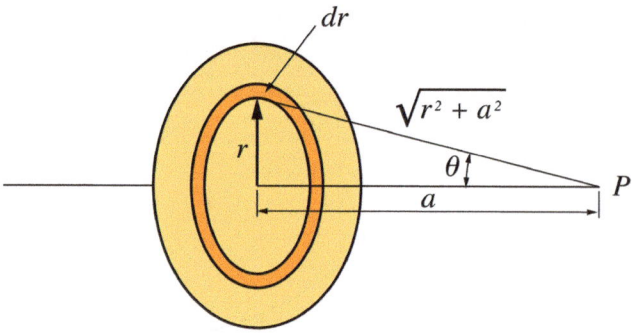

Fig. 9.12 The gravitational force exerted on a particle of mass m that is at a distance a from the center of a uniform solid disk of radius R and mass M

Integrating from a to $a + L$ gives

$$F = \frac{GmM}{L} \int_a^{a+L} \frac{dx}{x^2} = \frac{GmM}{L} \left[\frac{-1}{x} \right]_a^{a+L} = \frac{GmM}{L} \left[\frac{1}{a} - \frac{1}{a+L} \right] = \frac{GmM}{a(a+L)}$$

In vector form,

$$\mathbf{F} = \frac{GmM}{a(a+L)} \mathbf{i}$$

(b) if $a \gg L$, then

$$\mathbf{F} = \frac{GmM}{a^2} \mathbf{i}$$

That is, the rod can be considered as a particle of mass M that is at a distance a from m.

Example 9.4 Find the gravitational force exerted on a particle of mass m that is at a distance a from the center of a uniform solid disk of radius R and mass M as shown in Fig. 9.12.

Solution 9.4 Let us divide the disk into thin concentric rings of radius r and thickness dr. By symmetry, the resultant force on the particle is directed along the axis of the ring, since the y-components of the forces exerted by all particles of the ring will cancel out, where their x-components will add up. That is,

$$dF = \frac{G\, dMm \cos\theta}{r^2 + a^2}$$

Since the mass element dM is given by $dM = \sigma(2\pi r dr)$, we have

$$dF = \frac{G\sigma(2\pi r dr)m \cos\theta}{r^2 + a^2}$$

or

$$dF = \frac{G\sigma(2\pi r dr)ma}{(r^2 + a^2)^{3/2}}$$

The total force is

$$F = 2\pi G\sigma ma \int_{r=0}^R \frac{r dr}{(r^2 + a^2)^{3/2}} = \pi G\sigma ma \left[\frac{(r^2 + a^2)^{-1/2}}{-1/2} \right]_0^R$$

$$F = 2\pi G\sigma m \left[1 - \frac{a}{\sqrt{a^2 + R^2}} \right]$$

Example 9.5 A uniform solid sphere has a mass of 4.7 kg and a radius of 0.05 m. Find the magnitude of the gravitational force that the sphere exerts on a 0.02 kg particle located at (a) 0.5 m from the center of the sphere; (b) 0.03 m from the center of the sphere; (c) at the surface of the sphere; (d) at the center of the sphere.

Solution 9.5 (a)

$$F_{1s} = \frac{GmM}{r^2} = \frac{(6.67 \times 10^{-11}\ \text{Nm}^2/\text{kg}^2)(0.02\ \text{kg})(4.7\ \text{kg})}{(0.5\ \text{m})^2} = 2.5 \times 10^{-11}\ \text{N}$$

(b)

$$F_{1s} = \frac{GmMr}{R^3} = \frac{(6.67 \times 10^{-11}\ \text{Nm}^2/\text{kg}^2)(0.02\ \text{kg})(4.7\ \text{kg})(0.03\ \text{m})}{(0.05\ \text{m})^3} = 1.5 \times 10^{-9}\ \text{N}$$

(c)

$$F_{1s} = \frac{GmM}{R^2} = \frac{(6.67 \times 10^{-11}\ \text{Nm}^2/\text{kg}^2)(0.02\ \text{kg})(4.7\ \text{kg})}{(0.05\ \text{m}^2)} = 2.5 \times 10^{-9}\ \text{N}$$

(d)

$$F_{1s} = 0$$

Example 9.6 Three concentric spherical shells have masses of M_1, M_2, and M_3 and radius of R_1, R_2, and R_3, respectively, as in Fig. 9.13. Find the gravitational force exerted on a particle of mass m located at (a) $r = a$ (b) $r = b$ (c) $r = c$ (d) $r = d$.

Solution 9.6 (a)

$$F = 0$$

(b)

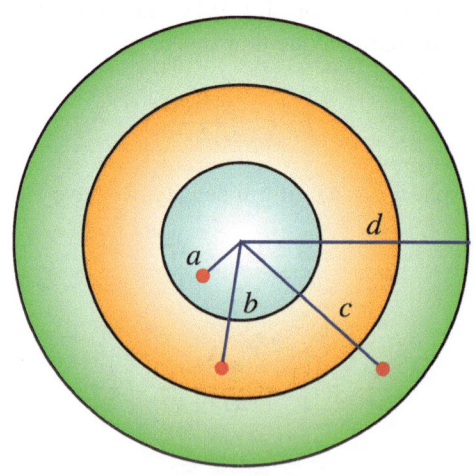

Fig. 9.13 Three concentric spherical shells

$$F = \frac{GM_1 m}{b^2}$$

(c)

$$F = \frac{GM_1 m}{c^2} + \frac{GM_2 m}{c^2} = \frac{Gm}{c^2}(M_1 + M_2)$$

(d)

$$F = \frac{Gm}{d^2}(M_1 + M_2 + M_3)$$

Example 9.7 A spaceship of mass m_1 is moving along a straight line path between the earth and the sun. At what distance from the center of the earth will the gravitational force of the sun balances that of the earth?

Solution 9.7 At that point, we have

$$F_{1E} = F_{1S}$$

$$\frac{Gm_1 M_E}{r^2} = \frac{Gm_1 M_S}{(d-r)^2}$$

or

$$\frac{(d-r)^2}{r^2} = \frac{M_S}{M_E}$$

$$r = \frac{d[M_E - (M_E M_S)^{1/2}]}{M_E - M_S}$$

Example 9.8 An artificial satellite is moving in a circular orbit about the earth at a distance of 1500 km above the earth's surface. Find its speed and period.

Solution 9.8

$$\frac{Gm_s M_E}{r^2} = \frac{m_s v^2}{r}$$

$$v = \sqrt{\frac{GM_E}{r}}$$

where r is the distance between the center of the earth and the satellite. That is,

$$r = (6.37 \times 10^6 \text{m}) + (1500 \times 10^3 \text{m}) = 7.9 \times 10^6 \text{m}$$

Hence,

$$v = \sqrt{\frac{GM_E}{r}} = \sqrt{\frac{(6.67 \times 10^{-11} \text{Nm}^2/\text{kg}^2)(5.98 \times 10^{24}\text{kg})}{(7.9 \times 10^6 \text{m})}} = 7.1 \times 10^3 \text{m/s}$$

$$T = \frac{2\pi r}{v} = \frac{2(3.1.4)(7.9 \times 10^6 \text{m})}{(71 \times 10^3 \text{m/s})} = 6968.8\text{s} = 116.15 \text{ min}$$

9.2.3 Weight and Gravitational Force

In Chap. 4, we've seen that the weight of an object is defined as the gravitational force exerted on the object by the earth (or any other astronomical object) and it is directed toward the center of the earth. The weight of an object is given by $\mathbf{w} = m\mathbf{g}$, where $\mathbf{g}$ is the free-falling acceleration and its value near the earth's surface is 9.8 m/s^2. The exact form of the gravitational force between any two objects was given earlier in this chapter by Newton's law of gravity In the case of an earth–particle system, the gravitational force that each one exerts on the other is

$$F_g = \frac{GM_E m}{r^2}$$

where M_E is the mass of the earth and m is the mass of the particle that is at a distance r from the center of the earth. Note that, it is assumed that the earth is a perfect sphere of uniform mass distribution, and therefore behaves as a particle. In reality, the earth is not a perfect sphere but rather an ellipsoid. Furthermore, the earth's density is not uniform since it varies with the radius of earth.

The earth's density also varies at the earth's surface from one region to another. In addition, if the earth's rotation is included, then the resultant force on an object will be its weight plus the centripetal force exerted on the object due to the rotation. However, these variations are often neglected. From the definition of weight, we have

$$w = mg = F_g = \frac{GM_E m}{r^2}$$

therefore

$$g = \frac{GM_E}{r^2} \tag{9.21}$$

As you can see the free-falling acceleration does not depend on the mass of the object as was predicted before. If the object is falling near the earth's surface, then distance r in Eq. 9.21 can be replaced by R_E which is the radius of the earth and we have

$$g = \frac{GM_E}{R_E^2}$$

If the object is at a distance h from the earth's surface, we may write

$$g = \frac{GM_E}{(R_E + h)^2}$$

Thus, the weight of an object decreases with increasing altitude. Table 9.1 shows the variation of g with altitude.

Table 9.1 Variation of g with altitude

Altitude h (km)	g (m/s^2)
1000	7.34
6000	2.6
10000	1.49
30000	0.3
60000	0.09

Example 9.9 A man can jump vertically upward from the earth's surface and reach an altitude of 0.2 m. Find the altitude the man can reach if he jumps with the same initial velocity on the surface of the moon.

Solution 9.9 Using the formula $y - y_0 = \frac{v^2 - v_0^2}{-2g}$ and by taking $y_0 = 0$ at the earth's surface and $y = h$ at the maximum height and that $v = 0$ there, we have

$$h = \frac{v_0^2}{2g}$$

Since the initial velocity of the man is the same on earth and on moon, we have

$$h_E g_E = h_m g_m$$

At the surface of the moon

$$g_m = \frac{GM_m}{R_m^2} = \frac{(6.67 \times 10^{-11}\,\text{N m}^2/\text{kg}^2)(7.36 \times 10^{22}\,\text{kg})}{(1.74 \times 10^6\,\text{m})^2} = 1.6\,\text{m/s}^2$$

$$h_m = h_E \frac{g_E}{g_m} = (0.2\,\text{m})\frac{(9.8\,\text{m/s}^2)}{(1.6\,\text{m/s}^2)} = 1.2\,\text{m}$$

That is, the maximum height reached by the man on the moon is six times the height reached on earth.

Example 9.10 A neutron star of radius of 12 km has a gravitational acceleration of 1×10^{12} m/s^2 at its surface. Calculate its average density.

Solution 9.10 The gravitational acceleration of a particle near the surface of the star is

$$g = \frac{GM_n}{R_n^2}$$

$$M_n = \frac{gR_n^2}{G} = \frac{(1 \times 10^{12}\,\text{m/s}^2)(12 \times 10^3\,\text{m})^2}{(6.67 \times 10^{-11}\,\text{N m}^2/\text{kg}^2)} = 2 \times 10^{30}\,\text{kg}$$

$$\rho = \frac{3M_n}{4\pi R_n^3} = \frac{3(2 \times 10^{30}\,\text{kg})}{4(3.14)(12 \times 10^3\,\text{m})^3} = 2.8 \times 10^{17}\,\text{kg/m}^3$$

Example 9.11 Find the free-fall acceleration of a body that is at a distance of $0.05R_E$ above the surface of the earth.

Solution 9.11

$$g = \frac{GM_E}{(R_E + h)^2} = \frac{GM_E}{(R_E + 0.05R_E)^2} = \frac{GM_E}{(1.05R_E)^2}$$

$$= \frac{(6.67 \times 10^{-11}\,\text{N m}^2/\text{kg}^2)(5.98 \times 10^{24}\,\text{kg})}{(6.7 \times 10^6\,\text{m})^2} = 8.9\,\text{m/s}^2$$

9.2.4 The Gravitational Field

As mentioned previously, the gravitational force is a field force that can act through empty space, i.e., physical contact between objects is not necessary for such a force to act. An alternative way in describing the gravitational attraction is by introducing the concept of the gravitational field. Suppose a test particle of mass m_0 is placed at different points from another mass M (which represents the center of the gravitational force). At each point, the test particle will experience a gravitational force that depends on its distance from M and is given by

$$\mathbf{F}_g = \frac{-GMm_0}{r^2}\mathbf{r}_1$$

where $\mathbf{r}_1$ is a unit vector that points radially outwards. Therefore, M may be considered as producing a gravitational field in the space around it. This field can be sensed by the force that the test particle experience when placed in the vicinity of M. The gravitational field produced by M at any point in space is thus given by

$$\mathbf{g} = \frac{\mathbf{F}_g}{m_0} = \frac{-GM}{r^2}\mathbf{r}_1$$

That is, the gravitational field at a point is defined as the gravitational force per unit mass at that point. A map of the field can be drawn showing the gravitational field at any point in space. Figure 9.14 shows the gravitational field vectors near the earth's surface and at large distances from the earth. Note that, the gravitational field is an example of a static field since the field at any point is constant with time.

Example 9.12 Find the magnitude and direction of the gravitational field at the point P in the arrangement shown in Fig. 9.15, where all particles have equal masses.

Solution 9.12 Since all masses are equal, the net gravitational force at P is due to the sum of the x-components of F_3 and F_2. That is,

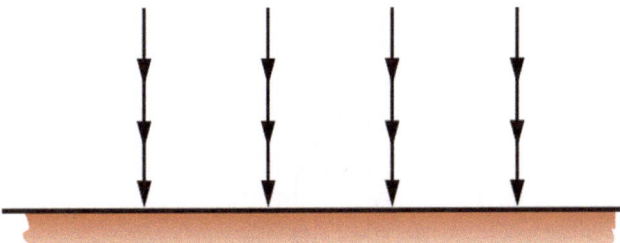

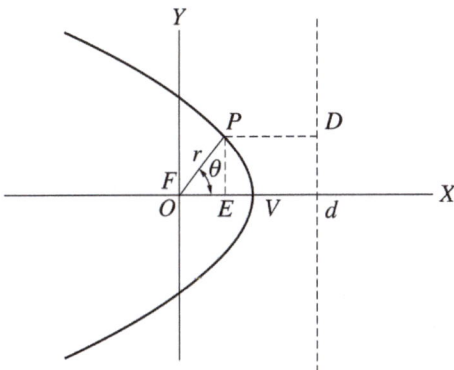

Fig. 9.16 A conic section has the property that the ratio e (called the eccentricity) of the distance between any point on the curve (for example point P) and another point called the focus (F) to the distance between P and a line called the directrix is equal to a constant

9.3.1 The Polar Equation of a Conic Section

A conic section has the property that the ratio e (called the eccentricity) of the distance between any point on the curve (for example point P) and another point called the focus (F) to the distance between P and a line called the directrix is equal to a constant (see Fig. 9.16). This constant differs from one conic section to another. Consider Fig. 9.16 where the focus F is at the origin O of the x and y coordinate system and the directrix is at $x = d$. Since the distance between P and F is

$$PF = r$$

then, the nearest distance between P and the directrix is

$$PD = d - FE = d - r \cos \theta$$

The eccentricity is therefore given by

$$e = \frac{PF}{PD} = \frac{r}{d - r \cos \theta}$$

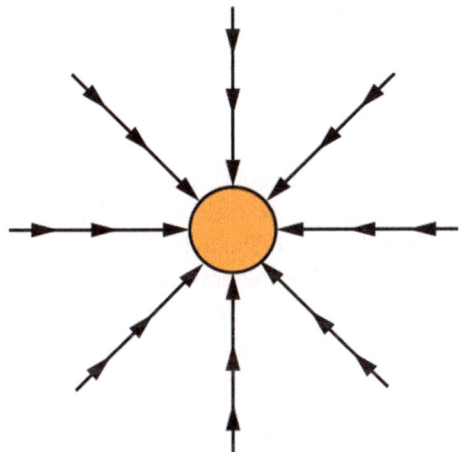

Fig. 9.14 The gravitational field vectors near the earth's surface and at large distances from the earth

Hence,

$$r = \frac{ed}{1 + e \cos \theta} \qquad (9.22)$$

This equation is the polar equation of a conic section.

1. **Ellipse:** $e < 1$ From Fig. 9.17, you can see that at $\theta = 0$, $r = OV$ and at $\theta = \pi$, $r = OV'$. Substituting this into Eq. 9.22 gives

$$OV = \frac{ed}{1 + e}$$

and

$$OV' = \frac{ed}{1 - e}$$

Since VV' is the length of the major axis which is equal to $2a$, (a is the length of the semimajor axis) we have

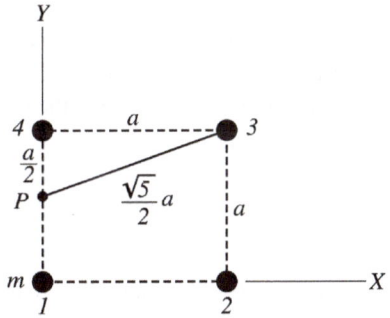

Fig. 9.15 Finding the magnitude and direction of the gravitational field at P

$$\mathbf{F} = 2F_3 \cos \theta \mathbf{i} = \frac{4Gmm_0}{5a^2} \cos \theta \mathbf{i} = \frac{4Gmm_0}{5a^2} \frac{2}{\sqrt{5}} \mathbf{i} = \frac{8Gmm_0}{5\sqrt{5}a^2} \mathbf{i}$$

$$\mathbf{g} = \frac{8Gm}{5\sqrt{5}a^2} \mathbf{i}$$

9.3 Conic Sections

Conic sections are produced if a double right circular cone intersects with a plane. It may be a circle, a parabola, an ellipse, or a hyperbola.

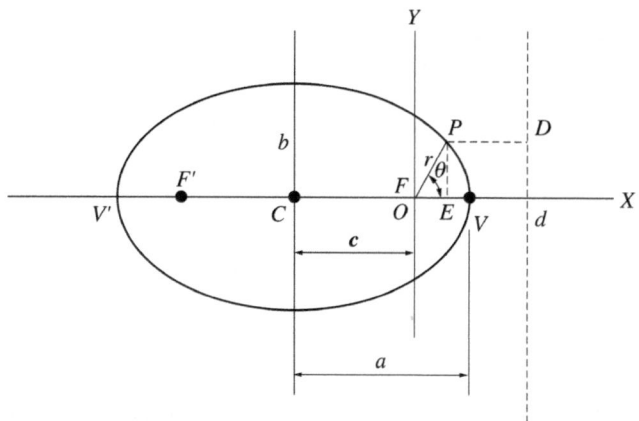

Fig. 9.17 In an ellipse, at $\theta = 0$, $r = OV$ and at $\theta = \pi$, $r = OV'$

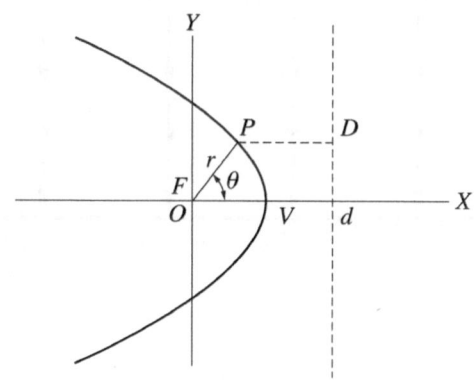

Fig. 9.18 In a parabola, as θ approaches π, r becomes infinite and hence $a \to \infty$

$$OV + OV' = 2a \qquad (9.23)$$

or

$$\frac{ed}{1+e} + \frac{ed}{1-e} = 2a$$

Hence,

$$a = \frac{ed}{1-e^2}$$

Or

$$ed = a(1 - e^2)$$

Substituting into Eq. 9.22, the polar equation of an ellipse is

$$r = \frac{a(1-e^2)}{1 + e\cos\theta}$$

That gives

$$OV = \frac{a(1-e^2)}{1+e} = a(1-e) \qquad (9.24)$$

and

$$OV' = \frac{a(1-e^2)}{1-e} = a(1+e) \qquad (9.25)$$

The distance C between the center of the ellipse and the focus is

$$C = CV - OV = a - a(1-e) = ae$$

Since from Fig. 9.17, we have $c < a$, i.e., the distance between the foci is less than that between the vertices, then $e < 1$. Furthermore, you can prove that $c = \sqrt{a^2 - b^2}$ or $b = a\sqrt{1 - e^2}$ where b is the length of the semiminor axis of the ellipse.

2. **Parabola**: $e = 1$ Since $e = 1$, Eq. 9.22 becomes

$$r = \frac{d}{1 + \cos\theta}$$

(Polar Equation of a Parabola) As θ approaches π, r becomes infinite and hence $a \to \infty$ (see Fig. 9.18).

3. **Hyperbola**: $e > 1$ The hyperbola has two branches as shown in Fig. 9.19. For the gravitational force, only the first branch (I) represents a possible motion of the particle since GM/h^2 must be positive. The polar equation of a hyperbola is given by

$$r = \frac{a(e^2 - 1)}{1 + e\cos\theta}$$

9.3.2 Motion in a Gravitational Force Field

The path of a particle in any central force field can be found by solving the equation of motion $(d^2u/d\theta^2 + u = -1/(mh^2u^2)f(1/u)$ (Eq. 9.9) if the form of the force is known. In the case of a gravitational force, we have

$$f(r) = \frac{-GMm}{r^2}$$

where M is assumed to be fixed and that it is attracting a particle of mass m and r is the distance between them. In terms of u, we have

$$f(1/u) = -GMmu^2$$

Substituting this into the equation of motion gives

$$\frac{d^2u}{d\theta^2} + u = \frac{-1}{mh^2u^2}(-GMmu^2)$$

or

$$\frac{d^2u}{d\theta^2} + u = \frac{GM}{h^2} \qquad (9.26)$$

This equation is a nonhomogeneous linear differential equation. Its solution may be given by

Fig. 9.19 The hyperbola

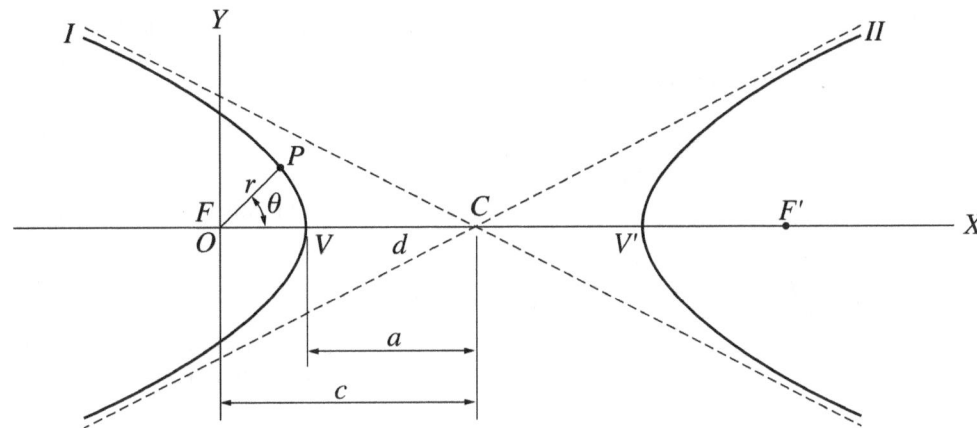

$$u = \frac{1}{r} = C\cos(\theta - \phi) + \frac{GM}{h^2}$$

where C and ϕ are integration constants. ϕ is known as the phase angle and it can be chosen to be $\phi = 0$ if the x-axis is chosen such that at $\theta = 0$, r is a minimum. That gives

$$u = \frac{1}{r} = C\cos\theta + \frac{GM}{h^2} \tag{9.27}$$

or

$$r = \frac{h^2/GM}{1 + \frac{Ch^2}{GM}\cos\theta} = \frac{ed}{1 + e\cos\theta}$$

Thus, the path of the particle under the influence of the gravitational force field is a conic with $ed = h^2/GM$ and $d = 1/C$ and $e = h^2C/GM$. If a planet is moving in elliptical orbit about the sun, then the maximum and minimum distances of the planet from the sun (OV and OV') are called the aphelion and perihelion respectively If a satellite is moving about a planet in an elliptical orbit, the maximum and minimum distances of the satellite from the planet are called the apogee and perigee respectively.

9.3.3 The Gravitational Potential Energy

Consider a particle of mass m moving under the influence of a larger particle of mass M ($M \gg m$). By using Eq. 9.10 ($\Delta U = U_f - U_i = -\int_{r_i}^{r_f} f(r)dr$) and noting that $f(r) = -GMm/r^2$, the change in the gravitational potential energy of the system as m moves from r_i to r_f in the field of M is

$$\Delta U_g = U_{gf} - U_{gi} = \int_{r_i}^{r_f} \frac{GMm}{r^2} dr = GMm \int_{r_i}^{r_f} \frac{dr}{r^2}$$

$$= GMm \left[\frac{-1}{r}\right]_{r_i}^{r_f} = GMm\left(\frac{1}{r_i} - \frac{1}{r_f}\right)$$

That is, as the particle of mass m moves toward or away from M, the potential energy of the system decreases and increases respectively Note that, the lighter particle (m) gains most of the kinetic energy as the potential energy changes. By choosing the reference point at infinity ($r_i = \infty$) then $U_i = 0$ and taking $r_f = r$ gives

$$U_g(r) = \frac{-GMm}{r}$$

For more than two-particle systems, there is more than one gravitational force (one for each pair of particles). Hence, there is more than one potential energy The total potential energy is the sum of the potential energies of each pair. For example if there are three particles, the total potential energy is

$$U_{tot} = U_{12} + U_{13} + U_{23} = -\left(\frac{Gm_1m_2}{r_{12}} + \frac{Gm_1m_3}{r_{13}} + \frac{Gm_2m_3}{r_{23}}\right)$$

Force from Potential Energy The gravitational force may be obtained from its corresponding potential energy. That is,

$$\mathbf{F}_g = -\frac{d}{dr}\left(\frac{-GMm}{r}\right)\mathbf{r}_1 = \frac{-GMm}{r^2}\mathbf{r}_1$$

Example 9.13 Find the potential energy of the system as shown in Fig. 9.20.

Solution 9.13

$$U = U_{12} + U_{13} + U_{23}$$

$$= -G\left(\frac{m_1m_2}{r_{12}} + \frac{m_1m_3}{r_{13}} + \frac{m_2m_3}{r_{23}}\right)$$

$$= -(6.67\times10^{-11}\,\mathrm{N\,m^2/kg^2})\left(\frac{(8\times10^4\,\mathrm{kg})}{(0.3\,\mathrm{m})} + \frac{(12\times10^4\,\mathrm{kg})}{(0.32\,\mathrm{m})} + \frac{(6\times10^4\,\mathrm{kg})}{(0.36\,\mathrm{m})}\right) = -5.4\times10^{-5}\,\mathrm{J}$$

Example 9.14 Two particles of equal masses (3 kg) are separated by a distance of 0.3 m : (a) Find the potential energy

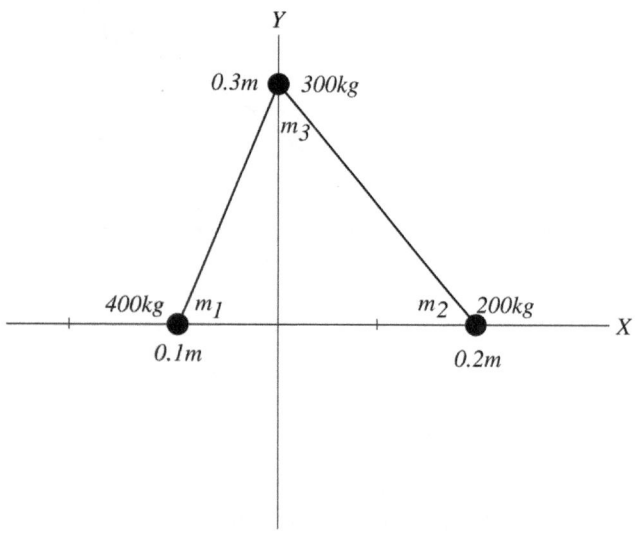

of the system; (b) how much work is required to reduce their separation to 0.1 m, (c) to increase their separation to 0.5 m.

Solution 9.14 (a)

$$U = \frac{-Gm^2}{r} = \frac{-(6.67 \times 10^{-11} \ \text{N m}^2/\text{kg}^2)(3 \ \text{kg})^2}{(0.3 \ \text{m})} = -2 \times 10^{-9} \ \text{J}$$

(b) The work done by the gravitational force is

$$W = -\Delta U = U_i - U_f = -Gm^2 \left(\frac{1}{r_i} - \frac{1}{r_f} \right)$$

$$= -(6.67 \times 10^{-11} \ \text{N m}^2/\text{kg}^2)(3 \ \text{kg})^2 \left(\frac{1}{(0.3 \ \text{m})} - \frac{1}{(0.1 \ \text{m})} \right)$$

that gives $W = 4 \times 10^{-9}$ J. The work done by an external agent is $W = -4 \times 10^{-9}$ J.

(c) The work done by the gravitational force is

$$W = -\Delta U = -Gm^2 \left(\frac{1}{r_i} - \frac{1}{r_f} \right) = -(6.67 \times 10^{-11} \ \text{N m}^2/\text{kg}^2)(3 \ \text{kg})^2 \left(\frac{1}{(0.3 \ \text{m})} - \frac{1}{(0.5 \ \text{m})} \right)$$

$$W = -8 \times 10^{-10} \ \text{J}$$

The work done by an external agent is $+8 \times 10^{-10}$ J.

9.3.4 Energy in a Gravitational Force Field

The equation of motion in terms of energy is given by Eq. 9.12:

$$\left(\frac{du}{d\theta} \right)^2 + u^2 = \frac{2(E-U)}{mh^2}$$

The gravitational potential energy of a two-particle system of masses M and m is given by

$$U_g(r) = \frac{-GMm}{r}$$

In terms of u we may write

$$U_g(1/u) = -GMmu \tag{9.28}$$

Furthermore, the solution of the equation (Eq. 9.26) of motion in the gravitational force field is

$$u = \frac{1}{r} = C \cos\theta + \frac{GM}{h^2} \tag{9.29}$$

Substituting Eqs. 9.28 and 9.29 into Eq. 9.12 gives

$$(C \sin\theta)^2 + \left(C \cos\theta + \frac{GM}{h^2} \right)^2 = \frac{2E}{mh^2} - \frac{2}{mh^2} \left(-GMm \left(C \cos\theta + \frac{GM}{h^2} \right) \right)$$

That gives

$$C^2 = \frac{2E}{mh^2} + \frac{G^2M^2}{h^4}$$

or

$$C = \sqrt{\frac{2E}{mh^2} + \frac{G^2M^2}{h^4}} \quad (assuming \ C > 0)$$

Substituting this value of C into Eq. 9.29 gives

$$u = \frac{GM}{h^2} + \sqrt{\frac{2E}{mh^2} + \frac{G^2M^2}{h^4}} \cos\theta$$

$$= \frac{GM}{h^2} + \frac{GM}{h^2} \sqrt{1 + \frac{2Eh^2}{G^2M^2m}} \cos\theta$$

or

$$u = \frac{GM}{h^2} \left[1 + \sqrt{1 + \frac{2Eh^2}{G^2M^2m}} \cos\theta \right] \tag{9.30}$$

Comparing this equation with the polar equation of a conic section (Eq. 9.22), we have

$$e = \sqrt{1 + \frac{2Eh^2}{G^2M^2m}}$$

Thus the trajectory of the particle is an ellipse if $e < 1$, that is if $E < 0$. Therefore, if the potential energy of the particle is greater than its kinetic energy the particle's path is an ellipse since it does not have enough energy to reach infinity. The trajectory of the particle is a parabola if $e = 1$ and hence

Fig. 9.21 Different paths

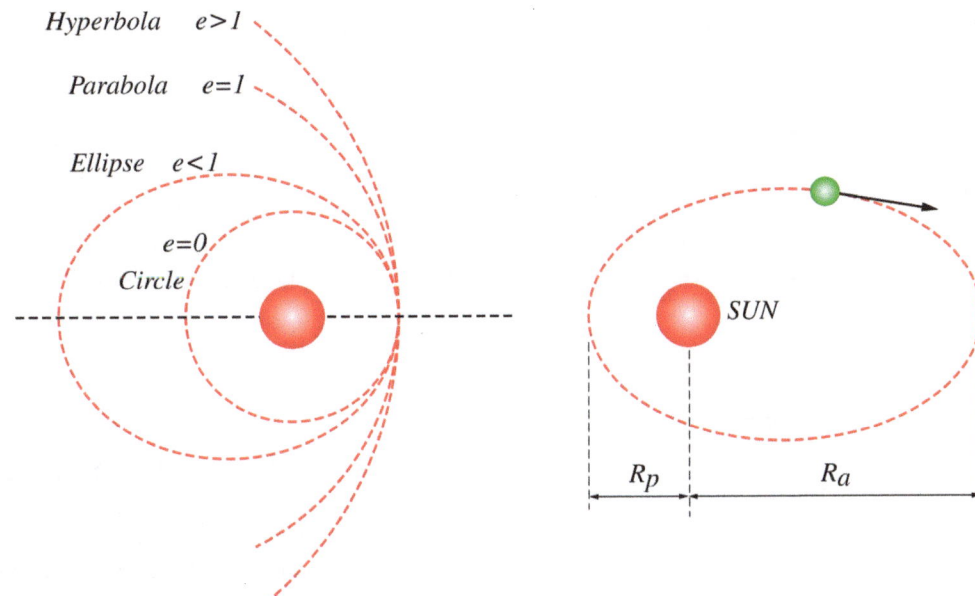

if $E = 0$. In that case, the kinetic energy of the particle is equal to its potential energy and thus it can reach infinity with zero kinetic energy. Finally, the trajectory of the particle is a hyperbola if $e > 1$ and therefore if $E > 0$. That is, if the kinetic energy of the particle is greater than its potential energy, then it will reach infinity with positive kinetic energy

- Elliptical Orbit $E < 0$
- Parabolic Orbit $E = 0$
- Hyperbolic Orbit $E > 0$

Different paths are shown in Fig. 9.21.

9.4 Kepler's Laws

After analyzing the astronomical data of the Danish astronomer Tycho Brahe, the German astronomer Johannes Kepler formulated his three laws of planetary motion.

9.4.1 Kepler's First Law

Every planet moves in an elliptical orbit with the sun at one focus as shown in Fig. 9.21.

Proof The gravitational force between the sun and a planet is

$$\mathbf{F} = \frac{-GM_SM_P}{r^2}\mathbf{r}_1$$

where M_S and M_P are the masses of the sun and the planet, respectively The acceleration of the planet is

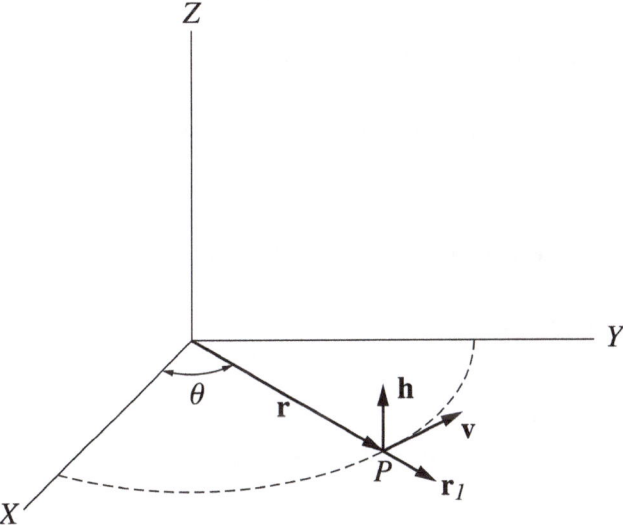

Fig. 9.22 From the first property of a central force we have $\mathbf{r} \times \mathbf{v} = \mathbf{h} =$ constant, where $\mathbf{h}$ is a constant vector perpendicular to the x-y plane

$$\mathbf{a} = \frac{-GM_S}{r^2}\mathbf{r}_1$$

From the first property of a central force, we have $\mathbf{r} \times \mathbf{v} = \mathbf{h} =$ constant, where $\mathbf{h}$ is a constant vector perpendicular to the x–y plane (see Fig. 9.22). Since $\mathbf{r} = r\mathbf{r}_1$ and $\mathbf{v} = d\mathbf{r}/dt = dr\mathbf{r}_1/dt = rd\mathbf{r}_1/dt + (dr/dt)\mathbf{r}_1$ we have

$$\mathbf{h} = r\mathbf{r}_1 \times \left(r\frac{d\mathbf{r}_1}{dt} + \frac{dr}{dt}\mathbf{r}_1 \right) = r^2\left(\mathbf{r}_1 \times \frac{d\mathbf{r}_1}{dt} \right) + r\frac{dr}{dt}\left(\mathbf{r}_1 \times \mathbf{r}_1 \right)$$

$$= r^2\left(\mathbf{r}_1 \times \frac{d\mathbf{r}_1}{dt} \right)$$

$$\mathbf{a} \times \mathbf{h} = \left(\frac{-GM_S}{r^2}\mathbf{r}_1\right) \times \left(r^2\left(\mathbf{r}_1 \times \frac{d\mathbf{r}_1}{dt}\right)\right) = -GM_S\left[\left(\mathbf{r}_1\frac{d\mathbf{r}_1}{dt}\right)\mathbf{r}_1 - (\mathbf{r}_1 \cdot \mathbf{r}_1)\frac{d\mathbf{r}_1}{dt}\right]$$

Using

$$\mathbf{A} \times (\mathbf{B} \times \mathbf{C}) = (\mathbf{A} \cdot \mathbf{C})\mathbf{B} - (\mathbf{A} \cdot \mathbf{B})\mathbf{C}$$

Since $\mathbf{r}_1 \cdot d\mathbf{r}_1/dt = 0$ and $\mathbf{r}_1 \cdot \mathbf{r}_1 = r_1^2 = 1$, we have

$$\mathbf{a} \times \mathbf{h} = GM_S\frac{d\mathbf{r}_1}{dt} = \frac{d}{dt}(GM_S\mathbf{r}_1)$$

Also we have

$$\mathbf{a} \times \mathbf{h} = \frac{d\mathbf{v}}{dt} \times \mathbf{h} = \frac{d}{dt}(\mathbf{v} \times \mathbf{h})$$

since $\mathbf{h}$ is a constant vector. That gives

$$\frac{d}{dt}(\mathbf{v} \times \mathbf{h}) = \frac{d}{dt}(GM_S\mathbf{r}_1)$$

or

$$\mathbf{v} \times \mathbf{h} = GM_S\mathbf{r}_1 + \mathbf{C}$$

where $\mathbf{C}$ is a constant vector. Since

$$h^2 = \mathbf{h} \cdot \mathbf{h} = (\mathbf{r} \times \mathbf{v}) \cdot \mathbf{h} = \mathbf{r} \cdot (\mathbf{v} \times \mathbf{h})$$

$$= (r\mathbf{r}_1) \cdot (GM_S\mathbf{r}_1 + \mathbf{C}) = rGM_S(\mathbf{r}_1 \cdot \mathbf{r}_1) + r(\mathbf{r}_1 \cdot \mathbf{C})$$

and since

$$\mathbf{r}_1 \cdot \mathbf{C} = C\cos\theta$$

we have

$$h^2 = rGM_S + rC\cos\theta$$

or

$$r = \frac{h^2}{GM_S + C\cos\theta} = \frac{h^2/GM_S}{1 + C/GM_S\cos\theta}$$

This equation is of a conic section and since the only closed conic section is an ellipse the law is proved.

9.4.2 Kepler's Second Law

The radius vector drawn from the sun to the planet sweeps out equal areas in equal periods of time.

Proof This was proved in Sect. 9.1 as a property of a central force, where we've seen that for any central force, the position vector r of the particle from the center of force O sweeps out equal areas in equal times. That is,

$$\frac{dA}{dt} = \frac{h}{2} = \text{constant}$$

or

$$\frac{dA}{dt} = \frac{L}{2m} = \text{constant}$$

Here, the center of force is the sun and the particle is the planet, hence we have

$$\frac{dA}{dt} = \frac{L}{2M_P}$$

9.4.3 Kepler's Third Law

The square of the period of revolution of any planet about the sun is proportional to the cube of the semimajor axis of its orbit.

Proof The area of an ellipse is given by $A = \pi ab$, where a and b are the semimajor and semiminor axis of the ellipse, respectively. From Kepler's second law, the areal velocity is a constant given by

$$\frac{dA}{dt} = \frac{h}{2} = \text{constant}$$

Therefore, the period of revolution may be considered as the time it takes the radius vector to sweep an area of πab

$$T = \frac{\pi ab}{h/2}$$

From Sect. 9.3, we have $b = a\sqrt{1 - e^2}$. That gives

$$T = \frac{\pi a^2\sqrt{1 - e^2}}{h/2}$$

Also, we've seen that the eccentricity for the gravitational force is given by $e = h^2C/GM$ or $e = h^2C/GM_S$ in the case of the planet–sun system. Since $ed = a(1 - e^2)$, we have

$$\frac{h^2}{GM_S} = a(1 - e^2)$$

or

$$\sqrt{1 - e^2} = \frac{h}{\sqrt{GM_Sa}}$$

Thus,

$$T = \frac{2\pi a^2 h}{h\sqrt{GM_Sa}} = \frac{2\pi}{\sqrt{GM_S}}a^{3/2}$$

or

$$T^2 = \left(\frac{4\pi^2}{GM_S}\right)a^3 = K_S a^3$$

where K_S is a constant that has a value given by

$$K_S = \frac{4\pi^2}{GM_S} = 2.97 \times 10^{-19} \text{ s}^2/\text{m}^3$$

This proves Kepler's third law. Note that, Kepler's laws apply also for satellites. In such cases, the mass of the sun in the previous equations is replaced by the earth or any other planet about which the satellite revolves.

9.5   Circular Orbits

The orbits of most planets in our solar system are almost circular. Next, we will find the total energy of a body of mass m moving in a circular orbit about a massive body of mass M that is assumed to be fixed (at rest) in an inertial frame of reference. From that energy, we will find the eccentricity and prove that the orbit is circular. The potential energy of such system is

$$U = \frac{-GMm}{r}$$

where r is the radius of the circular orbit. Applying Newton's second law to m gives

$$\frac{GMm}{r^2} = m\frac{v^2}{r} \qquad (9.31)$$

Therefore, the kinetic energy of the particle is

$$K = \frac{1}{2}mv^2 = \frac{GMm}{2r}$$

The total energy of m is therefore given by

$$E = K + U = \frac{GMm}{2r} - \frac{GMm}{r}$$

or

$$E = -\frac{GMm}{2r} \qquad (9.32)$$

In Sect. 9.4, the eccentricity of orbit in terms of energy was given by

$$e = \sqrt{1 + \frac{2Eh^2}{G^2M^2m}} \qquad (9.33)$$

Substituting Eq. 9.32 into Eq. 9.33 gives

$$e = \sqrt{1 + \left(\frac{-GMm}{2r}\frac{2h^2}{G^2M^2m}\right)}$$

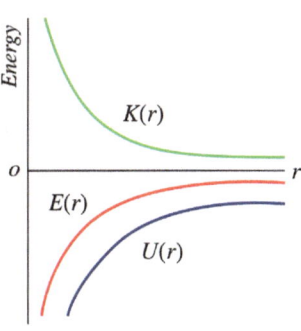

Fig. 9.23 The potential, kinetic and total energy as functions of r of an object in a circular orbit

Since $h = rv$ for a circular orbit and since $GMm/r^2 = mv^2/r$ and thus $v = \sqrt{GM/r}$, we have

$$h = \sqrt{rGM}$$

and

$$e = \sqrt{1 + \left(\frac{-GMm}{2r}\frac{2rGM}{G^2M^2m}\right)} = 0$$

Hence the orbit is circular. The potential, kinetic, and total energy as functions of r of an object in circular orbit are shown in Fig. 9.23.

Example 9.15 A satellite of mass of 1000 kg is in circular orbit about the earth at an altitude of $R_E/2$. What is the amount of work required to move the satellite to an altitude of $2R_E$.

Solution 9.15

$$W = \Delta E = E_f - E_i = GM_E m_s\left(\frac{-1}{2r_f} - \left(\frac{-1}{2r_i}\right)\right) = GM_E m_s\left(\frac{-1}{4R_E} + \frac{1}{R_E}\right)$$

$$= \frac{3GM_E m_s}{4R_E} = \frac{3(6.67 \times 10^{-11} \text{ Nm}^2/\text{kg}^2)(5.98 \times 10^{24} \text{ kg})(1000 \text{ kg})}{4(6.37 \times 10^6 \text{ m})} = 4.7 \times 10^{10} \text{ J}$$

9.6   Elliptical Orbits

For an elliptical orbit, we have

$$ed = a(1 - e^2) = \frac{h^2}{GM} \qquad (9.34)$$

Substituting Eq. 9.33 into Eq. 9.34 gives

$$a\left(1 - \left(1 + \frac{2Eh^2}{G^2M^2m}\right)\right) = \frac{h^2}{GM}$$

That gives

$$E = \frac{-GMm}{2a}$$

The speed of an object in an elliptical orbit can be found from

$$K = E - U$$

$$\frac{1}{2}mv^2 = \frac{-GmM}{2a} + \frac{GmM}{r}$$

$$v^2 = GM\left(\frac{2}{r} - \frac{1}{a}\right)$$

$$v = \sqrt{GM\left(\frac{2}{r} - \frac{1}{a}\right)}$$

9.7   The Escape Speed

The escape speed v_{esc} is the speed required for an object to escape from the influence of the gravitational field of an astronomical object or system. Suppose an object of mass m is projected from the surface of a planet of mass M. The minimum speed for the object to escape the gravitational field of the planet is that in which the object has zero total mechanical energy at infinity. From conservation of energy, we have

$$K_i + U_i = K_f + U_f$$

$$\frac{1}{2}mv_{esc}^2 + \left(\frac{-GMm}{R}\right) = 0$$

Hence

$$v_{esc} = \sqrt{\frac{2GM}{R}}$$

where R is the radius of the planet. If the object's initial speed is greater than the escape speed from that planet, then the object will still have some kinetic energy at infinity. Table.9.2 shows planetary data escape speeds

Example 9.16 What is the escape speed from the surface of: (a) Earth; (b) Mars; (c) Pluto.

Solution 9.16 (a)

$$v_{esc} = \sqrt{\frac{2GM_E}{R_E}} = \sqrt{\frac{2(6.67 \times 10^{-11} \text{ Nm}^2/\text{kg}^2)(5.98 \times 10^{24} \text{ kg})}{(6.37 \times 10^6 \text{ m})}} = 1.12 \times 10^4 \text{ m/s}$$

(b)

$$v_{esc} = \sqrt{\frac{2GM_M}{R_M}} = \sqrt{\frac{2(6.67 \times 10^{-11} \text{ Nm}^2/\text{kg}^2)(6.42 \times 10^{23} \text{ kg})}{(3.37 \times 10^6 \text{ m})}} = 5 \times 10^3 \text{ m/s}$$

(c)

$$v_{esc} = \sqrt{\frac{2GM_P}{R_P}} = \sqrt{\frac{2(6.67 \times 10^{-11} \text{ Nm}^2/\text{kg}^2)(1.4 \times 10^{22} \text{ kg})}{(1.5 \times 10^6 \text{ m})}} = 1.1 \times 10^3 \text{ m/s}$$

Example 9.17 What must be the minimum speed of a spacecraft that is at a distance of $3R_E$ from the center of the earth in order for it to escape the gravitational field of the earth?

Solution 9.17 The minimum speed is that in which the spacecraft has zero total mechanical energy at infinity,

$$K_i + U_i = K_f + U_f$$

$$\frac{1}{2}mv_{esc}^2 + \left(\frac{-GM_Em}{3R_E}\right) = 0$$

$$v_{esc} = \sqrt{\frac{2GM_E}{3R_E}} = \sqrt{\frac{2(6.67 \times 10^{-11} \text{ Nm}^2/\text{kg}^2)(5.98 \times 10^{24} \text{ kg})}{3(6.37 \times 10^6 \text{ m})}} = 6.46 \times 10^3 \text{ m/s}$$

Example 9.18 Given that the period of Mars in its orbit about the sun is 1.88 years and its semimajor axis of the orbit is 22.8×10^{10} m, find the mass of the sun.

Solution 9.18 The period in seconds is

$$T = 5.94 \times 10^7 \text{s}$$

From Kepler's second law, we have

$$M_S = \frac{4\pi^2 a^3}{GT^2} = \frac{4(3.14)^2(22.8 \times 10^{10} \text{ m})^3}{(6.67 \times 10^{-11} \text{ N m}^2/\text{kg}^2)(5.94 \times 10^7 \text{ s})^2} = 1.99 \times 10^{30} \text{ kg}$$

Example 9.19 Halley's Comet moves in an elliptical orbit about the sun. Its semimajor axis of orbit is 2.7×10^{12} m and its farthest distance ($OV' = R_a$) from the sun (the aphelion) is 5.3×10^{12} m. Find its period and its closest approach to the sun (the perihelion $OV = R_p$).

Solution 9.19 From Kepler's third law,

$$T^2 = K_S a^3 = (2.97 \times 10^{-19} \text{ s}^2/\text{m}^3)(2.7 \times 10^{12} \text{ m})^3$$

$$T = 2.4 \times 10^9 \text{ s} = 76 \text{ years}$$

From Eq. 9.23, we have

$$OV + OV' = 2a$$

or

$$R_p + R_a = 2a$$

$$R_p = 2a - R_a = 2(2.7 \times 10^{12} \text{ m}) - (5.3 \times 10^{12} \text{ m}) = 1 \times 10^{11} \text{ m}$$

Example 9.20 If Pluto's distance from the sun at perihelion is 4.43×10^{12} m, find (a) the ratio of its speed at perihelion to its speed at aphelion; (b) the eccentricity of orbit; (c) the total energy.

Table 9.2 Planetary data escape speeds

Body	Mass (kg)	Radius (m)	Semimajor axis a (m)	Escape speed (km/s)
Mercury	3.18×10^{23}	2.43×10^6	5.79×10^{10}	4.3
Venus	4.88×10^{24}	6.06×10^6	1.08×10^{11}	10.3
Earth	5.98×10^{24}	6.37×10^6	1.496×10^{11}	11.2
Mars	6.42×10^{23}	3.37×10^6	2.28×10^{11}	5
Jupiter	1.90×10^{27}	6.99×10^7	7.78×10^{11}	60
Saturn	5.68×10^{26}	5.85×10^7	1.43×10^{12}	36
Uranus	8.68×10^{25}	2.33×10^7	2.87×10^{12}	22
Neptune	1.03×10^{26}	2.21×10^7	4.5×10^{12}	24
Pluto	1.4×10^{22}	1.5×10^6	5.91×10^{12}	1.1
Moon	7.36×10^{22}	1.74×10^6		2.3
Sun	1.99×10^{30}	6.96×10^8		618

Solution 9.20 From Table. 9.2, we have $a = 5.9 \times 10^{12}$ m, therefore

$$R_a = 2a - R_p = 2(5.9 \times 10^{12} \text{ m}) - (4.43 \times 10^{12} \text{ m}) = 7.37 \times 10^{12} \text{ m}$$

From the conservation of angular momentum,

$$M_P v_a R_a = M_P v_p R_p$$

hence,

$$\frac{v_p}{v_a} = \frac{R_a}{R_p} = \frac{(7.37 \times 10^{12} \text{m})}{(4.43 \times 10^{12} \text{m})} = 1.7$$

(b) From Eq. 9.24 ($OV = R_p = a(1 - e)$), we have

$$e = 1 - \frac{R_p}{a} = 1 - \frac{(4.43 \times 10^{12} \text{ m})}{(5.9 \times 10^{12} \text{ m})} = 0.25$$

(c)

$$E = \frac{-GMm}{2a} = \frac{-(6.67 \times 10^{-11} \text{ Nm}^2/\text{kg}^2)(1.99 \times 10^{30} \text{ kg})(1.4 \times 10^{22} \text{ kg})}{2(5.9 \times 10^{12} \text{ m})} = -1.6 \times 10^{29} \text{ J}$$

Example 9.21 Two stars of equal mass M revolve about their center of mass with a speed v as shown in Fig. 9.24. Find the period of motion of each star.

Solution 9.21 The gravitational force that one star exerts on the other is

$$F = \frac{GM^2}{4r^2} = \frac{Mv^2}{r}$$

where r is the radius of orbit. Therefore,

$$v = \sqrt{\frac{GM}{4r}}$$

and

Fig. 9.24 Two stars of equal mass M revolve about their center of mass with a speed v

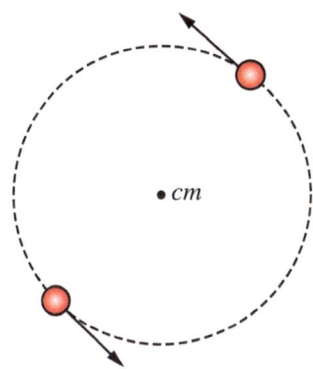

$$T = \frac{2\pi r}{v} = 2\pi r \sqrt{\frac{4r}{GM}} = 4\pi \sqrt{\frac{r^3}{GM}}$$

Example 9.22 A spaceship is fired from the surface of Mars with a speed of 12×10^3 m/s, find its speed at a very far distance from Mars.

Solution 9.22

$$K_i + U_i = K_f + U_f$$

$$\frac{1}{2}mv_i^2 - \left(\frac{GmM_M}{R_M}\right) = \frac{1}{2}mv_f^2 + 0$$

$$v_f^2 = v_i^2 - \frac{2GM_M}{R_M}$$

$$= (12 \times 10^3 \text{ m/s})^2 - \frac{2(6.67 \times 10^{-11} \text{ N m}^2/\text{kg}^2)(6.42 \times 10^{23} \text{ kg})}{(3.37 \times 10^6 \text{ m})}$$

That gives $v_f = 1.1 \times 10^4$ m/s.

Problems

1. Calculate the gravitational force between the earth and (a) the sun, (b) the moon.
2. Calculate the gravitational acceleration at the surface of Mars.
3. Three particles of masses $m_1 = 2$ kg, $m_2 = 6$ kg, and $m_3 = 3$ kg are located at the points $(0, 0)$, $(0, 5)$, and $(5, 0)$, respectively. Find magnitude and direction of the resultant gravitational force exerted on m_3.
4. The Geosynchronous satellites move in a circular orbit in the equatorial plane of the earth. They move in such a way that they always remain over the same point on the earth. Find the height and velocity of this satellite.
5. If the eccentricity of the orbit of Mercury about the sun is $e = 0.206$ and its semimajor axis is $a = 0.387$ AU, find (a) the distance of its farthest and closest approach to the sun (the aphelion and perihelion), (b) its period, (c) its total energy, (d) its angular momentum. (1 AU = 1.495×10^{11} m).
6. A body is released at a distance r from the center of the earth. Find its velocity just as it hits the surface of the earth.

7. Show that the speed of a satellite in an elliptical orbit about the earth at apogee and perigee are given by

$$v_p = \sqrt{\frac{GM}{a}} \sqrt{\frac{1+e}{1-e}} = \sqrt{\frac{GM}{a}} \sqrt{\frac{R_a}{R_p}}$$

and

$$v_a = \sqrt{\frac{GM}{a}} \sqrt{\frac{1-e}{1+e}} = \sqrt{\frac{GM}{a}} \sqrt{\frac{R_p}{R_a}}$$

8. An artificial satellite moves in an elliptical orbit about the earth. Its perigee and apogee altitudes are 1100 km and 4100 km respectively Find (a) the velocity of the satellite at perigee and apogee, (b) its semimajor axis, (c) its eccentricity, (d) the equation of its orbit, (e) its period, (f) its speed when it is at a distance of 3000 km above the earth's surface.
9. A satellite is at a distance of $1.2R_E$ from the center of the earth. Find the speed required for the satellite at this altitude (where it represents the orbit perigee) to be in (a) circular orbit, (b) parabolic orbit, (c) elliptical orbit of eccentricity of $e = 0.7$.
10. Suppose the earth suddenly stops moving about the sun, find the time it would take the earth to fall to the sun.

Open Access This chapter is licensed under the terms of the Creative Commons Attribution 4.0 International License (http://creativecommons.org/licenses/by/4.0/), which permits use, sharing, adaptation, distribution and reproduction in any medium or format, as long as you give appropriate credit to the original author(s) and the source, provide a link to the Creative Commons license and indicate if changes were made.

The images or other third party material in this chapter are included in the chapter's Creative Commons license, unless indicated otherwise in a credit line to the material. If material is not included in the chapter's Creative Commons license and your intended use is not permitted by statutory regulation or exceeds the permitted use, you will need to obtain permission directly from the copyright holder.

Oscillatory Motion

10.1 Oscillatory Motion

A motion repeating itself is referred to as periodic or oscillatory motion. An object in such motion oscillates about an equilibrium position due to a restoring force or torque. Such force or torque tends to restore (return) the system toward its equilibrium position no matter in which direction the system is displaced. This motion is important to study many phenomena including electromagnetic waves, alternating current circuits, and molecules. For a vibration to occur, two quantities are necessary to be present—stiffness and inertia.

10.2 Free Vibrations

When a system vibrates, a restoring force must be present. In addition to that force, there is always a retarding or damping force such as friction. If the effect of the damping force is small and can be neglected, then the motion is classified as free and undamped motion. Otherwise, the motion is classified as free damped motion. In both cases, the motion is known as free vibration since no forces other than the restoring and damping forces exist during vibration. If a driving force that does positive work on the system exists, the motion is classified as forced vibration.

This force may be applied externally to the system or sometimes is produced within the system. In this chapter, the case in which a restoring force is directly proportional to the displacement is considered. The resulting motion is then known as a harmonic vibration and the system is said to be linear. If the restoring force depends on the displacement in some other way, the resulting motion is known as anharmonic vibration and the system is said to be nonlinear.

10.3 Free Undamped Vibrations

This kind of motion is known as the simple harmonic motion. Next, we will examine examples of such motion in physics.

10.3.1 Mass Attached to a Spring

Consider a block of mass m attached to a light spring of spring constant k that is fixed at the other end (see Fig. 10.1). Suppose that the system lies on a frictionless horizontal surface. For small displacements, the restoring force acting on the block by the spring is given by Hook's law

$$F_s = -kx$$

As we've mentioned in Sect. 4.1, if the block is displaced slightly to the right (for example to $x = A$), the restoring spring force will accelerate the block to the left transferring its potential energy into kinetic energy As the block reaches its equilibrium position $x = 0$, all of its potential energy will be transformed into kinetic energy and it will overshoot to the other side. Again, as it moves left, the spring force decelerates the block to the right, transferring its kinetic energy into potential energy until all of its energy is potential at $x = -A$ where it comes to rest. At that point, it accelerates back to $x = 0$ and regains all of its kinetic energy where it overshoots again to $x = A$. Therefore, stiffness restores the mass where inertia is responsible for the mass to overshoot. From Newton's second law we, have

$$ma = -kx$$

or

$$m\frac{d^2x}{dt^2} + kx = 0$$

or

$$\frac{d^2x}{dt^2} + \omega_n^2 x = 0 \tag{10.1}$$

where $\omega_n = \sqrt{k/m}$ is called the natural angular frequency of the system. The general solution of this equation is of the form

$$x(t) = A_1 \cos \omega_n t + A_2 \sin \omega_n t \tag{10.2}$$

© The Author(s) 2019
S. Alrasheed, *Principles of Mechanics*, Advances in Science,
Technology & Innovation, https://doi.org/10.1007/978-3-030-15195-9_10

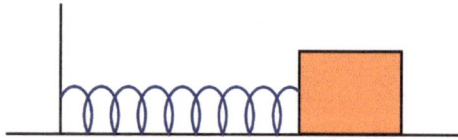

Fig. 10.1 A block of mass m attached to a light spring of spring constant k that is fixed at the other end

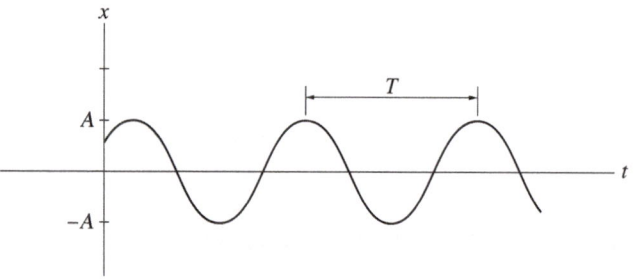

Fig. 10.2 Plot of x versus t for a simple harmonic oscillator

where A_1 and A_2 are arbitrary constants that can be found from the initial conditions. Therefore, there are many possible motions with the same angular frequency ω_n. By multiplying and dividing Eq. 10.2 by $\sqrt{A_1^2 + A_2^2}$, you can show that the solution may be written as

$$x(t) = A\cos(\omega_n t - \phi) \tag{10.3}$$

where $A = \sqrt{A_1^2 + A_2^2}$ is called the amplitude of motion and $\phi = \tan^{-1} A_2/A_1$ is called the phase constant. In general, ϕ is chosen such that $0 \le \phi \le \pi$. A and ϕ can be determined from the initial conditions, i.e., from the values of the displacement and velocity when the motion starts. The mass therefore oscillates between A and $-A$. The quantity $(\omega_n t - \phi)$ is called the phase angle. If this angle is increased by 2π, all physical quantities such as the displacement, velocity, and acceleration repeat themselves. The plot of x versus t is shown in Fig. 10.2. If A is fixed and ϕ is changed the motion will be the same except that the same physical quantities will appear either earlier or later than the preceding motion.

10.3.1.1 The Period and Frequency of Motion

The period of motion is the time required for one complete cycle or oscillation. Since the phase angle is changed by 2π after one complete cycle, we have for the mass–spring system,

$$\omega_n t + 2\pi = \omega_n(t + T)$$

or

$$T = \frac{2\pi}{\omega_n} = 2\pi\sqrt{\frac{m}{k}}$$

The frequency is defined as the number of complete cycles per unit time

$$f_n = \frac{1}{T} = \frac{\omega_n}{2\pi}$$

This frequency is called the natural frequency of the motion. The unit of the frequency is cycles/s or hertz (Hz).

10.3.1.2 The Phase Difference

The phase constant ϕ is important when comparing two or more oscillations of the same frequency Suppose a certain vibration has $\phi = 0$, this means that at $t = 0$ the displacement is maximum $x = A$. If a second vibration has also $\phi = 0$, then the two vibrations are said to be in phase (see Fig. 10.3 part a). Otherwise, the two vibrations are out of phase. If the phase constant of the second vibration is $\phi > 0$, then the second vibration is leading the first vibration in phase by ϕ. If $\phi < 0$, then the second vibration is lagging the first by ϕ. If $\phi = \pm\pi$, the two vibrations are said to be in antiphase with each other (see Fig. 10.3 part b).

10.3.1.3 The Velocity and Acceleration

The velocity of the mass is

$$v(t) = \frac{dx}{dt} = -\omega_n A\sin(\omega_n t - \phi) \tag{10.4}$$

This can also be written as

$$v(t) = \omega_n A\cos\left(\omega_n t - \phi + \frac{\pi}{2}\right) \tag{10.5}$$

The acceleration of the mass is

$$a(t) = \frac{dv}{dt} = -\omega_n^2 A\cos(\omega_n t - \phi) \tag{10.6}$$

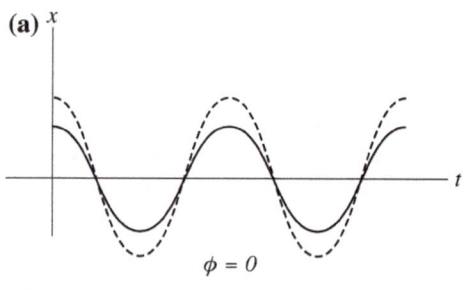

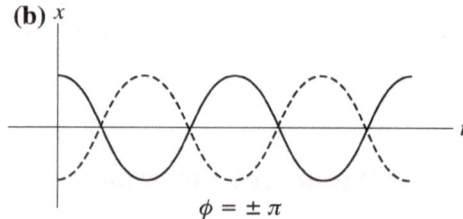

Fig. 10.3 **a** Two simple harmonic motions of the same frequency and same phase constant $\pi = 0$ but differing in amplitude. **b** Two simple harmonic motions of the same frequency and amplitude but differing in phase by $\phi = \pm\pi$

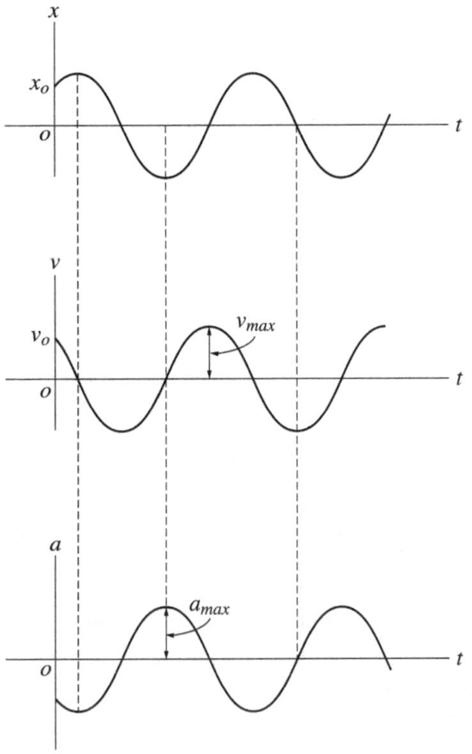

Fig. 10.4 The displacement, velocity and acceleration versus time

or

$$a(t) = \frac{dv}{dt} = \omega_n^2 A \cos(\omega_n t - \phi + \pi) \quad (10.7)$$

Hence, the velocity and acceleration also vary harmonically with time with amplitudes $\omega_n A$ and $\omega_n^2 A$, respectively, but they all have the same angular frequency From Eqs. 10.5 and 10.7 you can see that the velocity leads the displacement by $\pi/2$ or 90. The acceleration on the other hand leads the velocity by $\pi/2$ and the displacement by π or 180. Figure 10.4 shows the displacement, velocity, and acceleration versus time.

10.3.1.4 Boundary Conditions

Boundary conditions are used to find A and ϕ for a specific vibration. Suppose that the vibration is measured when the stopwatch is set to zero, i.e., at $t = 0$ and that at that instant the mass is released from rest at a distance of $x = A_1$ from its equilibrium position. Substituting these conditions into Eqs. 10.3 and 10.4, we have

$$x = A \cos \phi = A_1 \quad (10.8)$$

$$v = v_0 = -\omega_n A \sin \phi \quad (10.9)$$

Dividing Eq. 10.9 by Eq. 10.8 gives

$$\tan \phi = \frac{-v_0}{\omega_n A_1}$$

Squaring and adding Eqs. 10.9 and 10.8 gives

$$A_1^2 + \left(\frac{v_0}{\omega_n} \right)^2 = A^2 \cos^2 \phi + A^2 \sin^2 \phi$$

or

$$A = \sqrt{A_1^2 + \left(\frac{v_0}{\omega_n} \right)^2}$$

Example 10.1 An object oscillates in simple harmonic motion according to the expression $x = (3\text{m}) \cos(\pi t + \pi/3)$. Find (a) the amplitude, phase constant, period, and frequency of motion; (b) the displacement, velocity, and acceleration of the object at $t = 0.5\text{s}$ (c) the time when the object first reach $x = -1.5$ m.

Solution 10.1 (a)

$$A = 3 \text{ m}$$

$$\phi = \frac{\pi}{3}$$

$$T = \frac{2\pi}{\omega_n} = \frac{(2\pi)}{\pi} = 2 \text{ s}$$

and

$$f_n = \frac{1}{T} = \frac{1}{(2\text{s})} = 0.5 \text{ Hz}$$

(b) At $t = 0.5$ s

$$x = (3 \text{ m}) \cos \left(\pi (0.5 \text{ s}) + \frac{\pi}{3} \right) = -2.6 \text{ m}$$

$$v = -(3\pi \text{ m/s}) \sin \left(\pi t + \frac{\pi}{3} \right)$$

At $t = 0.5$ s

$$v = (-3\pi \text{ m/s}) \sin \left(\pi (0.5 \text{ s}) + \frac{\pi}{3} \right) = -4.7 \text{ m/s}$$

$$a = (-3\pi^2 \text{ m/s}^2) \cos \left(\pi t + \frac{\pi}{3} \right)$$

at $t = 0.5$ s

$$a = (-3\pi^2 \text{ m/s}^2) \cos \left(\pi (0.5 \text{ s}) + \frac{\pi}{3} \right) = 25.6 \text{ m/s}^2$$

(c) at $x = -1.5$ m

$$(-1.5 \text{ m}) = (3 \text{ m}) \cos \left(\pi t + \frac{\pi}{3} \right)$$

or

$$\frac{2\pi}{3} = \pi t + \frac{\pi}{3}$$

that gives $t = 0.3$ s.

Example 10.2 A 9 kg object is moving along the x-axis under the influence of a force given by $F = (-3x)$ N. Find (a) the equation of motion; (b) the displacement of the mass at any time if at $t = 0$, $x = 5$ m and $v = 0$.

Solution 10.2 (a)

$$F = -3x = ma = m\frac{d^2x}{dt^2}$$

hence,

$$\frac{d^2x}{dt^2} + 3x = 0$$

(b) The general solution of this equation is

$$x = A \cos \sqrt{3}t + B \sin \sqrt{3}t$$

Since at $t = 0$, $x = 5$ m, then $A = 5$ m and

$$x = (5\text{m}) \cos \sqrt{3}t + B \sin \sqrt{3}t$$

also we have at $t = 0$, $dx/dt = 0$, or

$$-5\sqrt{3} \sin \sqrt{3}t + \sqrt{3}B \cos \sqrt{3}t = 0$$

and therefore $B = 0$. Thus,

$$x = (5\text{m}) \cos \sqrt{3}t$$

Example 10.3 A 0.3 kg block is attached to a spring of force constant 20 N/m on a frictionless horizontal surface. If the initial displacement and velocity of the system is 0.02 m and 0.2 m/s, respectively, find the period, amplitude, and phase constant of motion.

Solution 10.3

$$\omega_n = \sqrt{\frac{k}{m}} = \sqrt{\frac{(20 \text{ N/m})}{(0.3 \text{ kg})}} = 8.2 \text{ rad/s}$$

$$A = \sqrt{A_1^2 + \left(\frac{v_0}{\omega_n}\right)^2} = \sqrt{(0.02 \text{ m})^2 + \left(\frac{(0.2 \text{ m/s})}{(82 \text{ rad/s})}\right)^2} = 0.03 \text{ m}$$

$$\tan \phi = \frac{-v_0}{\omega_n A_1} = \frac{-(0.2 \text{ m/.s})}{(8.2 \text{ rad/s})(0.03 \text{ m})} = -0.8$$

$$\phi = -38.7°$$

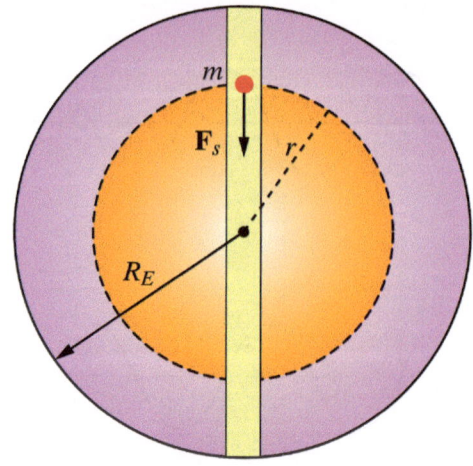

Fig. 10.5 A particle of mass m is dropped in a straight tunnel that is drilled through the earth and which passes through the center of earth

Example 10.4 A particle of mass m is dropped in a straight tunnel that is drilled through the earth and which passes through the center of earth as shown in Fig. 10.5. Show that the motion of the particle is simple harmonic motion and find its period.

Solution 10.4 Assuming that the earth is a perfect sphere of uniform density and since the particle is inside the earth, then from Sect. 9.2, the gravitational force exerted on the particle by the earth is

$$F = -\left(\frac{GmM_E}{R_E^3}\right) r = -kr$$

Because this force is directly proportional to the displacement and is opposite to it, then the particle will move in simple harmonic motion about the center of the earth. The equation of motion is

$$\frac{dr^2}{dt^2} + \left(\frac{GM_E}{R_E^3}\right) r = 0$$

hence,

$$\omega_n = \sqrt{\frac{GM_E}{R_E^3}} = \sqrt{\frac{(6.67 \times 10^{-11} \text{ Nm}^2/\text{kg}^2)(5.98 \times 10^{24} \text{ kg})}{(6.37 \times 10^6 \text{ m})^3}} = 1.24 \times 10^{-3} \text{ rad/s}$$

$$T = \frac{2\pi}{\omega_n} = \frac{2(3.14)}{(1.24 \times 10^{-3} \text{ rad/s})} = 5055.4 \text{ s} = 84.25 \text{ min}$$

Example 10.5 A 0.4 kg block is connected to two springs of force constants $k_1 = 20$ N/m and $k_2 = 50$ N/m as in

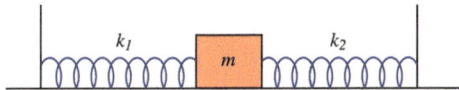

Fig. 10.6 A block connected to two springs

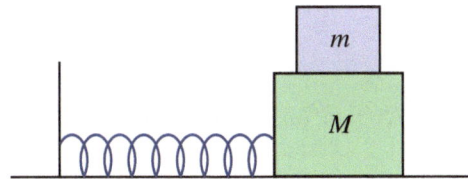

Fig. 10.7 A second block on top of a block connected to a spring

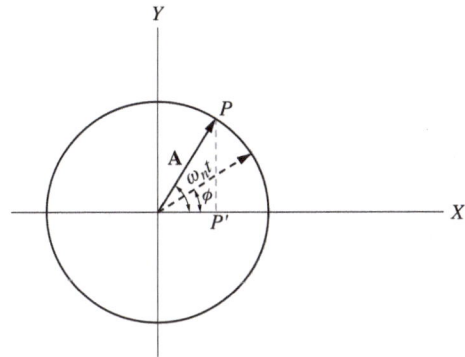

Fig. 10.8 A particle in uniform circular motion

Fig. 10.6. Find (a) the total force acting on the block; (b) the period of motion.

Solution 10.5 The force that each spring exerts on the block acts in the opposite direction of the displacement, therefore we have

$$\sum F = -k_1 x - k_2 x = -(k_1 + k_2)x = -(70 \text{ N/m})x$$

Thus the two springs can be considered as one spring of a force constant of $(k_1 + k_2)$. The period of motion is therefore

$$T = 2\pi \sqrt{\frac{m}{k_1 + k_2}} = 2(3.14)\sqrt{\frac{(0.4 \text{ kg})}{(70 \text{ N/m})}} = 0.5 \text{ s}$$

Example 10.6 A 6 kg block is connected to a light spring of force constant of 300 N/m on a frictionless horizontal surface. On top of it a second block of mass of 2 kg is placed. If the coefficient of static friction between the two blocks is 0.4 (see Fig. 10.7), find the maximum amplitude the system can have when it is in simple harmonic motion such that there is no slipping between the blocks.

Solution 10.6 The maximum acceleration of the lower block is $a_{max} = \omega_n^2 A$. In order for the upper block not to slip, the force of static friction between the two blocks must produce the same acceleration as the lower block. The maximum statistical frictional force that can be exerted on the upper block is $\mu_s mg$ and hence, the maximum acceleration that the force of static friction can produce is $\mu_s g$. Therefore, $\mu_s g = a_{max} = \omega_n^2 A$. Since

$$\omega_n = \sqrt{\frac{k}{(m + M)}}$$

we have

$$A = \frac{\mu_s g}{\omega_n^2} = \frac{\mu_s g (m + M)}{k} = \frac{(0.4)(9.8 \text{ m/s}^2)(8 \text{ kg})}{(300 \text{ N/m})} = 0.1 \text{ m}$$

10.3.2 Simple Harmonic Motion and Uniform Circular Motion

Consider a circle of radius A centered at the x and y axes as shown in Fig. 10.8. Let A be the position vector of a particle P rotating with a constant angular speed ω_n in the anticlockwise direction. The particle is thus in uniform circular motion. Suppose P starts the rotation at $t = 0$ at an angle of ϕ measured from the positive x-axis. At any time, the angular position of the particle is given by $(\omega_n t + \phi)$, therefore the vector position of the particle at any time is

$$\mathbf{A} = x\mathbf{i} + y\mathbf{j} = A\cos(\omega_n t + \phi)\mathbf{i} + A\sin(\omega_n t + \phi)\mathbf{j}$$

Hence,

$$x = A\cos(\omega_n t + \phi)$$

and

$$y = A\sin(\omega_n t + \phi)$$

That is, as P moves in uniform circular motion, its projection P′ on the x-axis moves in simple harmonic motion where the radius of the circle is equal to the amplitude of motion. The projection of P along the y-axis also undergoes simple harmonic motion. Thus, uniform circular motion may be considered as a combination of the simple harmonic motions of the projections of P on each axis. These two simple harmonic motions have equal amplitudes and angular frequencies but are in quadrature with each other (they differ in phase by $\pi/2$). The linear tangential velocity of the particle in this uniform circular motion is given by

$$v = A\omega_n$$

The x component of the velocity is from Fig. 10.9 given by

$$v_x = -\omega_n A\sin(\omega_n t + \phi)$$

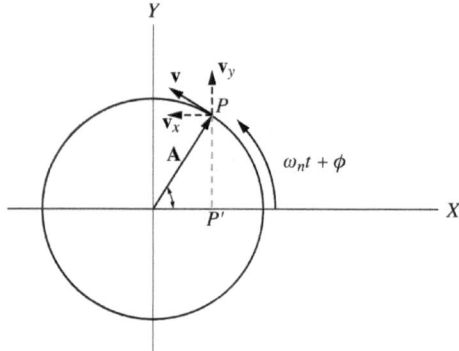

Fig. 10.9 The velocity components of the particle

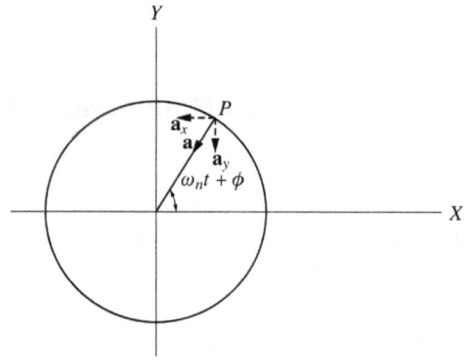

Fig. 10.10 The acceleration components of the particle

The acceleration of the particle in uniform circular motion is just the radial (centripetal) acceleration that is given by

$$a = \frac{v^2}{A} = A\omega_n^2$$

The x components of the acceleration (see Fig. 10.10) is

$$a_x = -\omega_n^2 A \cos(\omega_n t + \phi)$$

Hence as you can see, the displacement, velocity, and acceleration of the projection of P onto the x (or y axis) are the same as that of a simple harmonic motion. From this, we conclude that the simple harmonic motion can be represented as the projection of uniform circular motion along a diameter of the circle.

10.3.3 Energy of a Simple Harmonic Oscillator

Since in a simple harmonic oscillator, there aren't any dissipative forces, the total mechanical energy of the system is conserved and is equal to the sum of its kinetic and potential energies, that is

$$E = K + U$$

$$K = \frac{1}{2}mv^2 = \frac{1}{2}m\omega_n^2 A^2 \sin^2(\omega_n t + \phi)$$

$$U = \frac{1}{2}kx^2 = \frac{1}{2}kA^2 \cos^2(\omega_n t + \phi)$$

Thus,

$$E = \frac{1}{2}kA^2[\sin^2(\omega_n t + \phi) + \cos^2(\omega_n t + \phi)]$$

or

$$E = \frac{1}{2}kA^2 = \text{constant}$$

The equation of motion of a simple harmonic oscillator can be obtained from the total mechanical energy of the system as follows:

$$E = \frac{1}{2}m\dot{x}^2 + \frac{1}{2}kx^2 = \frac{1}{2}kA^2 \qquad (10.10)$$

$$\frac{dE}{dt} = m\dot{x}\ddot{x} + kx\dot{x} = 0$$

or

$$m\ddot{x} + kx = 0$$

Hence

$$\ddot{x} + \omega_n^2 x = 0$$

where $\omega_n = \sqrt{k/m}$. As the mass moves, its kinetic energy is transformed into potential energy and vice versa. Figure 10.11 shows the kinetic energy and potential energy of the system as a function of time and as a function of the displacement respectively Note that the variation of U and K with time is at twice the angular frequency of the variation of x, v, and a with time. This is because the potential energy is converted to kinetic energy twice in each cycle. The velocity of the simple harmonic oscillator can be obtained from the total energy of the system. From Eq. 10.10, we have

$$v = \pm\sqrt{\frac{k}{m}(A^2 - x^2)}$$

Hence, the maximum speed is at $x = 0$ and is zero at $x = \pm A$ which are called the turning points as discussed in Chap. chap444.

Example 10.7 A 0.3 kg mass is attached to a light spring. If the total energy of the system is 0.025 J and the amplitude of motion is 5 cm, find the period and frequency of motion.

Solution 10.7

$$E = (0.025\,\text{J}) = \frac{1}{2}kA^2 = \frac{1}{2}k(0.05\,\text{m})^2$$

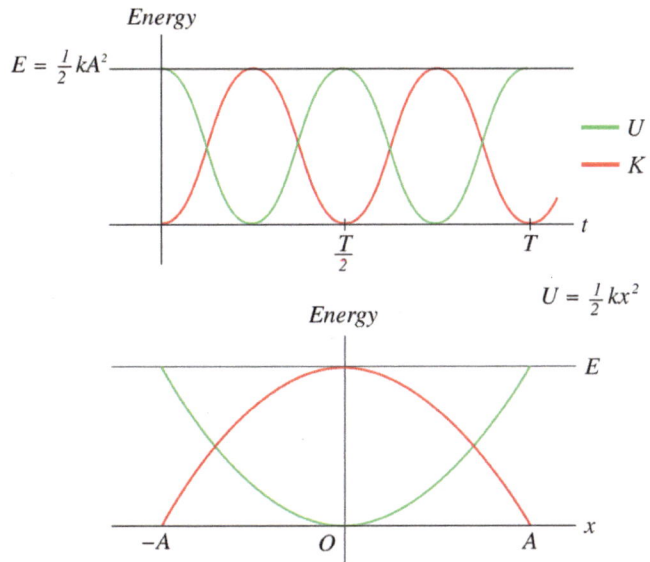

Fig. 10.11 As the mass moves, its kinetic energy is transformed into potential energy and vice versa

hence

$$k = 20 \text{ N/m}$$

The period of motion is therefore

$$T = 2\pi \sqrt{\frac{m}{k}} = 2(3.14) \sqrt{\frac{(0.3 \text{ kg})}{(20 \text{ N/m})}} = 0.8 \text{ s}$$

and the frequency is

$$f_n = \frac{1}{T} = \frac{1}{(0.8 \text{ s})} = 1.25 \text{ Hz}$$

Example 10.8 A 0.2 kg block is attached to a light spring of force constant of 11 N/m on a horizontal frictionless surface. If the block is displaced a distance of 8 cm from its equilibrium position, find (a) the amplitude, the angular frequency, the period and the frequency of motion when the block is released; (b) the maximum force exerted on the block; (c) the total mechanical energy of the system; (d) the maximum speed and maximum acceleration of the block; (e) the velocity of the block when its displacement is 2 cm; (f) the acceleration of the block when its displacement is 3 cm.

Solution 10.8 (a)

$$A = 8 \text{ cm}$$

$$\omega_n = \sqrt{\frac{k}{m}} = \sqrt{\frac{(11 \text{ N/m})}{(0.2 \text{ kg})}} = 7.4 \text{ rad/s}$$

$$T = \frac{2\pi}{\omega_n} = \frac{2(3.14)}{(7.4 \text{ rad/s})} = 0.85 \text{ s}$$

$$f_n = \frac{1}{T} = \frac{1}{(0.85 \text{ s})} = 1.2 \text{ Hz}$$

(b)

$$|F| = kA = (11 \text{ N/m})(0.08 \text{ m}) = 0.9 \text{ N}$$

(c)

$$E = \frac{1}{2}kA^2 = \frac{1}{2}(11 \text{ N/m})(0.08 \text{ m})^2 = 0.035 \text{ J}$$

(d)

$$v_{\text{max}} = \omega_n A = (7.4 \text{ rad/s})(0.08 \text{ m}) = 0.6 \text{ m/s}$$

$$a_{\text{max}} = \omega_n^2 A = (7.4 \text{ rad/s})^2(0.08 \text{ m}) = 4.4 \text{ m/s}^2$$

(e)

$$v = \pm\sqrt{\frac{k}{m}(A^2 - x^2)} = \sqrt{\frac{(11 \text{ N/m})}{(0.2 \text{ kg})}((0.08 \text{ m})^2 - (0.02 \text{ m})^2)} = 1.8 \text{ m/s}$$

(f)

$$a = -\omega_n^2 x = -(7.4 \text{ rad/s})^2(0.03 \text{ m}) = -1.6 \text{ m/s}^2$$

Example 10.9 An object connected to a spring is in simple harmonic motion on a frictionless surface. If the object's displacement when $(2v_{\text{max}}/3)$ is ± 0.015 m, find the amplitude of motion.

Solution 10.9

$$\frac{1}{2}kA^2 = \frac{1}{2}mv^2 + \frac{1}{2}kx^2 = \frac{1}{2}m\frac{4\omega_n^2 A^2}{9} + \frac{1}{2}kx^2$$

therefore

$$A^2 = \frac{9}{5}x^2 = \frac{9}{5}(0.015 \text{ m})^2$$

$$A = 0.02 \text{ m}$$

Example 10.10 A solid cylinder is connected to a light spring as in Fig. 10.12. If the cylinder rolls without slipping along the surface, show that the motion of the cylinder is simple harmonic motion and find its frequency.

Solution 10.10 At any instant the total mechanical energy is

$$E = \frac{1}{2}kx^2 + \frac{1}{2}I_{cm}\omega^2 + \frac{1}{2}Mv_{cm}^2 = \frac{1}{2}kx^2 + \frac{1}{2}I_{cm}\frac{v_{cm}^2}{R^2} + \frac{1}{2}Mv_{cm}^2$$

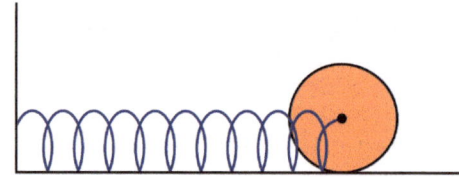

Fig. 10.12 A solid cylinder connected to a light spring

$$= \frac{1}{2}kx^2 + \frac{1}{2}\left(\frac{1}{2}MR^2\right)\frac{v_{cm}^2}{R^2} + \frac{1}{2}Mv_{cm}^2$$

Since the total mechanical energy is conserved

$$\frac{dE}{dt} = kv_{cm}x + \frac{1}{2}Mv_{cm}a_{cm} + Mv_{cm}a_{cm} = 0$$

$$kv_{cm}x = \frac{-3}{2}Mv_{cm}a_{cm}$$

or

$$a_{cm} = \frac{-2}{3}\frac{k}{M}x$$

$$\frac{d^2x}{dt^2} + \frac{2}{3}\frac{k}{M}x = 0$$

this equation is of a simple harmonic motion with

$$\omega_n = \sqrt{\frac{2}{3}\frac{k}{M}}$$

10.3.4   The Simple Pendulum

The simple pendulum is an example of an angular vibration in which the restoring effect is due to a restoring torque. A simple pendulum consists of a mass (called the bob) suspended by a light string of length L that is fixed at the other end (see Fig. 10.13). If the mass is pulled to the right or left from its equilibrium position and released, then the pendulum will swing in a vertical plane about an axis passing through O. The resulting motion is then a periodic or oscillatory motion. The restoring torque is due to gravity and is given by

$$\tau = -(mg\sin\theta)L$$

The minus sign indicates that the torque is a restoring torque, since it always tends to decrease θ. The moment of inertia of the bob about an axis passing through O is

$$I = mL^2$$

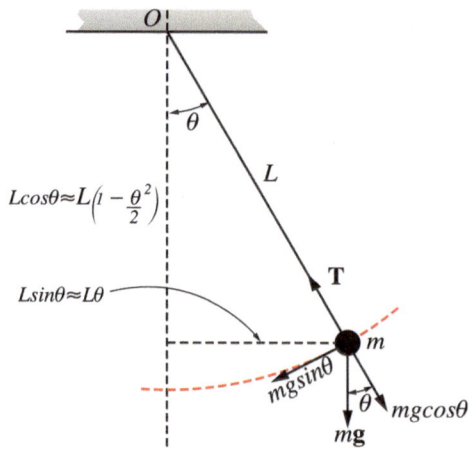

Fig. 10.13 The simple pendulum

From Newton's second law in angular form, we have

$$\tau = I\alpha = I\ddot{\theta}$$

Hence,

$$-mg\sin\theta L = mL^2\ddot{\theta}$$

or

$$\ddot{\theta} + \left(\frac{g}{L}\right)\sin\theta = 0 \qquad (10.11)$$

This equation does not represent a harmonic motion. That is because the torque is not directly proportional to the angular displacement. Thus, the system is nonlinear. However for small angular displacements, we have $\sin\theta \approx \theta$ (since $\sin\theta = \theta - \theta^3/3! + \theta^5/5!\ldots$) and Eq. 10.11 becomes

$$\ddot{\theta} + \left(\frac{g}{L}\right)\theta = 0$$

or

$$\ddot{\theta} + \omega_n^2\theta = 0 \qquad (10.12)$$

where $\omega_n = \sqrt{g/L}$. Hence for small angular displacements, the motion is a simple harmonic motion. The solution of Eq. 10.12 is of the form

$$\theta = \theta_m\cos(\omega_n t - \phi)$$

where θ_m is the maximum angular displacement and ϕ is the phase constant. The plot of this equation is shown in Fig. 10.14. The period of the simple pendulum is therefore given by

$$T = \frac{2\pi}{\omega_n} = 2\pi\sqrt{\frac{L}{g}}$$

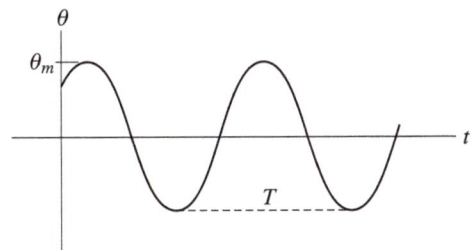

Fig. 10.14 The displacement versus time of a simple pendulum

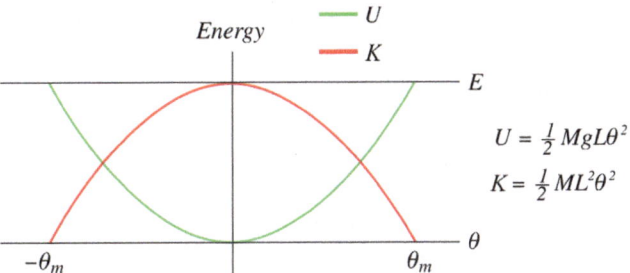

$$U = \frac{1}{2}MgL\theta^2$$

$$K = \frac{1}{2}ML^2\dot{\theta}^2$$

Fig. 10.15 The total energy of a simple pendulum

10.3.4.1 Energy

The kinetic energy of the simple pendulum is

$$K = \frac{1}{2}mv^2 = \frac{1}{2}mL^2\omega_n^2 = \frac{1}{2}mL\dot{\theta}^2$$

Taking the reference point of potential energy of the system to be zero when the bob is at the bottom, we have

$$U = MgL(1 - \cos\theta)$$

The total energy is therefore given by

$$E = K + U = \frac{1}{2}ML^2\dot{\theta}^2 + MgL(1 - \cos\theta)$$

For small θ, we have $\cos\theta \approx 1 - \frac{\theta^2}{2}$ since $\cos\theta = 1 - \theta^2/2! + \theta^4/4! \ldots$) thus

$$E = \frac{1}{2}ML^2\dot{\theta}^2 + \frac{1}{2}MgL\theta^2$$

Since

$$\dot{\theta} = -\theta_m\omega_n\sin(\omega_n t - \phi)$$

we have

$$E = \frac{1}{2}ML^2\theta_m^2\omega_n^2\sin^2(\omega_n t - \phi) + \frac{1}{2}MgL\theta_m^2\cos^2(\omega_n t - \phi)$$

or

$$E = \frac{1}{2}MgL\theta_m^2$$

Therefore, the total energy of the system is constant. Figure 10.15 shows the variation of the kinetic and potential energies with the displacement.

The equation of motion may also be obtained from energy as follows:

$$\frac{dE}{dt} = ML^2\dot{\theta}\ddot{\theta} + MgL\theta\dot{\theta} = 0$$

or

$$\ddot{\theta} + \left(\frac{g}{L}\right)\theta = 0$$

Example 10.11 A simple pendulum is 0.5 m long. Find its period at the surface of Mars and compare it to its period at the earth's surface.

Solution 10.11 At Mars's surface, the gravitational acceleration is

$$g_M = \frac{GM_M}{R_M^2} = \frac{(6.67 \times 10^{-11}\ \text{N } \textit{mathrmm}^2/\text{kg}^2)(6.42 \times 10^{23}\ \text{kg})}{(3.37 \times 10^6\ \text{m})^2} = 3.8\ \text{m/s}^2$$

The period at Mars is therefore

$$T_M = 2\pi\sqrt{\frac{L}{g_M}} = 2(3.14)\sqrt{\frac{(0.5\ \text{m})}{(3.8\ \text{m/s}^2)}} = 2.3\ \text{s}$$

At the earth's surface,

$$T_E = 2\pi\sqrt{\frac{L}{g_E}} = 2(3.14)\sqrt{\frac{(0.5\ \text{m})}{(9.8\ \text{m/s}^2)}} = 1.4\ \text{s}$$

Thus, $T_M = 1.6T_E$.

Example 10.12 A simple pendulum of length of 2 m is displaced through an angle of 12° and released. Find (a) the angular frequency of motion; (b) the maximum angular speed and maximum angular acceleration.

Solution 10.12 (a) The amplitude of motion is

$$\theta_{\text{max}} = (12°)\left(\frac{2\pi\ \text{rad}}{360°\ \text{deg}}\right) = 0.21\ \text{rad}$$

The angular frequency is

$$\omega_n = \sqrt{\frac{g}{L}} = \sqrt{\frac{(9.8\ \text{m/s}^2)}{(2\ \text{m})}} = 2.2\ \text{rad/s}$$

(b) The maximum angular speed is

$$\dot{\theta}_{max} = \omega_n A = (2.2 \text{ rad/s})(0.21 \text{ rad}) = 0.5 \text{ rad/s}$$

The maximum angular acceleration is

$$\ddot{\theta}_{max} = \omega_n^2 A = (2.2 \text{ rad/s})^2(0.21 \text{ rad}) = 1 \text{ rad/s}^2$$

Example 10.13 A simple pendulum 1.4 m in length is displaced through an angle of 10° and released. Find the velocity of the bob when it reaches the bottom.

Solution 10.13

$$\theta = (10°)\left(\frac{2\pi \text{ rad}}{360° \text{ deg}}\right) = 0.17 \text{ rad}$$

Taking the potential energy to be zero at the bottom, we have

$$mgL(1 - \cos\theta) = \frac{1}{2}mv^2$$

Since θ is small, $\cos\theta \approx 1 - \theta^2/2$ and therefore

$$mgL\frac{\theta^2}{2} = \frac{1}{2}mv^2$$

and

$$v = \sqrt{gL}\theta = \sqrt{(9.8 \text{ m/s}^2)(14 \text{ m})}(0.17 \text{ rad}) = 0.63\text{m/s}$$

10.3.5   The Physical Pendulum

The physical pendulum is a rigid body that oscillates about an axis passing through a point in the body other than its center of mass (the center of mass is assumed to be located at the center of gravity). Figure 10.16 shows a rigid body pivoted at point O that is at a distance d from the center of mass. The equilibrium position of the body is when its center of mass is directly below the pivot O. If the body is displaced either to the right or left from the equilibrium position, a restoring torque

due to gravity will act on it. As a result, the body will oscillate in a vertical plane where the axis of rotation is perpendicular to the page. The restoring torque is given by

$$\tau = -Mgd \ \sin\theta$$

where M is the mass of the body and d is the moment arm of the tangential component of the weight ($Mg \ \sin\theta$). From Newton's second law, we have

$$\tau = I\alpha$$

$$-Mgd \sin\theta = I\ddot{\theta}$$

For small angular displacements $\sin\theta \approx \theta$ and hence

$$\ddot{\theta} + \left(\frac{Mgd}{I}\right)\theta = 0$$

or

$$\ddot{\theta} + \omega_n^2\theta = 0$$

This equation is of a simple harmonic motion with an angular frequency of

$$\omega_n = \sqrt{\frac{Mgd}{I}}$$

and a period of motion of

$$T = \frac{2\pi}{\omega_n} = 2\pi\sqrt{\frac{I}{Mgd}}$$

Thus,

$$I = \frac{T^2Mgd}{4\pi^2}$$

Therefore, the moment of inertia of a body can be found by measuring its period when it is in simple harmonic motion as a physical pendulum. Note that, the simple pendulum is a special case of the physical pendulum since for a simple pendulum of mass m, the moment of inertia is

$$I = md^2$$

and thus, the angular frequency is

$$\omega_n = \sqrt{\frac{mgd}{md^2}} = \sqrt{\frac{g}{d}}$$

This angular frequency is of a simple pendulum where d represents the length of the string.

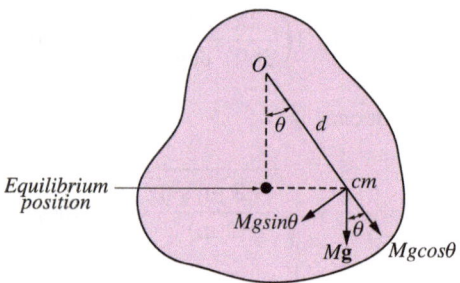

Fig. 10.16 The physical pendulum

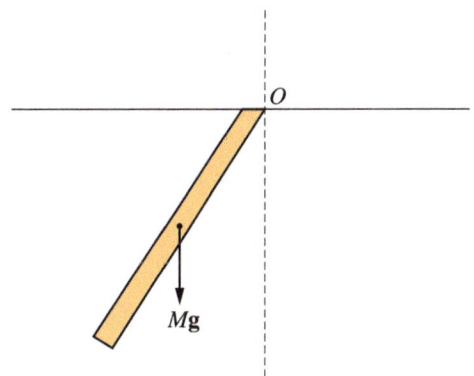

Fig. 10.17 A uniform rod suspended at one end oscillated with a small amplitude

Example 10.14 A uniform rod of length of 0.6 m that is suspended at one end oscillates with a small amplitude as in Fig. 10.17. Find the frequency of motion.

Solution 10.14

$$f_n = \frac{1}{2\pi}\sqrt{\frac{Mgd}{I}} = \frac{1}{2\pi}\sqrt{\frac{Mg(L/2)}{(1/3)ML^2}} = \frac{1}{2\pi}\sqrt{\frac{3g}{2L}} = \frac{1}{2(3.14)}\sqrt{\frac{3(9.8\text{m/s}^2)}{2(0.6\text{m})}} = 0.8\,\text{Hz}$$

Example 10.15 A uniform square plate of length a is pivoted at one of its corners and oscillates in a vertical plane as in Fig. 10.18. Find the period of motion if the amplitude is small.

Solution 10.15 The moment of inertia of a uniform rectangular plate about its center of mass is

$$I_{cm} = \frac{1}{12}M(a^2 + b^2)$$

Thus for a uniform square plate, we have

$$I_{cm} = \frac{1}{6}Ma^2$$

From the parallel axis theorem, the moment of inertia of the plate about an axis that is parallel to the center of mass axis

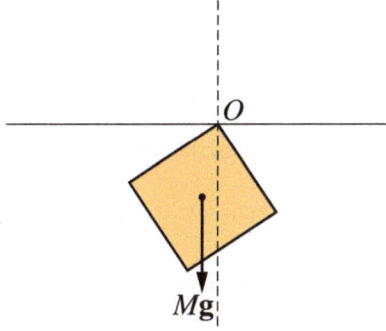

Fig. 10.18 A uniform square plate pivoted at one of its corners and oscillates in a vertical plane

and passing through one corner ($D = \sqrt{2}a$) is

$$I = I_{cm} + MD^2 = \frac{1}{6}Ma^2 + 2Ma^2 = \frac{13}{6}Ma^2$$

and hence

$$T = 2\pi\sqrt{\frac{I}{Mgd}} = 2\pi\sqrt{\frac{(13/6)Ma^2}{Mg\sqrt{2}a}} = 2\pi\sqrt{1.5\frac{a}{g}}$$

10.3.6 The Torsional Pendulum

The torsional pendulum consists of a rigid body suspended by a wire from its center of mass where the other end of the wire is fixed as shown in Fig. 10.19. The body is in equilibrium if the wire is untwisted. If the body is rotated through an angle θ it will oscillate about its equilibrium position (the line OP) due to a restoring torque exerted by the twisted wire on the body. This torque is found to be directly proportional to the angular displacement of the body. That is

$$\tau = -k\theta$$

where k is called the torsional constant. Its value depends on the property of the wire. Note that this equation is the rotational analogue of Hook's law in linear form ($F = -kx$). From Newton's second law, we have

$$\tau = I\alpha$$

or

$$-k\theta = I\ddot{\theta}$$

That gives

$$\ddot{\theta} + \left(\frac{k}{I}\right)\theta = 0$$

or

$$\ddot{\theta} + \omega_n^2\theta = 0$$

Fig. 10.19 The torsional pendulum

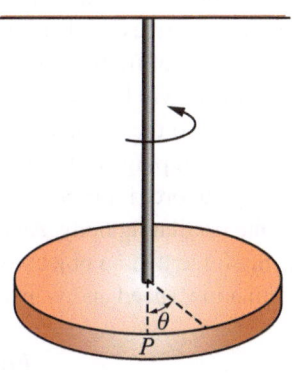

Fig. 10.20 A uniform solid sphere suspended at its midpoint by a light string

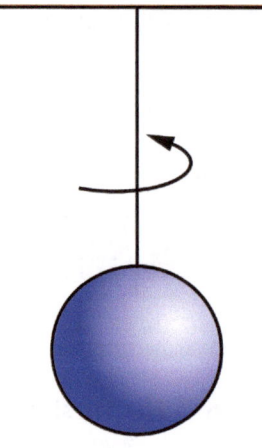

where $\omega_n = \sqrt{k/I}$ and the period is $T = 2\pi\sqrt{I/k}$.

Example 10.16 A uniform solid sphere of mass of 4.7 kg and radius of 5 cm is suspended at its midpoint by a light string (see Fig. 10.20) where it oscillates as a torsional pendulum. If the period of motion is 3.5 s, find the torsion constant.

Solution 10.16

$$T = 2\pi\sqrt{\frac{I}{k}}$$

for a uniform solid sphere

$$I_{cm} = \frac{2}{5}MR^2 = \frac{2}{5}(4.7 \text{ kg})(0.05 \text{ m})^2 = 4.7 \times 10^{-3} \text{ kg m}^2$$

hence,

$$k = \frac{4\pi^2 I_{cm}}{T} = \frac{4(3.14)^2(4.7 \times 10^{-3} \text{ kg m}^2)}{(3.5 \text{ s})} = 0.05 \text{ kg m}^2/\text{s}^2$$

10.4   Damped Free Vibrations

In this section, we will discuss the case in which the effect of damping that is due to a nonconservative force cannot be neglected. An example of such a force in mechanical systems is the force of friction. In this case, the mechanical energy of the system will be lost, the amplitude of motion will decrease to zero, and the oscillation dies out eventually. Here, we will discuss damping due to friction in the simplest case, where the frictional force is proportional to the first power of the velocity of the oscillating body. An example of such a frictional force is the force that an object experience when moving in a fluid with a low speed and is given by

$$F_D = -bv$$

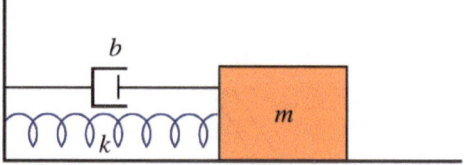

Fig. 10.21 A mass-spring system with damping

where b is a positive constant called the damping coefficient. Its SI units is $\text{N(m s}^{-1}) = \text{kg s}^{-1}$. The negative sign shows that the direction of the force is always opposite to the velocity. Now consider the spring–mass system as shown in Fig. 10.21, the cylinder shown in the figure contains a viscous fluid and a piston moving in it. Such device is known as the viscous damper. The net force on the oscillating body is

$$\sum F = F_s + F_D = -kx - bv$$

hence

$$m\ddot{x} + b\dot{x} + kx = 0$$

or

$$\ddot{x} + \gamma\dot{x} + \omega_n^2 x = 0 \qquad (10.13)$$

where $\gamma = b/m$ and $\omega_n = \sqrt{k/m}$. The units of γ is s^{-1}. This equation is a second order linear differential equation of constant coefficients. We may assume a solution of the form

$$x = Ce^{\lambda t}$$

Substituting this solution into Eq. 10.13 gives the characteristic (auxiliary) equation given by

$$\lambda^2 + \gamma\lambda + \omega_n^2 = 0$$

The roots of this equation are given by

$$\lambda_1 = -\frac{\gamma}{2} + \sqrt{\left(\frac{\gamma^2}{4} - \omega_n^2\right)}$$

and

$$\lambda_2 = -\frac{\gamma}{2} - \sqrt{\left(\frac{\gamma^2}{4} - \omega_n^2\right)}$$

From superposition, the general solution is given by

$$x = C_1 e^{\lambda_1 t} + C_2 e^{\lambda_2 t} \qquad (10.14)$$

Three possible solutions arise depending on whether the sign of the bracket $(\gamma^2/4 - \omega_n^2)$ is positive, negative or zero, i.e., depending on the size of the damping force. The roots λ_1 and λ_2 are either distinct real roots, equal real roots or a conjugate complex roots. Therefore, there are three possible motions of the system.

10.4.1   Light Damping (Under-Damped) ($\gamma < 2\omega_n$)

If $\gamma < 2\omega_n$ the resulting roots are complex roots given by

$$\lambda_1 = -\frac{\gamma}{2} + i\omega_D$$

and

$$\lambda_2 = -\frac{\gamma}{2} - i\omega_D$$

where

$$\omega_D = \left(\omega_n^2 - \frac{\gamma^2}{4}\right)^{1/2}$$

Hence, Eq. 10.14 may be written as

$$x = \left[C_1 e^{i\omega_D t} + C_2 e^{-i\omega_D t}\right] e^{\frac{-\gamma}{2}t}$$

Since $e^{\pm ix} = \cos x \pm i \sin x$ we have

$$x = [C_1(\cos\omega_D t + i\sin\omega_D t) + C_2(\cos\omega_D t - i\sin\omega_D t)]e^{\frac{-\gamma}{2}t}$$

$$= [(C_1 + C_2)\cos\omega_D t + i(C_1 - C_2)\sin\omega_D t]e^{\frac{-\gamma}{2}t}$$

$$= [A_1\cos\omega_D t + A_2\sin\omega_D t]e^{\frac{-\gamma}{2}t} \tag{10.15}$$

where $A_1 = C_1 + C_2$ and $A_2 = i(C_1 - C_2)$. As mentioned earlier Eq. 10.15 can be written as

$$x = A\cos(\omega_D t - \phi)e^{\frac{-\gamma}{2}t} \tag{10.16}$$

where A is the initial amplitude of motion. $Ae^{\frac{-\gamma}{2}t}$ is called the amplitude of motion and ϕ is the phase constant and ω_D is the angular frequency of the damped motion. This equation shows that the system oscillates in a decreasing harmonic motion where the amplitude of motion decreases exponentially with time until eventually the oscillation dies out (see Fig. 10.22). The dashed lines in Fig. 10.22 are called the envelope of the oscillation curve. The period of motion in light damping is therefore given by

$$\tau_D = \frac{2\pi}{\omega_D} = \frac{2\pi}{\sqrt{\omega_n^2 - \frac{\gamma^2}{4}}}$$

If $b = 0$ and thus $\gamma = 0$ the period of motion is reduced to that of a simple harmonic oscillator. If $\gamma \ll \omega_D$, the situation is referred to as very light damping and $\omega_D \approx \omega_n$. Furthermore

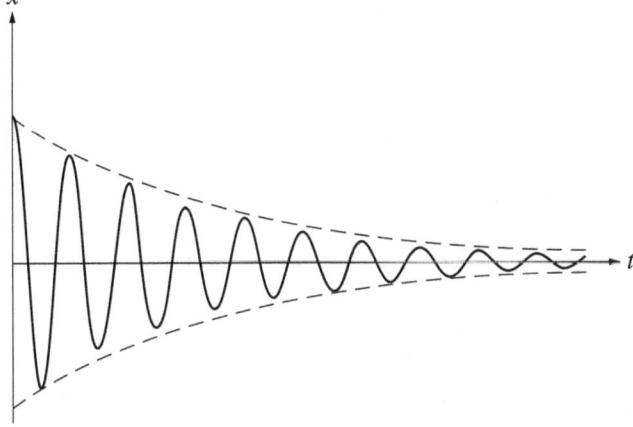

Fig. 10.22 In A lightly damped oscillator, the system oscillates in a decreasing harmonic motion where the amplitude of motion decreases exponentially with time until eventually the oscillation dies out

if there are two amplitudes A_a and A_b separated by the period of motion, then their ratio is given by

$$\frac{A_a}{A_b} = \frac{Ae^{-\frac{\gamma}{2}t_1}}{Ae^{-\frac{\gamma}{2}(t_1+\tau_D)}} = e^{\frac{\gamma}{2}\tau_D}$$

A quantity known as the logarithmic decrement is defined as

$$\delta = \ln\left(\frac{A_a}{A_b}\right) = \frac{\gamma}{2}\tau_D$$

Example 10.17 An 8 kg block is attached to a light spring and a light viscous damper. If at $t = 0, x = 0.12$ m and $v = 0$, find (a) the displacement at any time; (b) the logarithmic decrement. ($k = 30$ N/m, $b = 20$ N s/m).

Solution 10.17   (a)

$$\omega_n = \sqrt{\frac{k}{m}} = \sqrt{\frac{(30 \text{ N/m})}{(8 \text{ kg})}} = 1.9 \text{ rad/s}$$

$$\gamma = \frac{b}{m} = \frac{(20 \text{ N s/m})}{(8 \text{ kg})} = 2.5 \text{ s}^{-1}$$

and

$$\omega_D = \left(\omega_n^2 - \frac{\gamma^2}{4}\right)^{1/2} = ((1.9 \text{ rad/s})^2 - (2.5 \text{ Ns/m kg})^2 4)^{1/2} = 1.43 \text{ rad/s}$$

since $\gamma < 2\omega_n$, the damping is light. The displacement as a function of time is given by

$$x = A\cos(\omega_D t - \phi)e^{\frac{-\gamma}{2}t}$$

or

$$x = A\cos(1.43t - \phi)e^{-1.25t}$$

since at $t = 0$, $x = 0.12$ m, then

$$(0.12 \text{ m}) = A\cos\phi \tag{10.17}$$

the velocity of the block at any time is

$$\dot{x} = -1.43A\sin(1.43t - \phi)e^{-1.25t} - 1.25A\cos(1.43t - \phi)e^{-1.25t}$$

at $t = 0$, $v = 0$ and thus

$$0 = -1.43A\sin\phi - 1.25A\cos\phi \tag{10.18}$$

Solving Eqs. 10.17 and 10.18 for A and ϕ gives $\phi = -0.7$ rad and $A = 0.17$ m. Therefore,

$$x = 0.17\cos(1.43t - 0.7)e^{-1.25t}$$

(b)

$$\tau_D = \frac{2\pi}{\omega_D} = \frac{2\pi}{(1.43 \text{ rad/s})} = 4.4 \text{ s}$$

$$\delta = \frac{\gamma}{2}\tau_D = (1.25 \text{ s}^{-1})(4.4 \text{ s}) = 5.5$$

10.4.2   Critically Damped Motion ($\gamma = 2\omega_n$)

If $\gamma = 2\omega_n$, then the roots are equal real roots

$$\lambda_1 = \lambda_2 = -\frac{\gamma}{2} = -\omega_n$$

In that case, the motion decays without oscillation (see Fig. 10.23) and the general solution of Eq. 10.13 is

$$x = (C_1 + C_2\omega_n t)e^{-\omega_n t}$$

C_1 and C_2 are found from boundary conditions. If at $t = 0$, $x = A$, and $v = 0$, then

$$x(0) = C_1 = A$$

and

$$v(0) = \omega_n C_2 - \omega_n C_1 = 0$$

or

$$C_1 = C_2 = A$$

That gives

$$x = A(1 + \omega_n t)e^{-\omega_n t}$$

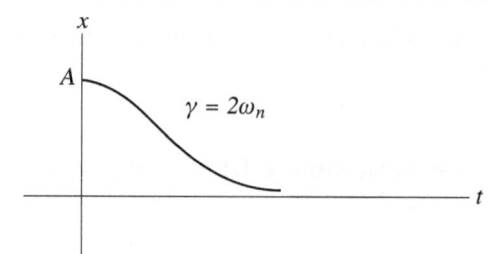

Fig. 10.23 In a critically damped motion, the motion decays without oscillation

10.4.3   Over Damped Motion (Heavy Damping) ($\gamma > 2\omega_n$)

If $\gamma > 2\omega_n$, the roots are distinct real roots given by

$$\lambda_1 = -\frac{\gamma}{2} + \sqrt{\left(\frac{\gamma^2}{4} - \omega_n^2\right)}$$

and

$$\lambda_2 = -\frac{\gamma}{2} - \sqrt{\left(\frac{\gamma^2}{4} - \omega_n^2\right)}$$

The general solution is given by

$$x = C_1 e^{\lambda_1 t} + C_2 e^{\lambda_2 t}$$

or

$$x = (C_1 e^{\alpha t} + C_2 e^{-\alpha t})e^{-\frac{\gamma}{2}t}$$

where

$$\alpha = \sqrt{\left(\frac{\gamma^2}{4} - \omega_n^2\right)}$$

C_1 and C_2 are found from boundary conditions. As critical damping, the resulting motion here is nonperiodic but the system returns to its equilibrium position at large values of t unlike critical damping (see Fig. 10.24).

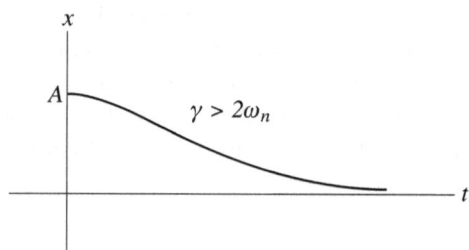

Fig. 10.24 As critical damping, the resulting motion here is nonperiodic but the system returns to its equilibrium position at large values of t unlike critical damping

Example 10.18 In Example 10.17, find the range of values of the damping coefficient for the system to be: (a) over damped; (b) critically damped.

Solution 10.18 (a) over damped if $\gamma > 2\omega_n$, i.e., if $\gamma > 3.8\text{s}^{-1}$ (b) critically damped if $\gamma = 3.8\text{s}^{-1}$.

10.4.4 Energy Decay

In damped free vibrations, the total mechanical energy is not constant since the damping force opposes the motion and dissipates the energy of the system. Now, consider the mass–spring system, the total mechanical energy of the system is

$$E = K + U = \frac{1}{2}m\dot{x}^2 + \frac{1}{2}kx^2$$

The rate of change of energy is

$$\frac{dE}{dt} = (m\ddot{x} + kx)\dot{x}$$

For damped vibrations in which the damping force is directly proportional to the velocity, we have

$$m\ddot{x} + kx = -b\dot{x}$$

Hence,

$$\frac{dE}{dt} = -b\dot{x}^2 \leq 0$$

Thus, the energy decreases with time in any damped motion and the rate in which it decreases is not uniform.

10.5 Forced Vibrations

In the previous sections, only free vibrations have been considered (i.e., vibrations in which only a restoring and damping force act within the system during motion). This section considers the case in which an external driving force is applied to the vibrator. This force is given as a function of time and we have

$$m\ddot{x} + b\dot{x} + kx = F(t) \tag{10.19}$$

Here, we will consider the case in which the force is a simple periodic force given by

$$F(t) = F_0 \cos \omega t \tag{10.20}$$

where F_0 is the amplitude and ω is the driving frequency. This force does positive work on the system to balance the energy loss due to damping. Substituting Eq. 10.20 into Eq. 10.19 gives

$$m\ddot{x} + b\dot{x} + kx = F_0 \cos \omega t \tag{10.21}$$

or

$$\ddot{x} + \gamma\dot{x} + \omega_n^2 x = \frac{F_0 \cos \omega t}{m}$$

Let us assume that the solution of Eq. 10.19 is given by

$$x = C_1 \cos \omega t + C_2 \sin \omega t$$

then, we have

$$\dot{x} = -\omega C_1 \sin \omega t + \omega C_2 \cos \omega t$$

and

$$\ddot{x} = -\omega^2 C_1 \cos \omega t - \omega^2 C_2 \sin \omega t$$

Substituting into Eq. 10.19 gives

$$(-\omega^2 C_1 \cos \omega t - \omega^2 C_2 \sin \omega t) + \gamma(-\omega C_1 \sin \omega t + \omega C_2 \cos \omega t)$$
$$+ \omega_n^2(C_1 \cos \omega t + C_2 \sin \omega t) = \frac{F_0 \cos \omega t}{m}$$

That gives

$$-\omega^2 C_1 + \gamma \omega C_2 + \omega_n^2 C_1 = \frac{F_0}{m}$$

and

$$-\omega^2 C_2 - \gamma \omega C_1 + \omega_n^2 C_2 = 0$$

Solving for C_1 and C_2 gives

$$C_1 = \frac{(F_0/m)(\omega_n^2 - \omega^2)}{(\omega^2 - \omega_n^2)^2 + \gamma^2 \omega^2}$$

and

$$C_2 = \frac{(F_0/m)\gamma\omega}{(\omega^2 - \omega_n^2)^2 + \gamma^2 \omega^2}$$

Hence,

$$x = \frac{(F_0/m)[(\omega_n^2 - \omega^2)\cos \omega t + \gamma\omega \sin \omega t]}{(\omega^2 - \omega_n^2)^2 + \gamma^2 \omega^2}$$

The term in brackets is of the form $A_1 \cos \omega t + A_2 \sin \omega t$ and thus it can be written as $A' \cos(\omega t - \phi)$ where

$$A' = \sqrt{A_1^2 + A_2^2}$$

i.e.,

$$A' = ((\omega_n^2 - \omega^2)^2 + \gamma^2 \omega^2)^{\frac{1}{2}}$$

and

$$\phi = \tan^{-1}\frac{A_2}{A_1} = \tan^{-1}\frac{\gamma\omega}{(\omega^2 - \omega_n^2)}$$

where $0 \leq \phi \leq \pi$. Hence,

$$x = \frac{(F_0/m)}{\sqrt{(\omega^2 - \omega_n^2)^2 + \gamma^2 \omega^2}} \cos(\omega t - \phi) \qquad (10.22)$$

If the driving force is applied for a long time compared with the time that the damped vibration dies out, then the system will eventually vibrate at the same frequency of the deriving force. Therefore, the general solution of Eq. 10.13 is called the transient solution since it approaches zero in a relativity short time whereas Eq. 10.21 is called the steady-state solution where the system oscillates with the same frequency as the deriving force. Therefore, the amplitude of a steady-state vibration is

$$A = \frac{(F_0/m)}{\sqrt{(\omega^2 - \omega_n^2)^2 + \gamma^2 \omega^2}}$$

When the deriving frequency ω approaches the natural frequency of the system ω_D, the amplitude of the resulting forced oscillation will increase. This is known as resonance. If the damping is very light, the amplitude reaches its peak when the deriving frequency is nearly equal to the natural frequency ω_n. As the damping becomes heavier, the maximum amplitude shifts to lower frequencies (see Fig. 10.25). In the case where there is no damping at all ($b = 0$), the amplitude of resonance is infinite at $\omega = \omega_n$.

Example 10.19 In Example 10.17, if a driving force of the form $F(t) = 5 \cos 4t$ is applied to the system, find the steady-state displacement as a function of time.

Solution 10.19

$$A = \frac{(F_0/m)}{\sqrt{(\omega^2 - \omega_n^2)^2 + \gamma^2 \omega^2}} = \frac{(5/8)}{\sqrt{((4)^2 - (1.9)^2)^2 + (2.5)^2(4)^2}} = 0.04 \text{ m}$$

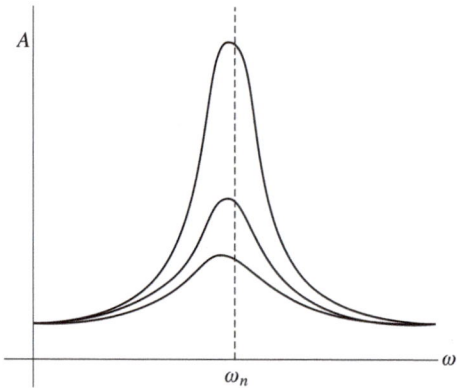

Fig. 10.25 When the deriving frequency ω approaches the natural frequency of the system ω_D, the amplitude of the resulting forced oscillation will increase. This is known as resonance. If the damping is very light the amplitude reaches its peak when the deriving frequency is nearly equal to the natural frequency ω_n. As the damping becomes heavier, the maximum amplitude shifts to lower frequencies

$$\phi = \tan^{-1} \frac{\gamma \omega}{(\omega^2 - \omega_n^2)} = \tan^{-1} \frac{(2.5)(4)}{((4)^2 - (1.9)^2)} = 0.8°$$

Hence,

$$x = 0.04 \cos(4t - 0.8)$$

Therefore, the forced vibration has the same frequency as the deriving force but lag in phase by $0.8°$

Example 10.20 In Example (10.17), find the steady-state displacement as a function of time if there is no damping.

Solution 10.20 The amplitude of the forced oscillation when the angular frequency ω of the deriving force is varied.

$$A = \frac{(F_0/m)}{\sqrt{(\omega^2 - \omega_n^2)^2 + \gamma^2 \omega^2}} = \frac{(5/8)}{\sqrt{((4)^2 - (1.9)^2)^2}} = 0.05 \text{ m}$$

$$x = 0.05 \cos 4t, \quad \phi = 0.$$

Problems

1. A 2 kg block is fastened to a spring of force constant 98 N/m on a horizontal frictionless surface. If the block is released a distance of 6 cm from its equilibrium position, find (a) the angular frequency, the frequency and the period of the resulting motion, (b) the time it takes the block to first reach $x = -5$ cm and its velocity at that time, (c) the maximum speed and maximum acceleration of the oscillating block, (d) the total mechanical energy of the oscillator.

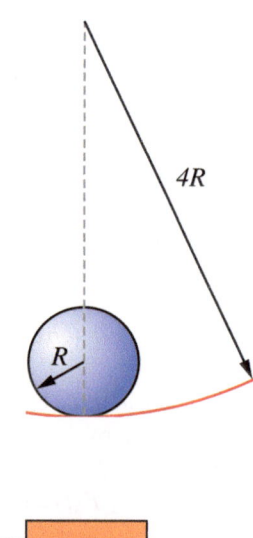

Fig. 10.26 A uniform solid cylinder of radius R and mass M rolls without slipping on a track of radius $4R$

4R

R

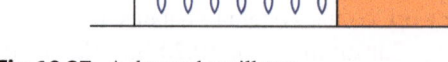

Fig. 10.27 A damped oscillator

2. A 10 kg block is attached to a light spring of force constant 200 N/m on a smooth horizontal surface. Find the amplitude of motion if at $x = 0.06$ m the velocity of the block is $v = 0.5$ m/s.

3. A particle rotate counterclockwise in a circle of radius 0.2 m with a constant angular speed of 2 rad/s. If at $t = 0$ the x-coordinate of the particle is 0.14 m, find the displacement, velocity and acceleration of the particle at any time.

4. If a simple pendulum has a period of 2 s, find its period when its length is increased by 20%.

5. A simple pendulum of length lm and mass of 0.4 kg oscillates in a region where $g = 9.8$ m/s^2. If the amplitude of oscillation is 10°, find (a) the angular displacement, angular velocity and angular acceleration of the pendulum as a function of time.

6. A uniform solid cylinder of radius R and mass M rolls without slipping on a track of radius $4R$ as shown in Fig. 10.26. Find the period of oscillation when the cylinder is displaced slightly from its equilibrium position.

7. A planer body of mass 3 kg oscillates as a physical pendulum. If the period of oscillation is 3 s and if the pivot

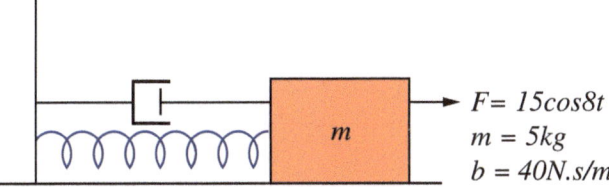

Fig. 10.28 A forced oscillator

point is at 0.2 m from the center of mass, find the moment of inertia of the body.

8. A uniform hollow cylinder of radius R and mass M is suspended at its midpoint from a wire and form a torsional pendulum. If the period of motion is T, find the torsion constant.

9. For the system shown in Fig. 10.27, determine the displacement of the block at any time if at $t = 0$, $x = 0$ and $v = 0$. ($k = 200$ N/m, $b = 200$ N s/m).

10. For the system shown in Fig. 10.28, find the steady-state displacement as a function of time.

In the figure: $F = 15\cos 8t$, $m = 5kg$, $b = 40N.s/m$

Open Access This chapter is licensed under the terms of the Creative Commons Attribution 4.0 International License (http://creativecommons.org/licenses/by/4.0/), which permits use, sharing, adaptation, distribution and reproduction in any medium or format, as long as you give appropriate credit to the original author(s) and the source, provide a link to the Creative Commons license and indicate if changes were made.

The images or other third party material in this chapter are included in the chapter's Creative Commons license, unless indicated otherwise in a credit line to the material. If material is not included in the chapter's Creative Commons license and your intended use is not permitted by statutory regulation or exceeds the permitted use, you will need to obtain permission directly from the copyright holder.

References

Alonso, M., Finn, E.J.: Fundamental University Physics: Volume 1 Mechanics. Addison-Wesley Publishing Company (1967)

Anton, H., Rorres, C.: Elementary Linear Algebra Application Version. Wiley (1994)

Arfken, G.B., Weber, H.J.: Mathematical Methods for Physicists. Academic Press (1995)

Beiser, A.: Concepts of Modern Physics. McGraw-Hill Inc. (1995)

Boas, M.L.: Mathematical Methods in the Physical Sciences. Wiley (1983)

Bueche, F.J., Hecht, E.: Schaum's Outline of Theory and Problems of College Physics. The McGraw-Hill Companies Inc. (1997)

Finney, R.L., Thomas, G.B.: Calculus. Addison-Wesley Publishing Company (1990)

Foulds, K.: Physics. John Murray Publishers Ltd. (2000)

Giancoli, D.C.: Physics Principles and Applications. Prentice-Hall International Inc. (1991)

Goldstein, H., Poole, C., Safko, J.: Classical Mechanics. Addison Wesley (2002)

Halliday, D., Resnick, R., Walker, J.: Fundamentals of Physics Extended. Wiley (1997)

Halliday, D., Resnick, R., Krane, K.S.: Physics: Volume 1. Wiley (2003)

King, A.R., Regev, O.: Physics with Answers. Cambridge University Press (1997)

Kittel, C., Knight, W.D., Ruderman, M.A., Helmholz, C.A., Moyer, B.J.: Mechanics. McGraw-Hill Inc. (1973)

Kleppner, D., Kolenkow, R.J.: An Introduction to Mechanics. McGraw-Hill Book Company (1973)

Main, I.G.: Vibrations and Waves in Physics. Cambridge University Press (1994)

Meriam, J.L., Kraige, L.G.: Engineering Mechanics: Volume 2 Dynamics. Wiley (1998)

Narula, G.K.: Physics for Class XI. Vikas Publishing House Pvt, Ltd (1996)

Serway, R.A.: Physics for Scientists and Engineers with Modern Physics. Saunders College Publishing (1996)

Seto, W.W.: Theory and Problems of Mechanical Vibrations. McGraw-Hill Book Company (1964)

Spiegel, M.R.: Theory and Problems of Theoretical Mechanics. McGraw-Hill Book Company (1982)

Swokowski, E.W., Olinick, M., Pence, D., Cole, J.A.: Calculus. PWS Publishing Company, Boston (1994)

Young, H.D., Freedman, R.A.: University Physics with Modern Physics. Addison Wesley Longman Inc. (2000)

Zill, D.G.: A First Course in Differential Equations. PWS-KENT Publishing Company (1993)

© The Editor(s) (if applicable) and The Author(s) 2019
S. Alrasheed, *Principles of Mechanics*, Advances in Science,
Technology & Innovation, https://doi.org/10.1007/978-3-030-15195-9